Scientific and Technical Writing

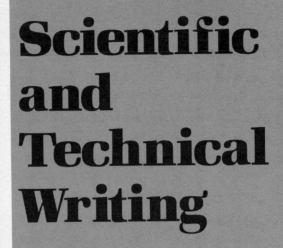

Scientific and Technical Writing

Peter M. Sandman
Rutgers—The State University of New Jersey,
New Brunswick
Carl S. Klompus
Rutgers—The State University of New Jersey,
New Brunswick
Betsy Greenleaf Yarrison
University of Baltimore

HOLT, RINEHART AND WINSTON
New York   Chicago   San Francisco
Philadelphia   Montreal   Toronto
London   Sydney   Tokyo   Mexico City
Rio de Janeiro   Madrid

Acquisitions Editor: Charlyce Jones Owen
Senior Project Editor: Lester A. Sheinis
Production Manager: Annette Mayeski
Design Supervisor: Gloria Gentile

To Patty and Jim;
Alison, Jennifer, and James

Library of Congress Cataloging in Publication Data

Sandman, Peter M.
  Scientific and technical writing.

  Includes index.
  1. English language—Rhetoric. 2. English language—
Technical English. 3. Technical writing. I. Klompus,
Carl S. II. Yarrison, Betsy Greenleaf. III. Title.
PE1475.S26   1985   808′.0666   84–19740
ISBN 0-03-041056-8

CBS COLLEGE PUBLISHING
Holt, Rinehart and Winston
The Dryden Press
Saunders College Publishing

Preface

Planning a technical career that doesn't include writing is like planning a basketball career that doesn't include dribbling. Writing is an intrinsic part of every scientific and technical occupation. Our purpose in producing this book is to help you make sure your writing skills are as fully developed as your other professional skills.

To achieve this purpose, *Scientific and Technical Writing* is organized systematically. Part One traces the steps in the technical writing process, from defining the problem to proofreading the manuscript. Part Two focuses on the varieties of technical content, such as description, definition, analysis, and illustration. In Parts Three and Four, finally, we detail how to put these components together into technical reports and correspondence. The fundamental premise of the book is that technical writing is a technical task, grounded in conscious choice among writing alternatives. Thus, we have tried to show you how to suit your content, organization, and language to your goals and your readers' needs.

We hope the book itself is an illustration of this emphasis on conscious choice to match the goal and audience. For example, we think you will find the style of *Scientific and Technical Writing* a little less formal, maybe even a little more fun, than the style of most technical documents. This suits an introductory text on writing, designed for college students. The same style would hardly be appropriate for a final report to management on a research project.

We have debts of gratitude to acknowledge to many people: to Alan Rauch, Lisa Tichauer, John Willard, and the rest of the technical writing staff at Cook College, Rutgers University; to Barbara Goff, Lois Kaufman, Tom Matro, Norma Reiss, and Adelaide Szukics, all of the Cook College Department of Humanities and Communications; to Richard Budd and David Sachsman of the Rutgers University School of Communication, Information, and Library Studies; to William Keach and Thomas Van Laan of the Rutgers University Department of English; to Roth Wilkofsky, Susan Katz, Nedah Abbott, Charlyce Jones Owen, Cynthia Burns, Lester Sheinis, and others at (or once at) Holt, Rinehart and Winston; to Catherine Ambos and her IBM P.C.; and to the following friends, colleagues, students, and ex-students who helped in special ways: Ralph Barthine, Elizabeth Bress, Deborah Cloeren, Mary Dudley Culbertson, William Barnett Culbertson, Dale Demy, Peter Demy, Jim Garry, Gerard Gilliam, Lisa Lofland Gould, B. J. Hance, Beverly Johnson, Robert Jordan, Sheri Krams, Joseph LoPorto, Patricia Castelli Machat, Jane Manetta, Mark Marcelli, Christopher Motta, David Salveson, Alice Sims-Gunzenhauser, Philip Spadaro, Dori Steinberg, and JoAnn Myer Valenti.

Colleagues at other institutions who took time to review various drafts of this book include: Theodore Andra, Utah State University; Rosemary Ascherl, Hartford State Technical College; Lynn Diane Beene, University of New Mexico; Virginia A. Book, The University of Nebraska-Lincoln; Donald H. Cunningham, Texas Tech University; L. W. Denton, Auburn University; Jone Goldstein, Wayne State University; Joyce Hicks, Valparaiso University; Paulino Lim, Jr., California State University, Long Beach; Susan H. McLeod, San Diego State University; David B. Merrell, Abilene Christian University; Carolyn Miller, North Carolina State University; Don Norton, Brigham Young University; Gary Poffenbarger, Texas Tech University; George Redman, University of South Carolina; William E. Rivers, University of South Carolina; John R. Rodman, Chesapeake College; Don W. Sieker, New Hampshire College; Emil F. Symonik, Milwaukee School of Engineering; John Zubritsky, Prince George's Community College. They are of course not responsible for the remaining defects.

We are grateful to Mary Paden for collaborating on an earlier version of this book. Although the book's thrust has changed substantially since then, we appreciate her contribution.

Our deepest thanks must be reserved for our students in Technical Writing, Advanced Technical Writing, Scientific and Technical Writing and Reporting, and Popular Writing on Technical Topics. They taught us how to teach them.

PMS
CSK
BGY

Contents

PART 3   Technical Reports   261

Scientific and Technical Writing

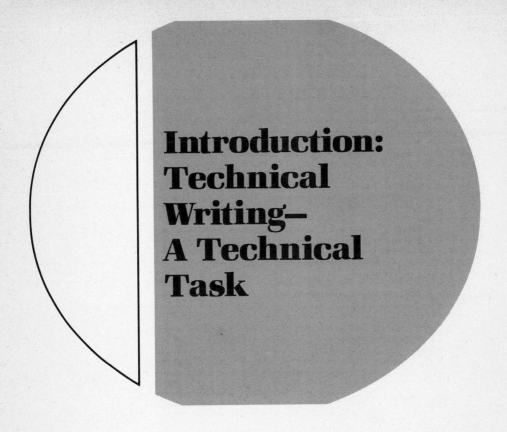

Introduction: Technical Writing— A Technical Task

The skills required for effective technical writing are essentially *technical* skills—planning, organization, clear thinking, and attention to detail. If you have training in any technical field—science, engineering, business, even the arts and humanities—you have learned numerous technical procedures as part of that training. Learning technical writing is no different from learning any other technical procedure such as gram staining, film developing, regression analysis, or poetry analysis. Throughout this book we will talk about the "technical procedure" of communicating effectively.

▶ The Need for Technical Writing

When most students preparing for careers in science and technology think about what they will be doing on the job, they tend to envision themselves in the laboratory, in the field, at the computer console, or at the drafting table—doing technical work rather than writing about it. Yet the reality is that technical professionals have no choice but to be technical writers too.

Two recent surveys demonstrate the importance of technical writing in scientific and technical careers. A 1977 survey of successful engineers revealed that 95 percent of the respondents considered the ability to write effectively

1

to be "very important" or of "critical importance" in their present jobs. Ninety-eight percent said their own writing skills or lack of them had affected their career advancement. And when they were asked how much of their working time was devoted to writing, the mean answer was 24 percent—well over a day a week.[1] Similarly, in a survey conducted by the American Society of Mammalogists, graduates of programs in the biological sciences reported that technical writing was the second most valuable "tool" course they had taken as preprofessionals, more valuable than computer science and second only to statistics.[2]

To understand why professionals find technical writing such an important skill, imagine for a moment that you work in the research and development division of a large company. You have an idea for improving the quality of the widgets your company makes. So far, however, it is only an idea; before you can test it out, you will have to get authorization. That means that you will have to write a proposal, perhaps two—an informal one to your supervisor, and if he or she likes your idea, a formal one to management. If the proposal is approved, you will have more writing to do: You will have to write out the testing procedures that you want your staff to follow; you will probably have to write to the machine shop explaining the special equipment you need to perform your tests; you may very well have to write to people either inside or outside the company who have particular expertise in an area of your project that you don't understand well enough; you will certainly have to write progress reports to management regularly, in order to keep them abreast of how your work is proceeding.

If your idea proves feasible, you will have to write the most important report of all—the one to convince management that your idea is worth implementing. And if *that* report is successful, it will generate others: You will have to write out the procedures for the production staff; you will have to write to the people in marketing so that they will know how to promote the new, improved widgets; you will have to work on the patent application to protect your innovation. In short, virtually every on-the-job activity you perform will require you to have competent writing skills.

▶ What Is Technical Writing?

The most obvious characteristic of technical writing is its technical subject matter. When we think about technical subjects, those that come to mind first are science and engineering, and most of the examples in this book will be chosen from those fields. But a technical subject is really *any* subject that is highly specialized, with its own theory and principles, its own vocabulary, and its own body of knowledge. If you have ever dipped into an advanced publication devoted to economics, football, literature, stamp collecting, or chess, you know that these, too, can be technical subjects.

A second distinguishing characteristic of technical writing is its audience. Although this audience may include scientists and engineers, it is not limited to them; the audience for a technical document may be machinists, managers, marketing experts, attorneys, or specialists in scores of fields who may or may not share an organizational affiliation with the technical writer. In fact, the audience may be any combination of these. Technical readers are a highly diversified group, and they don't always know as much about the subject as the writer. What unites all technical readers, however, is their *reason for wanting* to know about the subject. Unlike people who read for pleasure, technical readers are not passive consumers of information; they read because they intend to use the information to carry out their own specialized jobs.

Based on what we have said so far, the following definition of technical writing emerges. *Technical writing is writing that tries to inform people about a complex and specialized subject so that they can perform their own complex and specialized jobs more effectively.*

Technical subject matter, utilitarian purposes—these are characteristics of technical writing that nearly everyone agrees on. Most people agree, too, that technical writing has a characteristic tone, style, and point of view. Most technical audiences want to feel that what they are reading is accurate, objective, clear, impartial, concise, unemotional, restrained, meticulous, unambiguous, scientific, and straightforward.

At its best, technical writing is pleasing to the senses as well as to the mind. But its aesthetic is neither decorative nor evocative; it is functional. As W. Earl Britton has contended in a now-famous essay:

> The primary, though certainly not the sole, characteristic of technical and scientific writing lies in the effort of the author to convey one meaning and only one meaning in what he says. That one meaning must be sharp, clear, precise. And the reader must be given no choice of meanings; he must not be allowed to interpret a passage in any way but that intended by the writer. . . . Varied interpretations of a work of literature may add to its universality, whereas more than one interpretation of a piece of scientific and technical writing would render it useless.[3]

Technical writers attain this singleness of meaning by planning in advance what they want their writing to accomplish, then working and reworking their material until it means what they want it to—and nothing else. But in a sense, the goal of attaining this singleness of meaning is not limited to the task of writing. Technical professionals pursue it in their lab and field work as well, in the form of unambiguous and accurate research. In fact, if you think about the list of words we gave you earlier—accurate, objective, clear, impartial, concise, unemotional, restrained, meticulous, unambiguous, scientific, straightforward—you will see that these are not just the virtues of technical writing; they are the virtues of *all* technical work. It is not surprising, then, that technical experts often make the best technical writers.

▶ **Myths About Technical Writing**

If you have the makings of a good technical professional—in any field—then you can certainly master the skills required for sound functional prose. You must abandon, however, the following myths:

1. The myth that writing is ever fun and easy. Probably the most common complaint in technical writing courses is that every word is a struggle, proving presumably that the student just isn't a writer. Every word *is* a struggle—for professionals as well as beginners. Samuel Johnson made the point bluntly two hundred years ago: "No man but a blockhead ever wrote for anything but money." Writing is work, often agonizingly hard work.

2. The myth that good writing has to be inspired. Professional writers cannot afford to wait for inspiration; they must force themselves to write on schedule, often against a deadline, whether they feel like it or not.

3. The myth that good writers are born, not made. The very best writers are born *and* made; they train their inborn talent. The rest of us make do with training alone. It does not take genius to write an effective technical report; anyone in college can learn to write clearly and convincingly enough to communicate information to technical readers.

4. The myth that technical work and technical writing are unrelated processes. In many ways, technical writing is at the heart of all scientific and technical work. Scientists and engineers only rarely produce new technologies, new objects, or new procedures. What they usually produce is new *information*. This, in turn, is amalgamated with other information produced by other technical specialists to help technology inch forward. Thus, you will spend most of your career finding and sharing information with your professional associates. Communicating all that information *is* technical writing. It is absolutely central to all technological progress and one of the most important aspects of your job.

Good technical writing is not a product of natural talent or easy inspiration. It is *labor*. It is meticulously choosing words and phrases and sentences and paragraphs to achieve a specific goal. Learning technical writing is learning how to make the right choices.

▶ **Making Choices:
The Technical Writing Process**

The process of technical writing, like much science and engineering, is fundamentally a process of planning, testing, and refining. You wouldn't expect much of value from a researcher who cleared up a research problem by settling promptly and gratefully for the first possible solution that came to mind. That first possible solution might turn out to be totally impractical; it might have costs and side effects that would make another solution more appropriate; at best it would need to be refined, streamlined, tinkered with. Even if that first solution did turn out to be the ideal solution—a rarity, but it happens—the researcher wouldn't know that for sure until he or she had worked with it for a while, comparing it with several alternatives.

All this is equally true for a piece of writing. Occasionally, your first approach to a sentence will turn out to be the best one. But professional writers—creative and technical—always examine that first try to see if it is practical. They test it for costs and side effects; they refine it, streamline it, and tinker with it; they compare it with possible alternatives. Only poor writers ever settle for their first attempt.

Consider the preceding sentence for a minute. It first appeared in our manuscript this way:

| It is only the poor writers who are ever satisfied to settle for their first attempt.

That sentence is not too bad for a first draft, but it is flabby. "Satisfied to settle" is redundant; either you are satisfied with your first attempt or you settle for it. So:

| It is only the poor writers who ever settle for their first attempt.

That sentence is better, but the beginning still seems wordy, indirect, and needlessly complicated. "It is only the poor writers who" is a slow start. So:

| Only poor writers ever settle for their first attempt.

There are literally an infinite number of ways to express the idea contained in that sentence. For instance, you can use the personal approach:

| Your writing will invariably be poor if you ever let yourself settle for your first attempt.

You can turn the sentence around to make it positive:

| Good writers always refuse to settle for their first attempt.

You can use the singular instead of the plural:

| Only a poor writer ever settles for his or her first attempt.

You can hedge cautiously:

| As a rule, it is only poor writers who ever settle for their first attempt.

You can turn the verb into the subject:

| Settling for their first attempt is something that only poor writers ever do.

You can qualify the statement to make it less blunt:

| Only poor writers (and very insecure ones) ever settle for their first attempt.

You can turn the sentence into a command:

| If you want to be a good writer, don't ever settle for your first attempt.

Or you can abandon the structure and vocabulary of the original entirely:

| What most distinguishes good writers from poor ones is effort and choice; good writers think a sentence through instead of settling immediately for the first version that comes to mind.

The word "effort" here suggests still another possibility:

| The apparently effortless prose of good writers comes precisely from *effort*—that is, from conscious choice.

We have now shown you twelve ways of expressing the sentence. After we created all twelve, we looked them over and chose this one to end the paragraph:

| Only poor writers ever settle for their first attempt.

This is what writing is all about.

In all fairness, we must admit that only the most compulsive writers actually consider twelve alternatives for every word, phrase, sentence, and paragraph. An experienced writer rejects most alternatives without writing them down, almost without thinking about them. Only the serious contenders get careful consideration. For inexperienced writers, the process of choice involves two contradictory dangers. The first danger is the one we have been focusing on so far—that you won't consider enough alternatives and will settle for your first attempt. The second danger is that you will become so preoccupied with the alternatives that you won't be able to choose at all.

Writing courses and writing texts often have a paralyzing effect on students, especially at the beginning. You walk in as a cheerfully mediocre writer who composes each sentence only once; two weeks later twelve or thirteen versions of each sentence are competing with one another in your head, and you cannot get anything down on paper. The antidote is fairly simple. Bear in mind that writing and editing are *different* processes or, at least, different stages in the same process. Professionals do a good deal of editing in their heads as they write (and later do more on paper). Rank amateurs don't edit at all. As an apprentice professional, your best bet is to write first, edit later.

 Overview

Editing or revising is the last step in the technical writing process. First come careful analysis of goal and audience, collection and organization of the relevant information, and preparation of an effective first draft. In the next four chapters, we walk you through this process one step at a time.

In Part 2 we introduce you to the essential components of technical documents: definitions, analyses, descriptions, and illustrations. Virtually all technical documents require at least some of these components. In fact, technical manuals, among the most common output of technical writers, are often nothing more than a compilation of these components headed by an explanatory introduction.

In Parts 3 and 4 we discuss the special design considerations of specific types of technical reports and correspondence.

Notes

1. R. M. Davis, "How Important Is Technical Writing?—A Survey of the Opinions of Successful Engineers," *The Technical Writing Teacher* 4 (Spring 1977), p. 83.
2. G. W. Barrett and G. N. Cameron, *Career Trends and Graduate Education in Mammalogy* (New York: American Society of Mammalogists, 1981).
3. W. Earl Britton, "What Is Technical Writing?" *College Composition and Communication*, 16 (May 1965).

The Technical Writing Process

PART 1

1 Analyzing Goal and Audience

**Making Choices Based on Goal and Audience:
   A Case in Point**

Choice of Vocabulary
Degree of Detail
Tone
Emphasis

The most basic principle of effective communication is that what you say (both its content and its style) hinges on whom you are addressing and why you are addressing them. Competent technical writers know that they can't even begin a rough draft until they have a clear idea of what they want their writing to accomplish—their *goals*—and who their readers are—their *audience*. Once they have these two pieces of information, they can determine how to present their material.

 # Clarifying Your Goals

In the Introduction, we defined technical writing as writing designed to convey information on a specialized subject to readers who want or need the information so that they can perform their own jobs more effectively. Actually, we were being deliberately overgeneral when we said that the goal of technical writing is "to convey information." The goal of any *single* piece of technical writing is usually a great deal more specific and immediate than that—to describe a new piece of equipment that your company has developed so that potential customers can understand how it works, to instruct your company's technicians on how to perform a new procedure, to examine the pros and cons of a proposed design change to executives or to your supervisors.

Informing and Persuading

All technical writing sets out to inform its readers, and sometimes that is its only major goal. But much of the time it also undertakes to influence their decisions or actions. For example, if you were writing a proposal to undertake a new research project, a major part of your goal would be to convince company officials that the project is worth doing and that you are the person to do it. Or if, after a yearlong study, you had some recommendations to make about ways to streamline your company's production, you would surely want to convince readers that your recommendations should be implemented. This sort of rational persuasion is an important goal of most technical reports and proposals.

Whether your goal is to inform or to convince—or both—you should formulate it as explicitly as possible and *write it down*. Some version of this goal will probably appear as the statement of purpose when the piece you are writing takes its final form. Knowing what you want your writing to accomplish before you begin a draft will greatly simplify the task of deciding what to say and how to say it.

Articulating Hidden Goals

You should write down not only your overt goals, the ones that will actually appear in your final draft, but also those goals that probably won't ever be expressed directly in any draft. Start with *organizational goals*—to show that your project deserves the budget increase you have asked for, to demonstrate that the two extra lab technicians you talked management into hiring really do increase your effectiveness, to prove that your research group is making satisfactory progress on the project you began six months ago.

Next add to your list any *personal goals* you might have—to make your supervisor look good, to get a raise, to show your older colleagues that you are as knowledgeable as they are, to hint that you would be perfect for that new position opening up in Hawaii. Be as honest and forthright as you can; these goals are entirely legitimate, and having them in front of you as you write will help you achieve them.

Coping with Multiple Goals

Any single piece of technical writing is likely to have several goals. Suppose, for example, that you have decided that your company should change the way it manufactures a particular product. To build the case for this recommendation, you may need to show the shop managers that the change will cut their down time; the quality control department will want to see your evidence that it will improve product reliability; the finance people will need convincing that it won't raise manufacturing costs. Thus you have derived three explicit goals from just one recommendation—not counting personal or organizational goals.

Elaborate formal reports and proposals may have literally dozens of goals; short, informal reports and letters, usually much more modest in scope, may have only one or two. As we take up the various types of technical writing in subsequent chapters, we'll have more to say about the likely goals of each. Most of the time, one or another of these goals will determine whether you discuss a certain piece of information in detail, mention it briefly, or leave it out altogether; whether you choose one particular way of saying something or opt for a jazzier (or more dignified, more sophisticated, simpler, and so on) alternative; whether you choose a method of organization that emphasizes one specific conclusion or pick one that gives equal weight to several. If you force yourself to be clearheaded about your goals, even if you don't admit most of them to anyone but yourself, it will reduce the likelihood that you will make such choices unintentionally—and badly!

▶ Matching Each Goal with Its Audience

As you generate your list of goals, be sure to tie each one to a specific audience:

> to explain the new federal regulations to the *production staff*
>
> to inform *company executives* that your pilot project has been completed
>
> to convince your *supervisor* that your research should be continued

Linking every goal to an audience helps ensure that everything you write will be of technical interest to *somebody*. Obviously, you can't achieve your goals if you don't address them to the readers who can act on them.

Moreover, how much you need to explain, to describe, to persuade, and so on depends entirely on the audience you are addressing. No matter what you are writing or to whom you are writing it, your most basic and immediate goal is always to produce a document that your readers can *use*—that is, a document that your readers will find comprehensible, relevant, and reliable. To determine what your readers are likely to find comprehensible, relevant, and reliable, you must carefully analyze the readers on your list.

▶ Analyzing Your Audience

Since not all technical readers want, know, or need the same information, no piece of technical writing is *universally* useful. There is, in fact, no such thing as "the technical audience." Instead, the audience for any given piece of technical writing may consist of one reader or a dozen or 8,000. It may be composed entirely of engineers or scientists, of administrators, or of lab technicians—or it may be a mixture of people with many different backgrounds and areas of expertise. The most frequent error committed by novice technical writers—and the most serious—is to act as though all audiences were the same.

The Audience Profile

Granting that there is a unique audience for everything you write and that you must reach this audience if your report or letter or memo is to be successful, you are faced with the problem of determining as precisely as you can who this audience is. We have devised a series of questions to help you do this (see Figure 1.1); the answers should tell you a good deal about your readers. Study this *Audience Profile* now, and use it as a guide whenever you write a technical paper.

If you have few goals and few audiences (as in most letters, for instance), filling out the profile should be fairly easy. However, if you are going to write

AUDIENCE (or audiences) _____

_____ Individual? _____ Group?

What are the major common denominators among members of the group?

_____

What are the areas of greatest disparity among them? _____

_____

What do _I_ want from this audience? (What do I want them to do?) _____

_____

NEEDS:

What use will my audience be making of this information? _____

_____

KNOWLEDGE:

How much technical expertise do my readers have in my specific _topic_? Is it

_____ Profound? _____ Thoroughgoing? _____ Working?

_____ Fair? _____ Minimal? _____ Nonexistent?

How much technical expertise do my readers have in my _field?_

_____ Much less than I have   _____ Less than I have

_____ About as much as I have   _____ More than I have

_____ Much more than I have

How did they acquire their knowledge?   _____ Through reading?

_____ Through hands-on experience?

What is their educational level? _____

STATUS:

Are my readers professionals? _____

What position(s) do they hold? _____

Within my organization? _____ Outside my organization? _____

In the organizational hierarchy, are they _____ At my level?

_____ Slightly above me?   _____ Slightly below me?

_____ Considerably above me?   _____ Considerably below me?

ATTITUDES:

How would I describe the attitude of this audience toward me? _____

_____

How would I describe the attitude of this audience toward the material? _____

_____

FIGURE 1.1   The Audience Profile.

to disparate groups of readers, each with widely differing needs, knowledge, status, and attitudes, you will be much better off profiling each group separately; otherwise, you will find yourself checking off all the boxes on the profile.

Assessing Readers' Needs. The sections of the profile marked "NEEDS" and "KNOWLEDGE" are the most critical. To answer the question, "What use will my audience be making of this information?" you have to think less about your goals and more about your audience's needs. If you think for a moment about the sort of work your readers do—marketing, purchasing, finance management, administration, research and development, and so on—you should have little trouble deducing the sort of information that they need. A description of a new piece of equipment written for technicians whose job it will be to use it will have to be different from the specifications written for the company that manufactures it.

If figuring out what each audience needs is fairly easy, failing to bother is often disastrous. Many a solid report has been sent back for more work— or worse yet, filed away and forgotten—simply because the author neglected to include the information most relevant to the audience receiving it. Whether you are writing a one-page memo or a three-hundred-page report, make sure that it deals explicitly with the special needs of your readers; make sure that they can put it to use.

Assessing Readers' Knowledge. The section on "KNOWLEDGE" is just as important. The questions here are designed to determine two related but distinct pieces of information: how much background your readers have on the specific topic of your paper and how much background they have in your field. As for the first, the answer is usually very little. Even your immediate supervisor is likely to know far less about your specific topic than you do; otherwise, he or she wouldn't need you to write about it at all. And, of course, if you are writing about any type of original research, what you tell your readers will be *all* they know; your paper will be their first contact with the subject.

On the other hand, some readers may know a good deal—more than you even—about your field in general or about your little corner of it in particular, whereas others are bound to be specialists in some other field entirely. This presents problems. A great deal rides on the technical background of your readers—which terms you have to define, which concepts you have to explain, which jargon you have to avoid.

Technical readers will make the effort to understand what you are telling them. But sometimes readers will think they understand when they don't— and make mistakes in their own work as a result. Sometimes they will send for clarification, delaying their own work and creating additional work for you. And sometimes, in desperation, they will simply give up, dismiss what you have written as incomprehensible, and move on to their next task.

Readers with advanced technical educations, meanwhile, are quick to dismiss writing that they feel is beneath their level of understanding. They may skim too quickly and miss vital information, or they may complain that they need more detail before they can get on with their work.

These problems are difficult to solve even if you have thought about the background of your readers. If you haven't even asked yourself what they know, the solution is unattainable.

Bridging the Gap Between Needs and Knowledge. Fortunately, what your readers need to know and what they already know are usually connected— audiences generally have the strongest background in those areas where they need the most detail. The research and development people who genuinely need the technical details of a proposed project are equipped to understand those details; the financial managers who couldn't possibly understand them don't really need them, either. This relationship between an audience's background and its information needs will frequently make your writing task easier. But don't rely on it too heavily. It is inevitable that on certain occasions you are going to have to fill in the background for an audience that knows little and needs to know a lot.

How do you know what your readers know and don't know about the material, what terminology they will feel most comfortable with, what tone to use with them, and—most important—what they need in the way of evidence, examples, statistics, graphs, proofs, and so on, so that they can make the best possible use of the information? You may actually know your readers. Almost all technical writing is put together on demand. Either you want somebody specific to read it or somebody specific has asked you to write it. And even when you cannot identify every single reader, you will be able to identify the *kinds* of readers you are writing to—to group them into categories.

Groups of Readers

The criteria you should use to group your readers are, of course, their information needs and their backgrounds. Using both criteria at once, you will find that most of your audiences fall into one of the following categories.

Specialists in Your Field. These are your most immediate colleagues in terms of background and interests. They share your vocabulary, your foundation in the basics of the discipline, and probably most of your biases. As a result, you can present them with highly technical and complex information—details about methodology, mathematical formulas, underlying theory—with confidence that you will be understood. In fact, specialists in your field usually *demand* this kind of detail; if it is not there, they assume that *you* don't know it.

Specialists in Other Fields. These are your counterparts in other technical disciplines. You share with them a basic technical education, but at some point your training diverged from theirs; beyond that point you have to explain terms and concepts as though you were writing to an educated layperson.

Consider the following sentence, for instance: "The 96-hour LC50 for *Salmo gairdneri* was 22.8 mg/l." All technical people know the accepted abbreviations for milligram and liter and know what quantities these terms measure. All biologists (but not all technical people) know that the 96-hour LC50 is the concentration that kills half the population in four days. All marine biologists (but not all biologists) know that *Salmo gairdneri* is the rainbow trout.

Like all specialists, specialists in other fields are interested almost exclusively in their own disciplines. If you are writing to them, evidently there is some overlap between your work and their discipline. The overlap determines what they need to know. Explaining *why* they need to know—showing what you offer that they can use, or what you need that they have to provide—is your job, not theirs. Of course, you don't share their expertise any more than they share yours, but you have more than half the responsibility for building bridges between the two. If you can do it, use *their* vocabulary and *their* key concepts. At least be explicit about where you think the overlap is.

Managers. We are using the word "managers" rather broadly here to mean nontechnical decision-makers. They are specialists in their own right, in fields like finance, law, public relations, and management itself, though they may well be scientific illiterates.

Unlike the first two audiences, managers don't expect to learn anything about their fields from you—and they don't want to learn any more than they have to about *your* field. Their reasons for reading are entirely practical. They care little about your theory and methods, or even about your findings. But they care a great deal about how your recommendations could further the organization's goals—increase profits, streamline production, improve the company's image, and so on. And they care equally about what they would have to do to implement those recommendations. Very likely they represent a variety of departments with different responsibilities and, therefore, have different information needs. They should be profiled carefully.

What unites all managers is their need to get their information fast. Thus it is tempting to rely on technical language and other techniques to pack as much information as possible into as little space as possible. Resist the temptation. Most won't have the background to understand you. Extremely dense writing, moreover, is hard to follow even for readers who know all the words. The typical scientific abstract, for example, is slower to read and harder to understand than the introduction that usually follows it.

The trick in writing for managers isn't condensation; it is *selection*. Decide carefully what this audience really needs to know; then take all the space you need to explain what you have to say thoroughly and interestingly.

In 1962, Westinghouse surveyed its managers, asking what they looked for in a technical report.[1] Not surprisingly, managers reported that they wanted to know (1) what the report was about, (2) the significance and implications of the work, and (3) the action called for in the report. They were also asked what parts of a report they read; their response is shown in Figure 1.2. Clearly, technical writers who lavish all their attention on the body of a report to be

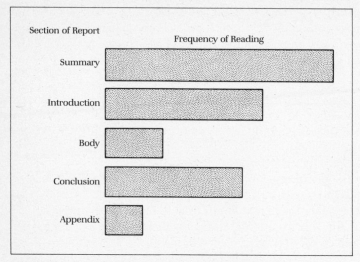

FIGURE 1.2   How Managers Read Reports.
(Reprinted by permission of the Westinghouse Electric Corporation.)

read by managers may be wasting the effort. The bodies of most technical reports take more time to read than many managers have to spare. Because managers are unlikely to change their reading habits, writers must organize their reports so that the most important information is in sections that the audience is certain to read.

Managers read technical writing with their heads full of nontechnical reservations—Will it cost too much? Can we find the necessary personnel? Will the Zoning Board oppose it? Will the Board of Directors approve it? Anticipating these concerns will help you include the information you need to alleviate them.

Technicians. Most technicians are experts, too, not in an area of knowledge, but in an area of skill. Like managers, technicians are not particularly interested in theory and methodology. Unlike managers, however, they are profoundly interested in the technical details of how to get a job done.

Like most specialists, technicians tend to be a bit contemptuous of people who don't understand the basics of their specialty. If you are going to suggest how they should do a job, you had better know the exact technical details of how it is done. In addition, technicians are more likely to *want* to get the job done if you explain clearly why the job is worth doing—and why it is worth doing your way. And they are more likely to be able to get the job done if you provide enough background so that they can do their own troubleshooting when things don't happen exactly as you predicted.

Common sense will tell you how much background technicians need. The lab technicians who will be analyzing water samples for chlorine obviously need to know what the new standard is and how to test for it. They would most likely want to know why the standard has changed as well. They would not, however, want the legislative history of the change and the details of the toxicity studies leading to the change.

Laypeople. In a sense, everyone outside your field is a layperson. But most readers have a specialized interest in one aspect or another of what you are writing about, the aspect that impinges on their own field. In the "laypeople" category, we are talking about the people who know almost nothing and need to know almost everything.

Documents that go all the way to the top of an organization or outside it are especially likely to have lay audiences, particularly if they make recommendations that might affect the community. Many reports prepared by or for the government—environmental impact statements, for instance—regularly have lay audiences. You may want—or need—to communicate with laypeople on other occasions as well: to promote good public relations, to sell them on a product or an idea, to share a useful or interesting piece of knowledge with them. And, of course, owner's and operator's manuals are usually aimed at laypeople.

It is extremely difficult to do justice to the needs of laypeople. You must keep your language simple and nontechnical, explain all specialized concepts and assumptions, and never allow the technical details to overshadow your major points.

Multiple Audiences

Life would be simpler if every member of the audience for a given piece of writing belonged to the same group—if all your readers were colleagues or managers or technicians or specialists in a related field. And often this is, in fact, the case. Virtually all letters have a single reader, and many reports are aimed at a single *group* of readers. But for many formal reports and proposals, you will wind up writing a single document for a multiple audience, addressing specialists, managers, and technicians all at once. Addressing a multiple audience is cost-efficient, certainly more efficient than writing four or five versions of a report, each tailored to the specific needs of a single group of readers. But it can also be a real chore, because with every sentence you run the risk of boring one portion of your audience while writing over the heads of another.

The solution is to organize these long reports into sections, each of which is aimed at the readers who need and can understand it. Information that all readers need goes in the front end of the report, and the specialized sections are listed in a table of contents so that readers who need them can find them quickly. We discuss the structure of long, complex reports in Chapter 13, "Technical Report Format."

Exercise 1.1

The passages that follow are adapted from a variety of technical reports. On the basis of the content and language of each passage, decide which of our five audience groups would find that passage suitable and which would not. (Of course, the passage may be clearly suitable or unsuitable for more than one group.) Here again are the five groups: (a) specialists in your field, (b) specialists in other fields, (c) managers, (d) technicians, and (e) laypeople. Write the letters of the groups you think should or should not see each passage.

1. Good for _____

   Bad for _____

   Hat sections of steel should be formed and welded to the back of the housing for stiffening and to provide support for collector suspension on the support structure. Figures 3.2 and 3.3 show the proposed housing design details.

2. Good for _____

   Bad for _____

   Early erosion measurement in Ocean City, New Jersey, was based on sand samples taken at various points: along the berm crest in December 1929; along the mid-tide line in August 1932 and July 1935; along the mid-tide line and offshore in March and October 1949; along the high-water line in July 1950; and at 12-, 18-, 24-, and 30-foot depths in June 1953.

3. Good for _____

   Bad for _____

   In addition to the technical reasons we have noted for proceeding slowly in the adoption of solar heating for in-plant use, several nontechnical arguments also suggest caution. The main concern is that federal tax policy is now under reconsideration as it applies to deductions for energy-saving modifications, and delay may well lead to increased savings under forthcoming tax laws.

4. Good for _____

   These findings concerning the adsorption of methyl mercury are

Bad for _____   consistent with the models proposed by Berglund and Berlin, confirming their finding that adsorption is an important factor in the toxicity of alkyl mercurials.

5. Good for _____   Despite its significant advantages, the proposed modification will

   Bad for _____   seriously increase the hazard of explosion in manufacturing, storage, and transportation. Chapter 5 assesses these safety risks and proposes ways to minimize them.

6. Good for _____   The pressure relief valve (PRV) on this apparatus functions like the

   Bad for _____   cock on an old-fashioned tea kettle, literally "blowing off steam" whenever the pressure exceeds a predetermined level. The appropriate level for the PRV depends on the size of the vessel (see Table 2.1).

Verifying Your Audience Analysis

Despite all the suggestions we have made, you may find it difficult to profile your audience accurately. Even if you are reasonably certain that you now know who your readers are, you can never be certain that you know *exactly* and *absolutely* what they want or need. Thus, we have two additional suggestions.

Consult with Your Audience. Most technical people welcome the opportunity to sketch out in advance the questions they need answered. Delineating the questions they need answers for is much easier and much more productive than doing without the answers. So if you are a little vague about the information needs of a particular audience, by all means *ask*.

Ask Test Readers to Examine Your Rough Draft. After you have started writing, try to find a few typical audience members—test readers—to read the relevant portions of your draft. Here again, most technical writers are too diffident; if they ask anyone for comments, it is usually a close colleague who shares their biases and blind spots. Pick test readers who share your *audience's* biases and blind spots instead. Stress that you are not looking for comments on technical accuracy (unless you are), but rather for advice on organization and style. It helps to ask your readers specific questions like these:

1. Is the purpose clear at all times? Please mark any sections where it is not clear *why* I am talking about a particular topic.
2. Is the information itself clear? Please mark any concepts or explanations that need clarification.
3. Is the content relevant? Please mark any sections that deal with information you don't need to know.
4. Is the organization helpful? Please indicate where the structure seems unclear or disorganized and where you have trouble.
5. Is anything missing? Please tell me what you need to know that isn't here.

Make sure your test readers recognize that *you*—not they—are being tested. "Read this and see if you understand it" puts the readers on trial, and most will respond by pretending to understand more than they do. "Show me where this gets fuzzy" prompts a much more helpful response.

▶ Making Choices Based on Goal and Audience: A Case in Point

By now you shouldn't need convincing that your goals and your audience determine what ends up in your report, letter, or memo. Writers who can adapt their material to fit different information needs or to achieve different goals are the ones who succeed. Such adaptations are not intuitive; they are *technical*. In the next few pages, we will analyze in some detail just *how* such adaptations were made by a professional writer—in this case, an eminent psychologist. Read through the following selections, both taken from articles written by Stanley Milgram, a professor at Yale. The topics of the two articles are the same. The goals of the two articles, however, are *not* the same, because the intended audiences are quite different. The first passage, which originally appeared in an article prepared for the *Journal of Abnormal and Social Psychology* in 1963, is intended for an audience of professional psychologists. The second, which appeared ten years later in an article for *Harper's Magazine*, aims at the general, college-educated reader.

A.   This article describes a procedure for the study of destructive obedience in the laboratory. It consists of ordering a naive S to administer increasingly more severe punishment to a victim in the context of a learning experiment. Punishment is administered by means of a shock generator with 30 graded switches ranging from *Slight Shock* to *Danger: Severe Shock*. The victim is a confederate of the E. The primary dependent variable is the maximum shock the S is willing to administer before he refuses to continue further. 26 Ss obeyed the experimental commands fully, and administered the highest shock on the generator. 14 Ss broke off the experiment at some point after the victim protested and refused to provide further answers. The procedure created extreme levels of nervous tension in the Ss. Profuse sweating, trembling, and stuttering were typical expressions of this emotional disturbance. . . . The variety of interesting behavioral dynamics observed in the experiment, the reality of the situation for the *S*, and the possibility of parametric variation within the framework of the procedure point to the fruitfulness of further study.

. . . As the experiment proceeds the naive subject is commanded to administer increasingly more intense shocks to the victim, even to the point of reaching the level marked *Danger: Severe Shock.* Internal resistances become stronger, and at a certain point the subject refuses to go on with the experiment. Behavior prior to this rupture is considered "obedient," in that the subject complies with the commands of the experimenter. The point of rupture is the act of disobedience. A quantitative value is assigned to the subject's performance based on the maximum intensity shock he is willing to administer before he refuses to participate further. Thus, for any particular subject and for any particular experimental condition, the degree of obedience may be specified with a numerical value. The crux of the study is to systematically vary the factors believed to alter the degree of obedience to the experimental commands.[2]

B.   I set up a simple experiment at Yale University to test how much pain an ordinary citizen would inflict on another person simply because he was ordered to by an experimental scientist. Stark authority was pitted against the subjects' strongest moral imperatives against hurting others, and, with the subjects' ears ringing with the screams of the victims, authority won more often than not. The extreme willingness of adults to go to almost any lengths on the command of an authority constitutes the chief finding of the study and the fact most urgently demanding explanation.

In the basic experimental design, two people come to a psychology laboratory to take part in a study of memory and learning. One of them is designated as a "teacher" and the other a "learner." The experimenter explains that the study is concerned with the effects of punishment on learning. The learner is conducted into a room, seated in a kind of miniature electric chair. . . . He is told that he will be read lists of simple word pairs, and that he will then be tested on his ability to remember the second word of a pair when he hears the first one again. Whenever he makes an error, he will receive electric shocks of increasing intensity.

The real focus of the experiment is the teacher. After watching the learner being strapped into place, he is seated before an impressive shock generator. The instrument panel consists of thirty lever switches set in a horizontal line. Each switch is clearly labeled with a voltage designation ranging from 15 to 450 volts. . . .

The teacher is a genuinely naive subject who has come to the laboratory for the experiment. The learner, or victim, is actually an actor who receives no shock at all. The point of the experiment is to see how far a person will proceed in a concrete and measurable situation in which he is ordered to inflict increasing pain on a protesting victim.

Conflict arises when the man receiving the shock begins to show that he is experiencing discomfort. At 75 volts, he grunts; at 120 volts, he complains loudly; at 150, he demands to be released from the experiment. As the voltage increases, his protests become more vehement and emotional. At 285 volts, his response can be described only as an agonized scream. Soon thereafter, he makes no sound at all.

For the teacher, the situation quickly becomes one of gripping tension. It is not a game for him; conflict is intense and obvious. The manifest suffering of the learner presses him to quit; but each time he hesitates to administer a shock, the experimenter orders him to continue. To extricate himself from this plight the subject must make a clear break with authority.[3]

In both these passages, Milgram explains the basic procedure of his experiment and summarizes his results. But the first passage sounds more "scientific," more "technical." Milgram has made conscious choices to achieve this difference in sound. He has made changes in his choice of vocabulary, in degree of detail, in tone, and in emphasis.

Choice of Vocabulary

Probably the most visible difference between the two articles is their vocabulary. The journal article is filled with the technical vocabulary expected by its technical audience; for instance, "S" and "E" replace "subject" and "experimenter." Terms like "primary dependent variable," "parametric variation," and "behavioral dynamics" would probably be unclear to most general readers, but other psychologists know what they mean. This isn't merely a case of using big words where little ones will do: These technical terms are a kind of professional shorthand that assumes an understanding of underlying theories of behavioral psychology and psychological experimentation. Indeed, other psychologists are so accustomed to such jargon that they would probably be surprised and suspicious if Milgram *didn't* use it.

Highly technical vocabulary would, however, be out of place in the second passage. Some readers of *Harper's* are conversant with the jargon of behavioral psychology, but most are not. Knowing that his shorthand might produce confusion rather than understanding, Milgram has left it out. Thus, where the professional reads:

The primary dependent variable is the maximum shock the S is willing to administer before he refuses to continue further

the *Harper's* audience reads:

The point of the experiment is to see how far a person will proceed in a concrete and measurable situation in which he is ordered to inflict increasing pain on a protesting victim.

It takes thirty-one words to tell the general readers what the professional reader can understand in twenty. In general, explaining a technical procedure, device, or idea to a lay audience takes longer than explaining the same to a technical audience.

Moreover, when writing for *Harper's*, Milgram is careful to remove not only the easily recognizable technical terms like "parametric" but also the technical terms that a general audience might not realize *are* technical. These include words like "confederate," "rupture," and "condition," all of which have specific meanings in psychology that are not the same as their everyday meanings. In common usage, for instance, to say that someone is a "confederate" may mean only that the person is a friend, but it may also carry the suggestion that the person is involved in some shady enterprise. In psychology, the term means a person working for the experimenter who pretends to be a subject. Similarly, a lay audience might not realize that the word "condition," which has a welter of meanings in everyday speech, refers in psychology to the experimental technique of making quantifiable modifications in the way a procedure is set up. To avoid possible confusion or misinterpretation, Milgram replaces the word "confederate" with the phrase "an actor who receives no shock at all." Again, conciseness must yield to clarity.

Degree of Detail

Technical terminology not only is more dense and economical than workaday prose, but it is also a great deal more precise. "More economical" need not mean "fewer words." Look again at the sentence we just quoted from the *Harper's* article. Once Milgram says that the experiment provides a "concrete and measurable situation" in which to observe obedience behavior, he never again mentions the actual data-gathering process. He tells his audience that subject responses were "measurable" and leaves it at that. For his colleagues in psychology, he is much more explicit:

> A quantitative value is assigned to the subject's performance based on the maximum intensity shock he is willing to administer before he refuses to participate further. Thus, for any particular subject and for any particular experimental condition, the degree of obedience may be specified with a numerical value.

In this case, what the general audience gets in four words, the professional audience gets in *forty-six*. Of course, they aren't getting the same amount or kind of information; as we have seen, the general audience is being told only that measurements were taken whereas the professionals are given a detailed methodology. Nevertheless, each audience is getting the information it needs. The general audience is interested in the *outcome* of the experiment, in how many people were willing to follow orders to hurt others and why.

General readers aren't likely to rush out and duplicate Milgram's experiment. The chances are that Milgram's fellow psychologists won't retire to their own laboratories and duplicate it either. But they probably *do* want to duplicate it in their heads. They want enough information to allow them to visualize the experiment all the way through, so they can decide for themselves if Milgram's work is logically airtight, statistically valid, and properly executed. Lay audiences usually take procedural matters on faith; professional audiences cannot afford to. Knowing this, Milgram gives them the details they need.

Tone

As you can see, Milgram has made appropriate choices concerning both his vocabulary and the kind of information to include. He has also made choices about his tone. Compare these two sentences:

> 26 *S*s obeyed the experimental commands fully, and administered the highest shock on the generator.

> Stark authority was pitted against the subjects' strongest moral imperatives against hurting others, and, with the subjects' ears ringing with the screams of the victims, authority won more often than not.

The first sentence is both more precise and more concise than the second. In half the number of words, it says exactly how many subjects were fully obedient whereas the second says only that a majority of the subjects obeyed. The second sentence, however, is infinitely more exciting to read; it is action-packed, vividly descriptive, charged with emotion. We are not merely told that the subjects obeyed; rather, we are shown a nightmarish picture of suffering that emphasizes the subjects' moral dilemma. The second passage abounds with such strong, vivid language. Clearly, Milgram wants to engage his lay readers' *emotions* as well as their minds. He wants them to identify with his subjects; he wants them to feel the tension that his subjects felt as they poised their hands over his electric shock console; he wants them to consider how they would respond if they were in his subjects' shoes.

This sort of descriptive, emotional language is noticeably absent in the article written for his fellow psychologists. In the *Journal of Abnormal and Social Psychology*, Milgram's writing is straightforward and emotionless, presenting the facts, the results, and the conclusions unadorned. If he wants the readers of the *Harper's* piece to identify with the subjects, he wants his colleagues to identify with the experimenter, the emotionless observer.

Professional readers often regard subjective qualities in the style as transgressions, as failures to maintain scientific detachment and, consequently, as evidence of bias. Thus, Milgram is careful to keep his tone unemotional. Look again at the sentence in which he describes his subjects' extreme anxiety:

Profuse sweating, trembling, and stuttering were typical expressions of this emotional disturbance.

He might have written the sentence in other ways:

Many of the subjects sweated profusely, trembled, or even stuttered.

Many of the subjects showed their emotional tension by sweating profusely, trembling, or stuttering.

So great was their emotional tension that many of the subjects began to sweat profusely, to tremble, even to stutter.

The unrelenting tension generated by the experience reduced many of the subjects to sweating, trembling, stuttering wrecks.

As they agonized over their situation, many of the subjects broke out in cold sweats, trembled visibly, or stuttered incoherently.

All these sentences contain the same factual information as Milgram's, so why does his version sound more "objective"? One of the major differences between his original and our revisions is that in his sentence there is no mention of *people*. The grammatical subject of his sentence is "sweating, trembling, and stuttering," and this subject is followed with the lackluster verb "were." Not people, but *behavior patterns*, are emphasized. In its barest form, Milgram's sentence tells its readers that "certain behaviors were typical."

In each of our versions, the focus is on the people who exhibited these behaviors. In fact, in four of the five versions, the phrase "many of the subjects" is the subject of the sentence. And in all our versions, the verbs are forceful and concrete. Unlike "were," which is emotionally neutral, verbs like "sweated," "trembled," and "agonized" carry an emotional charge. In their barest form, then, our versions show their readers that "people suffered." What Milgram has done, in other words, is select a subject-verb relationship that effectively neutralizes the emotional charge of his sentence. The technique that he uses here is called "smothering the verb," and we discuss its use—and *misuse*—in Chapter 12.

Emphasis

We have pointed out to you that our rewrites of Milgram's sentence sound less objective than his original version because ours emphasize people whereas his emphasizes certain behavioral conditions. A similar difference in emphasis exists throughout the two articles. You can see this difference clearly by comparing the following two lists. What we have done is reduce the first ten sentences of each article to their grammatical essentials—their subjects and verbs.

Journal of Abnormal and Social Psychology

1. This article describes
2. It consists
3. Punishment is administered

4. The victim is
5. The primary dependent variable is
6. 26 *Ss* obeyed
7. 14 *Ss* broke off
8. The procedure created
9. Profuse sweating, trembling, and stuttering were
10. The variety of interesting behavioral dynamics . . . and the possibility of parametric variation . . . point

Harper's Magazine

1. I set up
2. Stark authority was pitted
3. The extreme willingness constitutes
4. Two people come
5. One is designated

6. The experimenter explains
7. The learner is conducted
8. He is told
9. He will receive

10. The real focus is

In the first list, the emphasis is clearly on the design, procedure, and results of the experiment. The structure of the sentences draws the reader's attention to what was done, how it was done, and why it was worth doing. In the second list, the reader's attention is focused on people *doing* things. Note the number of sentences that have a human subject. Moreover, by reading just these subjects and their verbs, you can get a hint of suspense, the thread of a story about two people who come into a lab and are manipulated by an experimenter. Even the title, "The Perils of Obedience," is designed to pique readers' curiosity—it sounds vaguely ominous. The title of the journal article—"A Behavioral Study of Obedience"—lacks "punch," but, as we have seen, Milgram's colleagues don't want punch; they just want the facts.

Our examination of the many choices that Milgram has made in preparing these two articles shows how careful and deliberate he has been to put his writing in phase with his audiences' differing backgrounds and needs. This phasing together of the readers' needs and the writer's goals is what produces relevant, reliable, and comprehensible writing. Our students are always astonished to learn that Milgram wrote both of these articles himself. But it is not really so surprising; we told you before that technical people make the best technical writers. No one knows his own experiments better than Milgram, and no one cares about their implications more deeply. To disseminate his information, he has learned to design his prose as exactly as he designs his experiments. Once you have figured out, as Milgram has, who will be reading what you write, why they will be reading it, and what you want them to learn from it, you can learn to adapt it successfully.

Review Questions

1. Name several overt goals that a report might have.
2. What are organizational goals and personal goals, and how do they differ from overt goals?
3. Why is it important to link each of your goals to a specific audience?
4. What do *all* technical readers look for in a technical document?
5. How do you determine what a given audience *needs*?
6. What is the relationship between most readers' needs and knowledge?
7. What distinguishes managers from specialists? What types of information are of most interest to each?
8. How do technicians differ from managers? From specialists in your field?
9. What is a multiple audience, and how do you deal with one?
10. If you aren't certain that you understand the information needs of your audience even after you have analyzed it carefully, what should you do?
11. Why is technical writing usually more concise than nontechnical writing?
12. In what sense is technical writing usually more precise than nontechnical writing?
13. Explain how the style and tone of a piece of writing are affected by its goal and audience.

Assignments

1. Find a technical article that you admire from a publication for nontechnical readers, such as *Discover*, *Science 85*, or the science section of *The New York Times*. Then find at least one article on the same topic in a technical journal. Give three examples each of differences in vocabulary choice, degree of detail, tone, and emphasis; then add any other important differences you find.

2. You must write a quarterly progress report on your team's research project, but unfortunately little progress has been made, chiefly because the purchasing department ordered the wrong equipment. The report will be read by your direct supervisor, who understands the delay; by the purchasing manager, who has already apologized and ordered the correct equipment; and by the section head (the boss of both supervisors), who knows nothing yet of the delay. List your goals for the report, with each goal connected to the appropriate audience or audiences.

3. Imagine that you have been assigned to write a description of a complex piece of technical equipment. For each of the following audiences, identify the goals of the description, the relevant audience characteristics, and the implications of both for the description you would write: (a) a technician who must use the equipment for the first time; (b) a company you have hired to pick up the equipment and deliver it to another address; (c) a manager reviewing your recommendation to purchase the equipment in preference to

its competitors; (d) the department that will repair the equipment whenever problems arise; (e) colleagues reading your report of research results obtained in using the equipment.

Notes

1. Richard W. Dodd, "What to Report," *Westinghouse Engineer* 22 (July–September 1962), pp. 108–111.
2. Stanley Milgram, "A Behavioral Study of Obedience," *Journal of Abnormal and Social Psychology* 67:4 (1963), pp. 371–379.
3. Stanley Milgram, "The Perils of Obedience," *Harper's Magazine* 247 (December 1973), pp. 62–66. Abridged and adapted from *Obedience to Authority*, by Stanley Milgram, copyright © 1974 by Stanley Milgram, by permission of Harper & Row, Publishers, Inc.

2 Gathering and Organizing Information

Evaluating Information

Evaluating the Reliability of Your Sources
Evaluating the Reliability of Your Information
>   Facts
>   Inferences
>   Opinions

Organizing Information

Formulating Your Main Point

The Thesis Statement
The Statement of Content
The Thesis Statement as Unifying Device

Identifying Your Readers

Grouping Your Data

Listing the Major Points
Sorting Your Notes
Making Groups
Labeling the Groups
Combining the Groups
Double-checking Your Groups

Putting the Groups in Order

Natural Orders of Presentation
>   Chronological Order
>   Spatial Order

Logical Orders of Presentation
>   Order of Decreasing Importance
>   Order of Increasing Importance
>   Other Hierarchical Orders
>   General-to-Specific Order
>   Specific-to-General Order
>   Comparative Order

The Matrix Problem
Organizing for Multiple Audiences

Outlining

By now you should be convinced of the cardinal importance of analyzing your audience and your goals before you begin writing. Many novice writers skip that vital *pre*writing step in their rush to begin accumulating information, and they end up lavishing hours on irrelevant, superfluous, or misdirected research. Instead, you should approach the universe of information on your topic selectively, with a clear idea of who your readers are and what they need to know. When you run across a fact or an idea in a source, you should be able to ask yourself: Do my readers need to know this? Will knowing this help my readers to perform their complex and specialized jobs more effectively? Do I want them to know it? Of course, you cannot answer those questions until you know who your readers are and what their task is. Therefore, analysis of goal and audience is a necessary prelude to all technical writing.

Technical writing itself begins, however, with data collection. Like good scientists, good writers go through this laborious process before they ever venture a hypothesis or draw a conclusion. The information-gathering process may be a simple matter of thinking and jotting down notes. At the other extreme, information gathering for a technical report may require elaborate and extensive research that could take years. Formal documents on technical topics require formal information-gathering procedures, and even an informal document like an interoffice memo isn't created in a vacuum; it starts as a list of the points you want to make and the facts you want to mention.

The more systematically you go about assembling your material in the first place, the easier it will be to bring order out of chaos when you sit down to write. In this chapter, we will discuss both these steps in the technical writing process—first how to gather information efficiently, then how to organize it coherently.

▶ Gathering Information: Where to Start, When to Stop

Two questions pervade the information-gathering process: "Where do I start?" and "When do I stop?" You may think that you cannot answer the second question until you are well into the process, but in fact you must answer both in part before you begin—that is, you must *set an agenda* for collecting your data.

First of all, you must decide realistically how long it will take you to write the report: first draft, revised draft, final copy. Then you should automatically double your estimate (writing always takes much longer than you expect) and set the first day of the report-writing period as the absolute deadline for information gathering. Then select a scope that suits the time frame.

Setting the Scope

When you set the scope of an inquiry, you decide what sources of information you will and will not consult. On any topic, more information exists than you could possibly unearth, let alone ingest, in a few days or a few weeks; therefore, you must concentrate your investigation on those sources whose information is most relevant to your readers' needs. Begin by examining your list of goals. For each goal, list the kinds of information that will best achieve it. Will facts, eyewitness accounts, or authoritative opinions most strongly sway your readers? Will you need to go to the library for journal articles and government documents? Will you need to interview authorities in the field, to take a survey, to run a series of laboratory experiments, to examine a site firsthand?

Ask yourself how rigorous your audience will expect you to be in your pursuit of information. Will you need to check all the existing sources or just a representative sample? Will you need to get all the facts, most of them, or a select few? Must you avoid bias and present both sides of the issue, or can you admit to bias and present only information that supports your point of view? Must you be precise or just close? Decide *how much* information you will need to achieve each goal. You want to convince your readers, but you don't want to inundate them. If you are writing a research report or a journal article, your readers will expect your review of the relevant literature to be exhaustive. If your goals are more modest and your audience is less demanding, a meticulous analysis of the pertinent data, complete with table after table of exact numerical breakdowns, might bore them, confuse them, and—worse yet—irritate them. If your readers want only ball-park figures and rough estimates, that's all you need supply. You will, no doubt, revise your scope as you gather your information—perhaps narrowing it, perhaps broadening it— so do not feel wedded to these preliminary estimates. You will need them, however, if you are to set a realistic agenda for getting your report completed on time.

Setting an Agenda

In order to use your information-gathering time efficiently, you are going to have to set yourself a schedule and force yourself to abide by it. Take your preliminary list of potential information sources and make two lists: a list of prospective sources in order of importance and a list of sources in order of accessibility (from the least to the most accessible). You'll want to begin with the most important sources, so that if you reach your deadline to begin writing without having considered every source that you wanted to consider, you will at least have considered all the sources with the highest priority. As you gather information, important sources will turn up that you didn't know about before you started, so continue to revise this list as you collect your data, always adding in *more* important sources ahead of *less* important ones.

As you establish a schedule for tracking down sources, you'll have to consider not only their importance but also their accessibility. Make arrangements early to obtain hard-to-locate materials; while you are waiting for them to arrive, you can peruse the sources close at hand. You may also want to build some personal priorities into your agenda: You may want to start with the most up-to-date sources and work your way back, or to start with less technical sources and work up to the most highly technical ones, or to start with the most general sources and move toward more specialized sources as you gain knowledge.

In the next several sections of this chapter, we will provide you with a general orientation to the kinds of information available to technical writers and to the places where that information is most likely to be found. We will begin with an information source you probably undervalue—yourself—and with an information-gathering technique known as "brainstorming."

 ## Brainstorming

Brainstorming is thinking out loud and on paper. Every writing project should begin with it, so start the information-gathering process by getting out a clean notepad and emptying your mind of everything you can think of that ought to be included in your report. If you are writing a progress report, for example, jot down what you have done since the last report and what you plan to do before the next one. If you are working on a more ambitious document—for instance, a final report to corporate management on the results of a three-year study—list the important subject areas that you think ought to be covered in the report; note down the conclusions you reached; set down in writing your reservations and your unanswered questions; record the results that you'd like to highlight, your recommendations, your suggestions for future research. Don't try to edit your thoughts; just write down whatever comes into your mind.

Don't try to separate specific details from broad generalities, and don't leave *anything* out, no matter how trivial it seems or how chaotic your list begins to look. Brainstorming is a debriefing process, a transfer of ideas from mind to paper, and it works best when you relax and free-associate, letting one idea lead to another and recording them as they come along. Later on, you can sift through your ideas and decide which ones are worth using. Now the important thing is for you to remove your knowledge from storage in your memory and place it in written storage, from which it can more easily be retrieved. If you stall temporarily, read over your list; that usually suggests new items. When you can think of nothing more, put the list in an accessible place so you can add to it later.

Ideally, brainstorming is a collective process in which ideas are spontaneously generated through intensive group discussion. Ask interested colleagues what they think ought to be included in your report. If they cannot think of anything right away or if the discussion starts to bog down, show them your list, and ask them what they think ought to be added. Make appointments to discuss the contents of your report with people who may have ideas to contribute. Set up meetings of interested parties and encourage freewheeling discussion. (For advice on how to conduct such meetings, see Chapter 18, "Oral Reports and Group Meetings.") You will find that, in a work environment, almost every project report begins with a discussion session headed by the project leader. Obviously, the more complex the project, the more general the discussion (and your notes) will tend to be. Don't worry— you will be able to fill in the details from documents that are available to you firsthand, and it is to those documents that you should turn next.

Consulting Pertinent Documents

Almost no final report ever gets written that wasn't preceded by a preliminary and a final proposal, a series of progress reports, an exchange of correspondence among participants on the project, and various interim reports on the data. Get out the existing files and sort through them. Copy down relevant details from earlier reports on the project. Cull out sentences and paragraphs that you can use in your final report more or less as they stand.

If you are writing an evaluation or a recommendation report, you will also want to consult any original documents that will help you make your evaluation and provide evidence to support it. If you are awarding a contract, you will want to examine and compare the written bids submitted by each competitor. If you are hiring a new associate, you will want to examine and compare dossiers and personnel files. Where no documents exist, solicit them. If you are doing a feasibility study, ask for comments and recommendations from all interested parties. Ask people within your organization to put their ideas and opinions in writing for you. Write to people outside your organization who might have worthwhile information to contribute. Then organize all the materials that you have used into a new file that will serve as a repository of information on the project and a resource for others working on it after you.

Conducting Primary Research

Much of the information for your technical writing will naturally come from your own firsthand research. You will be writing about what you found in your laboratory experiments or field investigations. Most of this primary research (that is, research based on your own observations) will already be recorded in laboratory notebooks, progress reports, and similar documents. Thus, when you gather the documentary sources relevant to your topic, you will be gathering both your own primary research and the research of others.

Sometimes it is appropriate to do additional primary research, especially for a technical report or other piece of technical writing. For example, you may want to interview experts to tap their firsthand knowledge and authoritative opinions on your subject matter—or to solicit their bibliographic suggestions. You may even want to conduct an informal survey, asking the same questions of all respondents and tabulating their answers. (A *formal* survey is a substantial undertaking, requiring as much expertise as any other scientific methodology.)

If you decide to conduct any additional primary research before you write, decide early and begin early. Surveys are extremely time-consuming, and even a single interview must be arranged well in advance. If you expect the interview to open up new paths for library research, all the more reason to do it early.

Conducting Library Research

The principal information sources for most technical writing are primary research and documents from the files. But technical writers must often use the library to nail down an elusive statistic, the derivation of a key formula, the source for a borrowed methodology, and so on. And some kinds of technical writing—especially formal proposals, journal articles, and research and development reports on technical topics—require substantial research in the technical literature. Sometimes they require a formal literature review, sometimes only an informal discussion of previous work on the topic, but they all require that you have more than a passing acquaintance with the library. Your employer will expect you to have learned library research skills in school. A course in technical writing is an excellent place to hone those skills.

Compiling a Bibliography

Your college or university library has an array of sophisticated accessing tools for locating original materials; these tools are the ones that professionals use when they compile a bibliography on a research topic. In the professional world, the card catalog is rarely useful because it contains only books, and published bibliographies are only occasionally useful because they are inevitably out-of-date. Professional bibliographic work begins with indexes and abstracts.

Indexes and Abstracts. Looking through indexes and abstracts is called "searching the literature," and it is standard procedure in all professional quality research. Some disciplines even publish guides to their fields, such as the American Chemical Society's *Searching the Chemical Literature*. When you go to the reference room in your library, you will see right away why such guides are necessary. You will find wall after wall of comprehensive indexes to every imaginable subject area. There are two kinds: broad indexes that cover an entire discipline and specialized indexes that address themselves to a specific topic. Following is a partial list (there are hundreds):

General Indexes and Abstracts

Applied Science and Technology Index
Bibliography of Agriculture
Biological Abstracts
Biological and Agricultural Index
Bioresearch Index
Business Periodicals Index
Chemical Abstracts
Computer and Control Abstracts
Current Abstracts of Chemistry and Index Chemicus
Current Index to Conference Papers in Chemistry
Current Index to Conference Papers in Engineering

Current Index to Conference Papers in Life Sciences
Current Index to Conference Papers: Science and Technology
Current Papers in Electrical and Electronic Engineering
Current Papers in Physics
Current Papers on Computers and Control
Education Index
Electrical and Electronics Abstracts
Engineering Index
Index Medicus
Pandex Current Index of Scientific and Technical Literature
Physics Abstracts
Psychological Abstracts

Topical Abstracts

Animal Breeding Abstracts
Apicultural Abstracts
Corrosion Abstracts
Environmental Abstracts
Gas Chromatography Abstracts
Genetics Abstracts
Institute of Petroleum Abstracts
Marine Fisheries Abstracts
Metals Abstracts
Nuclear Science Abstracts
Pollution Abstracts
Solid State Abstracts

Searching through the indexes and abstracts for prospective sources is a tedious process because of the sheer volume of material, much of it irrelevant. Nearly all indexes and abstracts are now prepared by computer, so they come out quite frequently (most are monthly with annual cumulations), and they are remarkably complete. This means that although there may be no dearth of information on your topic, there may also be far more information than you feel equipped to handle. Given the enormous increase in scientific and technical information in the last several decades and the strong likelihood that such information will continue to increase exponentially, you will have to limit your information-gathering efforts. Unless you further narrow your topic, the most sensible way to do that is to restrict the number of entries in your initial bibliography to a manageable number. The most reputable way to restrict is on the basis of currency. Choose journal articles over books and very recent work over earlier work.

Even if you limit in advance the number of sources that you will include in your bibliography, sifting through the indexes and abstracts for just those sources will still be time-consuming. Set aside several days with long, uninterrupted spans of time. Make yourself a list of words or phrases that you think you will be able to use as subject headings. (The *Library of Congress*

Subject Headings can help suggest alternatives.) The narrower and the more technical you make these "key terms," the more likely it is that you will not waste time with irrelevant sources. When you have arrived at key terms that you think are narrow enough to begin with, you are ready to start "searching the literature."

Searching the Literature. If your topic is one that enabled you to come up with very narrow key terms, you can afford to begin your search in the broader indexes and abstracts; they won't have a paralyzingly large number of entries under your key terms. But often your key terms are necessarily broad, even after you have narrowed them as much as possible. To avoid information overload, then, search in the most specialized index or abstract you can locate that is likely to include material on your topic.

Suppose you are investigating the relationship between cholesterol and heart disease as part of your company's proposal for a new diet food. Pass up *Index Medicus* and *Biological Abstracts* for now. *Biological Abstracts* comes out every two weeks, and each volume is about the size of the Manhattan telephone book. There is a great deal more there than you need, so look for an index with a narrower scope—*Nutrition Abstracts and Reviews*, for instance. Look up "cholesterol," and make a bibliography card for every book or article that looks promising.

One or more of the entries under "cholesterol" ought to tell you which diseases of the heart or the circulatory system or both (for example, athero-sclerosis) have been linked to cholesterol levels in the blood. That's good, because "heart disease" would have been too general a term to look up, especially in a computer-generated index. Such indexes are prepared according to "descriptors" or "search terms." Authors indicate several key words (usually four) under which they want their work to be subject-indexed. Researchers must use these same words to retrieve the names of articles from computer storage. In the case of articles and books written before the era of computer retrieval, programmers take descriptors from the titles. The problem, of course, is that if you want to look anything up, you have to use the same term for it that the authors did. Generally speaking, this means you should use scientific terminology. *Biological Abstracts* even includes a genus/species listing in the back so you can locate information under both *Odocoileus virginianus* and "white-tailed deer."

"Cholesterol," luckily, is a scientific term, and once it has led you to "atherosclerosis," you can start to breathe more easily. The narrower the key word, the fewer the number of entries. With "atherosclerosis," you might even be able to go back to *Index Medicus* to look for articles without having to spend days reading and copying the entries in one volume. If you check under *both* "atherosclerosis" and "cholesterol" in several indexes and abstracts for the last several years, you should pick up most of what has been published relevant to your paper. You may even be able to save yourself some time if your library has *Science Citation Index*. This index is based on *two* key words;

you look up articles that have *both* "atherosclerosis" and "cholesterol" in the title, not just one or the other. If you have to do it by hand, look for "cholesterol" under "atherosclerosis" and vice versa, and keep backtracking. If you find only a few sources, go back a number of years; if you find several the first place you look, you may need to go back only four or five. When you have accumulated ten or so recent sources, read through them, and check their bibliographies. These recent sources will cite most of the key older ones, and you can locate them more quickly through citation than you can by working through abstracts. Moreover, you will learn which of the older sources are considered authoritative by recent researchers. Unless (and this is unlikely) the sources peter out entirely, you will have to decide arbitrarily when to stop collecting sources and start collecting data.

Assembling a preliminary bibliography is a painstaking process, and there is only one way we know to simplify it: Use a computer yourself. Many large university libraries now have tie-ins to national data bases on scientific and technical topics. MEDLARS (at the National Library of Medicine), AGRICOLA (from the National Library of Agriculture), and SCISEARCH (from the Institute for Scientific Information) are among the data bases now being used by both students and professionals. Computer searches are the most efficient means of obtaining the very latest bibliographical information, but libraries do charge for them.

Using Abstracts. A publication like *Nuclear Science Abstracts* is actually an index, but it is called an "abstract" because, in addition to bibliographical information, it provides a short, highly concentrated summary of each book and article that it names. Abstracts are extremely useful to anyone gathering material from library sources, because titles alone can be misleading. Instead of running from stack to stack in the library, you can read through compilations of abstracts like *Solid State Abstracts* and *Biological Abstracts* and then seek out only those articles that sound most promising. Even if you decide not to seek out an article or if you cannot find it, you may still use the abstract. Certainly, you cannot evaluate the work on the basis of the abstract alone, but at least you have *some* information about the study—its authors, the date, the hypothesis, the methodology, the results. In your bibliography, cite the abstract, not the study you never found.

Guides to Government Documents. The United States Government Printing Office is the world's largest publisher, and U.S. government documents are a major information source used extensively by professional researchers. Because many university libraries are government repositories (institutions to which government documents are sent automatically upon publication), university students frequently have access to a wealth of information published under government auspices. The U.S. Government Printing Office publishes two comprehensive indexes to government documents. One is *A Monthly Guide to U.S. Government Publications*, whose detailed subject index can give you access to such documents as testimony at federal hearings, reports by cabinet departments and independent federal agencies, and technical reports by government scientists. A corollary publication, *U.S. Government Research Reports*, provides a subject index to all reports by individuals and corporations holding government contracts. Sponsored research is made public through such annual reports, and scientists working under federal grants report their results to their sponsors as well as to professional journals. In fact, if you pore through the government documents, you can sometimes pick up on work in progress before it hits the journals. Every library has its own way of subject-indexing government documents. Some libraries include them in the card catalog, and some have a separate subject listing for government documents alone. Ask your reference librarian.

This last statement bears repeating. Reference librarians are the most underutilized of all library personnel. They are experts in information retrieval, and they know exactly what the library they work in has and does not have. Ask them—they will be glad to help you.

Collecting Data

After all this work, you will be ready to *start* doing research. We aren't going to teach you how to take notes, but we do want to point out the three major kinds of notes: verbatim quotations, summaries, and records of your own ideas and opinions. Each of these three types serves a different purpose.

Quoting. When you take a single specific fact or inference from a source, copy it down word for word. In most technical reports you will use virtually no verbatim quotes, but will paraphrase your sources instead. But you should do your paraphrasing *while you are writing the paper* because taking down information and writing it up are different intellectual processes. If you take notes in your sources' own words, two weeks later you will know exactly what they said, not just what you *thought* they said.

Summarizing. In addition to recording the information in a source item by item, you will sometimes want to summarize part or all of a source on a card. For instance, you will want to record the central idea or main point of each of your sources. Even when you have transcribed all the information in an article onto dozens of separate cards, you will have no record of what the article *as a whole* was about unless you summarize it for yourself.

Recording Your Own Thoughts. As you conduct library research, you will constantly be forming opinions and making judgments about what you read. You should set those thoughts down on paper while the research is going on, before you forget them. As you read, jot down on a card any thought that enters your mind, and keep those cards interfiled with the others.

Evaluating Information

As you amass information on your topic, you should continually be asking yourself two questions: How reliable is the source of this information? How reliable is this information itself?

Evaluating the Reliability of Your Sources

One of the principal criteria for source reliability is citation by subsequent researchers. You will notice immediately as you begin to take notes on your sources that some references will turn up over and over in everyone's discussion and in everyone's bibliography. You will want to track down and study the publications of any scholar who appears to be treated in the literature as an authority in the field. *American Men and Women of Science* and similar biographical reference books will give you a fairly complete list of someone's important publications. You will find all of a scholar's book-length studies listed in the *National Union Catalogue*, a wall-sized reproduction of the entire card catalog of the Library of Congress. Every university library has these basic references, so check them to see how long a list of publications this much-mentioned scholar has to his or her credit. People who have published extensively on a topic are frequently regarded as experts; the very fact that their work continues to get published is an important indication that that work is highly regarded.

Try to examine the work of "name" scholars early in the research process. You will need to understand the seminal research on a topic in order to put subsequent research in the proper context. You will be expected to mention the seminal research (as others did), and, if you wish to disagree with widely accepted findings, do so with the utmost caution.

Not only can you learn from experience who the celebrities are, but you can also find out by looking in *Science Citation Index* and *Social Science Citation Index*. If you look up a specific book or article in either of these two reference works, you will find a list of every subsequent publication in which that book or article was mentioned. Work that has spawned other work can usually be considered sound, but, of course, you cannot tell from a citation index alone whether colleagues are citing someone's work to commend it or to criticize it.

Reliability varies among journals as much as it does among authors. Try to get a feel for the reputation of the journals you are using. Refereed journals (those whose articles are chosen by expert consultants) are considered more reliable than those whose articles are not refereed. Older, long-established journals can afford to pick and choose among manuscripts; they won't publish any but the most rigorously conducted and meticulously documented studies. Every academic discipline and profession has its major journals, and the data you take from them are likely to be welcomed as sound. Find out if the journal you are using is one of the top ones or a satellite. You might also examine the editorial policy of the journal—you will find it on the editorial page along with the directions for submitting manuscripts. Journals committed to a certain philosophy—pro-business, pro-conservation, pro- or anti-anything—are considered less reliable than those that maintain strict neutrality on political and ethical issues. This does not mean that you should not take information from them; it simply means that you should conduct some reliability tests on the information itself.

Evaluating the Reliability of Your Information

Broadly speaking, there are three kinds of information: facts, inferences, and opinions. Facts are the most reliable because what makes an item of information a fact is the overwhelming body of evidence that supports its accuracy. Take a statement like "Penicillin is bactericidal." It is more than highly probable that the statement is true, given the quantitative evidence we now possess. Bacteria have been treated billions of times with *Penicillium notatum*, and in almost every instance the bacteria were killed. You should keep in mind, however, that although it is now known that penicillin kills bacteria, that information was not always considered factual. When Sir Alexander Fleming initially reported in 1929 that penicillin was antibiotic, his idea was considered a promising speculation. Only when the theory had been exhaustively tested under laboratory conditions did it weigh in as a fact.

The difference between a fact and an inference is in the quantity of the supporting evidence; essentially, a fact is an inference that is universally accepted. Likely conclusions are extrapolated from empirical evidence; then those conclusions are tested deductively under experimental conditions. If the experimental evidence strongly supports the hypothetical conclusion being tested, researchers begin to regard that conclusion as a likely inference. (The technical definition of an inference is "a logical conclusion drawn from clear, solid evidence.") The stronger the experimental evidence, the stronger the inference. When the evidence has piled up to such a degree that the conclusion seems virtually irrefutable, an inference becomes a fact; that is the principle of validation by replication through which all "truth" in science is finally arrived at. Of course, even facts may turn out to be temporary; many of the "facts" of nineteenth-century science have been disproved by experimental evidence gathered over the course of the last hundred years, and today's facts, too, may turn out to be only partial truths based on an incomplete understanding of reality. Therefore, it is unwise for you to regard any information in your sources as the absolute truth. Maintain a healthy skepticism toward what you read, and carefully evaluate every bit of information in terms of the evidence supporting it. If there is a powerful relationship, but no evidence that is really conclusive, you are in the realm of inference. If only a possible or promising relationship exists, you can consider the information to be a matter of opinion, depending for its reliability on the authoritativeness of the person or persons holding the opinion and the number of knowledgeable persons sharing it.

As you conduct your research, it will be necessary for you to differentiate among facts, inferences, and opinions. You don't want to present opinions to your readers as facts, nor do you want to disregard the facts when they don't support your opinion. At the beginning of the research process, it's often difficult to distinguish opinion from fact; you know from experience that a persuasively worded opinion may be perceived by an uninformed audience as a statement of fact. In the process of gathering information, you develop from an interested but uninformed reader to a very well-informed one. At the center of that development process is an increase in your ability to discern which information in a source is factual, which is inferential, and which is informed opinion.

Facts. The universe of facts includes both quantifiable data and inferences that are accepted as true by virtually every expert. Facts are considered public property—that is, when you include a fact in a technical report, you need not cite its source unless it is an obscure fact and would be difficult for readers to confirm. One way you will be able to identify facts when you conduct research is by noting which pieces of information in documented sources are left undocumented. Anything left undocumented in a document that is other-wise scrupulously footnoted is either an original contribution or a fact. (You should be able to tell which.) Another way you will learn the facts on your topic is by reading them over and over again in source after source.

Inferences. Inferences are the property of their originators until they turn into facts. The first person to identify the characteristics common to all of the drugs we now classify as narcotics got credit for that idea for a time—until the information passed into the public domain and its originator was forgotten. The theory of relativity is now regarded as fundamentally factual and the intellectual property of all educated people, but at one time it belonged exclusively to Albert Einstein and had to be credited to him whenever it was cited. The theory of relativity began as one man's opinion, grew to be regarded as a reputable inference as it was accepted by more and more physicists, then drifted over into fact when almost no one challenged its accuracy.

Of all the kinds of information you will gather for a technical report, inferences will be the hardest for you to identify. It is difficult even for experts to tell when enough evidence exists to validate an inference conclusively. As a general rule, you should regard as inference any ideas that are shared by several people knowledgeable about a topic but not by everyone.

Often the best way to identify inferences in your sources is by examining the language in which they are couched. When scientists make inferences, they frequently express them with a kind of deliberate understatement: "It seems likely that sodium nitrite is itself carcinogenic." When you read a sentence like "One may readily conclude from these data that erosion will continue to decrease at about the same rate through 1988," you can be fairly sure that the person making the assertion considers it an inference rather than a fact. The impersonal introduction "One may readily conclude . . ." is not an expression of false modesty; rather, it is a means of keeping the researcher off the hook. Phrases like "the data suggest . . ." or "it is probable that . . ." or "it can be inferred from these results that . . ." sound overcautious and a little fussy, but they signal to readers that the data are not conclusive.

One of your major concerns as a researcher will be to avoid treating inferences as facts. You may agree with them or think them sound, but that in itself doesn't make them correct. When they are presented to you in cautious language, accept them cautiously and treat them cautiously when you pass them along to your own readers. When you transcribe an inference from a source, be sure to note that it is an inference and to record the names of all the sources that make that inference and treat it as valid. They constitute a school of thought, and you will want to cite all of them when you present the inference for your readers' consideration.

Opinions. Opinions are inferences that may well not be valid. Opinions propose a relationship between data and conclusions that is tenuous or hard to prove, or that requires a leap of faith, or that simply hasn't had enough time to spread through the field. Opinions can range from unreliable through marginally reliable all the way up to highly reliable, depending on how many people share them and who those people are. Not infrequently, the number of people sharing an opinion is directly proportional to the amount of evidence there is to back it up. Legal opinions and medical opinions—considered very reliable—are firmly grounded in evidence.

Should you use opinion in technical writing? Certainly. Ideas that are good but unsubstantiated are still good, and you should not necessarily reject a testable opinion simply because it has not yet been tested. Good intellectual work is valuable even if it lacks empirical support; in fact, the opinions of recognized authorities are considered perfectly legitimate evidence, provided those authorities are proffering opinion in subject areas where they have actual expertise. When you cite the opinions of recognized authorities, be sure to name them and list their credentials, so that your readers can decide how much trust to accord them. Also, because opinions are often unique to the people who hold them, you must exercise great care when transcribing them. Make absolutely sure that you have the name of the source and the page number correct and that you have copied the opinion verbatim. Wherever possible, quote opinions directly, so that there is no confusion in a report about who is talking. If you decide that, in your report, you want to paraphrase the opinion, to interpret it in your own words, identify your source clearly. Otherwise, the opinion will be perceived by readers as your own. In short, be sure that a source's opinion is presented as just that—not as your opinion and not as fact.

In Chapter 1, we warned you not to rely too heavily on the judgment of a reader who shares your biases and blind spots. The same warning applies to your own judgment in dealing with sources—don't restrict yourself to those that you agree with. You will write a better report if you can include and successfully refute counterevidence than you will if you refuse to confront such counterevidence by simply omitting it. And evidence you cannot refute deserves to be acknowledged.

 # Organizing Information

For many writers, the feeling of satisfaction that comes from completing the information-gathering phase of a project quickly fades as they realize that the mountainous pile of notes arrayed before them is not yet a report. All those 3-by-5-inch cards, legal pads, computer sheets, and envelope backs have to be carefully sifted for tidbits of information, and those tidbits have to be sorted out, arranged in logical and useful groups, and finally turned into sentences and paragraphs. The whole process can be intimidating. Indeed, surveys have shown that organizing a report is the problem that most exasperates technical writers.

The problem is not with formats—with where you put the abstract or the list of recommendations. (We discuss report formats in Chapter 13.) The problem is how to organize the *information* that forms the body of anything you write—the "meat" of what you are writing, rather than the "trimmings."

 # Formulating Your Main Point

The reason that you began to gather information in the first place was that you were facing a question that needed an answer. Now that your data are all in, you can answer the question. The first step in organizing your report is, quite simply, to write this answer out clearly and accurately in sentence form. Putting it in writing now, before you begin to sort out your note cards or struggle with an outline, will give you a clearly defined goal to aim for. You will be able to see from the outset exactly where all the points you will be making in your report must inevitably lead. This written statement of your main idea, usually called the thesis statement, encapsulates what you want your readers to learn from your report. If you take the time now to write it carefully, you will greatly reduce the risk of wandering off the mark as you draft your report.

The Thesis Statement

If your original question was very specific and focused, writing the thesis statement will be relatively easy. Here are two such questions and their corresponding thesis statements:

Question: Will the efficiency of the water heater be significantly increased by encasing the tank in 4 inches of insulation?

Thesis Statement:	Wrapping the tank in a 4-inch blanket of insulation will increase the water heater's efficiency by about 30%.
Question:	Will the incidence of mastitis in dairy cows be affected if peanut husks are substituted for wood shavings as bedding material?
Thesis Statement:	The incidence of mastitis in dairy cows will not be significantly affected by the change in bedding material.

Each of these thesis statements presents a specific conclusion that the rest of the report will validate with facts, inferences, and reasoned opinions. And each of these conclusions is brief enough to reside comfortably in a one-sentence thesis statement. Often, however, the question you are answering will be too broad and open-ended to admit of a brief thesis statement. Consider this question, for instance:

What is currently being done to improve the quality of optical fibers?

The question may have half a dozen answers or more. The best way to deal with multiple answers is to list them:

Current research in fiber optics is centering on five main areas:

1. raising the refractive index of the silica
2. obtaining low-loss silica by new extrusion techniques
3. . . . .

The advantage of listing specific points here is that in enumerating them, you will also be developing a plan of organization. If each point is equally important, chances are good that your report will devote a section to each one. Furthermore, seeing them here will give you the opportunity to decide if some are, in fact, more important than others.

Whether your thesis statement takes the form of a sentence or a list, make sure that it answers the question specifically. That is, make sure first of all that what you write down actually says something: "This is a report on the major effect of oil shale extraction on the environment" is not a thesis statement; it is just another way of asking the original question, "What is the major environmental problem associated with oil shale extraction?" On the other hand, "Disposing of the spent shale is the most serious environmental problem associated with oil shale extraction" *is* a thesis statement.

Second, make sure that your thesis statement answers the question that you set out to answer rather than some other question you raised along the way. You may, for instance, have decided halfway through your investigation that extracting oil from shale is not economically feasible; although this is an interesting thesis for a future report, it is irrelevant to the report you are supposed to be writing now.

The Statement of Content

In many of the reports you write, you will be analyzing data and drawing conclusions based on them. The thesis statements for these reports will present your opinion on what the data suggest—you will recommend X over Y; conclude that A, B, and C caused D; argue that P is the most important result of Q—and you will devote the report to marshalling evidence in support of your opinion. In some reports, however, you will be presenting information rather than analyzing it: giving directions, describing a process or mechanism, and so forth. For these informational reports, you won't be defending opinions or drawing conclusions, so your thesis statement, here called a *statement of content*, will be merely descriptive. The statement of content for a report detailing the operation of a four-stroke engine, for instance, might look like this: "One complete cycle of the piston involves four distinct phases or 'strokes': the intake stroke, the compression stroke, the power stroke, and the exhaust stroke. Each will be described in detail."

The Thesis Statement as Unifying Device

The primary purpose of the thesis statement (or the statement of content) is to keep *you* on track as you organize and write your report, to ensure that your report is unified. In the course of your research, you undoubtedly gathered at least some information that is interesting and provocative, but ultimately irrelevant to the subject at hand. Indeed, accumulating irrelevant material is almost inevitable, especially in the early stages of an investigation. Before an answer begins to take shape, all promising leads have to be run down. The tendency among inexperienced and undisciplined writers, painfully aware of how much time it took to track down and record this tangential material, is to find some place in the report to squeeze it in. A well-formulated, specific thesis statement will, therefore, do more than just remind you of what your report is about; it will remind you of what your report is *not* about. It is the key to a focused and unified report. Any idea, fact, or opinion, no matter how arresting, that does not in some way support or develop the thesis statement must be scrapped.

So valuable is the thesis statement in guaranteeing a report's unity that we recommend that you keep it prominently displayed as you organize and write your report. Most reports, in fact, include the thesis statement as part of the introduction. Just as it keeps writers oriented as they write, it keeps readers oriented as they read.

▶ Identifying Your Readers

Before you begin to deal with all the information you have gathered for your report, pull out the Audience Profile you have developed (pp. 14–17), and remind yourself of who your readers will be and what they will want from your report. Keeping your readers in mind will help you to develop an organization that emphasizes what is most important to them.

Here is an example. An engineer was asked to analyze three types of packaging material for durability and to recommend one for a particular commercial application. In conducting her experiment, she found that in order to test the durability of the materials reliably, she would have to make radical modifications in the existing laboratory apparatus. In fact, she spent three weeks redesigning the equipment and only two days running the actual tests. Her report recounted at length her inability to use the existing apparatus and her success at developing a new one. The bulk of her report was devoted to a description of the redesigned apparatus; the results of the tests—and her recommendation—received relatively little attention. She was unaware that her report had given disproportionate attention to her procedures until her supervisor called her in and told her to rewrite it with the emphasis on her conclusions rather than on how she arrived at them. Her problem wasn't exactly that she had strayed from her thesis, for she did present her conclusions eventually; rather, her problem was that she had neglected the needs of her audience, mistakenly assuming that their interest would be the same as hers.

▶ Grouping Your Data

With your thesis statement and audience clearly defined, you are ready to begin organizing the body of your report. The process of organizing involves two major phases: (1) grouping your random notes into logical and meaningful categories and (2) arranging the groups in an appropriate order of presentation. In this section, we detail the steps in the grouping process; in the next, we will discuss orders of presentation.

As you become more comfortable with the task of organizing, you will undoubtedly modify the process we outline here, omitting or combining some steps, performing others in your head. Indeed, if you have gathered your information methodically and logically, organizing it now will be greatly simplified.

Listing the Major Points

Make a list of all your major points and ideas, numbering this list as you go along. These are the points that you will obviously have to cover to prove your thesis. They may support your thesis directly:

Thesis: Plant security can be improved in five ways.
Major points: Window and door guards can be upgraded.
                Hallway monitors can be installed on level A.
                And so on.

Thesis: The company should purchase Robot X rather than Robot
                Y.
Major points: Robot Y costs twice as much to operate.
                Robot X is easier to program.
                Robot X needs servicing less frequently.
                And so on.

Major points and ideas may, however, support the thesis only indirectly. They may cover methodology or important background information—anything, in fact, that your readers will need to know before they can understand and accept your thesis.

Don't worry about supporting evidence yet; details will come after you have developed your list of points. You should be able to generate this preliminary list without looking at your notes. Indeed, many of the items on this list will have been on your mind ever since you began to gather information to support your thesis.

Sorting Your Notes

Now read through your notes carefully, looking for evidence—facts, quotations, data, and so on—to support each of your points. Put the number of the relevant point next to each note. If you have kept your notes on note cards, you can put all the cards with the same number in the same pile.

If you have notes that don't support any point, decide whether the note is relevant to your thesis. If it is, you need to add a point to your list. If it isn't, put it aside for now. You may discover its relevance as you continue to organize; more likely, you will eventually have to discard it.

Making Groups

When you have finished sorting out your notes, start looking for items in each of your piles that seem to belong together. Items may fit well together because they are aspects of the same issue, because they are opposites that deserve to be contrasted, or for any other reason that seems logical. Many items will go together quite naturally. An opinion stated in one of your sources may be supported by the test results published in another source; facts and figures that you have obtained from your own laboratory research may suggest something interesting about the conclusions of someone else's study. But do not force this grouping process. If some items do not fit with anything else, leave them as single-entry "groups."

Labeling the Groups

As your groups begin to take shape, give each one a heading that identifies its controlling idea. If, for instance, you are preparing a report on the pros and cons of using the insecticide carbaryl to control the gypsy moth population in your area, you might have groups of data with headings like "potential dangers to humans," "potential effects on birds," "potential effects on natural gypsy moth parasites and predators," "short-term effectiveness in controlling gypsy moths," "long-term effectiveness," and so on. These headings will facilitate the organizing process—you will be able to see at a glance what each of the groups is about. And later, when you begin to write the report, they will become the basis for paragraph and section divisions.

Combining the Groups

At this point, you have a list of major points with the supporting evidence for each broken down into small, labeled groups. Now check to see if some of these groups can be grouped together. Do two or more groups have similar headings? Are several groups devoted to the same aspect of the major point? If so, try to combine them; see if they will form a larger group. In the carbaryl example, for instance, several of the headings refer to effects on *nontarget species*; perhaps these could all be grouped under that larger heading.

At the same time, see if any of your major points seem to belong together. Suppose, for example, that your report deals with the likely effects of setting up a subsidized municipal bus system. A partial list of your important points includes the following:

Effects on use of downtown stores and thus on their profits
Effects on mobility, especially for elderly, handicapped, and poor people
Effects on use of private cars and thus on traffic congestion and air pollution
Effects on gasoline consumption
Effects on time spent in transit by the average user
Effects on individual and corporate attitudes toward staying in the community
Effects on taxi use and viability of taxi companies
Effects on user costs versus existing alternatives
Effects on municipal budget and tax rate

You could keep these all as discrete groups, but that would not help organize the content. And if you lumped them all together now under the heading of "effects," you would only be postponing trouble, because you would just have to divide later. But with a modest amount of squeezing and trimming, you can tighten up this list: You can fit all your effects into three broad categories—economic effects, social effects, and environmental effects—each of which can then be subdivided:

Economic effects
   on the city budget
   on taxi companies
   on downtown merchants
   and so on
Social effects
   on the elderly
   on the handicapped
   on the poor
   and so on
Environmental effects
   on gasoline consumption
   on use of private cars and thus on traffic and air pollution
   and so on

This process of grouping information into categories based on similar characteristics is called *classifying*. Classifications are not solely tools for organizing the elements of a report; systems of classification are designed whenever there is a need to impose order on an undifferentiated mass of things or ideas. The ability to classify is so necessary in the technical world that we have devoted an entire chapter (see Chapter 6) to classification and its corollary activity, partition.

You should keep in mind that any given assortment of raw data can be classified in any number of ways. The information for a report on the pollution caused by synthetic fuel production, for instance, might all fit comfortably into three large groups, one for each type of fuel source: coal, tar sand, and oil shale. The same information, however, might also be grouped according to the type of pollution caused by synthetic fuel production: water, air, noise, land and biota, and so on. How you group will depend, of course, on your goals and audiences. If your main goal is to catalog the damage that might be done to the environment by syn-fuel production, then the second set of groupings might be more emphatic and useful than the first. If, on the other hand, your goal is to assess the possibility that one syn-fuel method is less polluting than the others, then the first set of groupings might be better. Always group in the direction of your thesis statement. If your thesis is very specific, it will suggest the major groupings. In a report supporting the thesis that "five factors contributed to the failure of the project," you are bound to have five major groupings.

Double-checking Your Groups

When you have taken this grouping process as far as you can, check for completeness and unity. Is the information in each of your groups sufficient to justify the controlling idea of that group? If not, determine what you need to add, and then go back through your notes—or back to the laboratory or library—and find it. Is there anything in the group that does *not* help to justify the controlling idea? If so, remove it. Once you have checked each group, check the whole. Are the various controlling ideas, taken together, sufficient to prove your thesis statement? Are they all relevant to your thesis, either directly or indirectly? Some, as we suggested earlier, may support your thesis only indirectly (the methods section of a lab report, for instance). Any point or detail that your readers will need to see before they can accept your thesis is relevant.

What you are after can be represented by the diagram shown in Figure 2.1. Each of the major headings (the controlling ideas) should address an aspect of your thesis, and each of the subgroups under that heading should address a specific aspect of that heading.

▶ Putting the Groups in Order

At this point in the organizing procedure, you should have a work sheet with a thesis statement and a series of headings and subheadings representing the major ideas and supporting data of your report. The next step is to decide on an order of presentation. If you have six major points to make—and therefore six groups of supporting information—you must decide which you are going to discuss first, which second, and so on. Then you must decide on an order of presentation for the information *within* each supporting group.

Orders of presentation can be divided into two large categories. *Natural orders*, in which the material itself suggests the order of presentation, tend to be sequential. *Logical orders*, in which the writer imposes an order on the material, are hierarchical. As we discuss the varieties of natural and logical orders, bear in mind that an ordering principle typically governs only the body of the report (or of the section or subsection). Introductory and concluding sections are usually organized separately and then appended at either end.

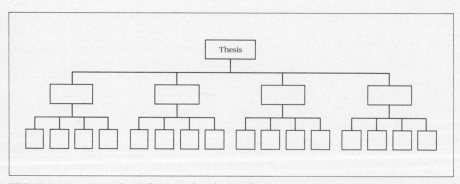

FIGURE 2.1   Grouping Ideas and Information.

Natural Orders of Presentation

Natural order is most often used in descriptive or informational reports that present data rather than analyze them. The two basic types of natural order are chronological and spatial.

Chronological Order. Chronological order is the obvious choice when your focus is a sequence of actions, events, or ideas unfolding over time. Instructions are written in exact chronological order (except, of course, for the introduction and conclusion). The same is true of most process descriptions (how a microprocessor works, for instance) and many progress reports (what happened in March, in April, in May, in June). You will use chronological order often.

But chronological order has three major disadvantages. First, it becomes monotonous in long passages. Second, because chronological order treats every item in a sequence with equal emphasis, it tends to bury important items that happen to fall in the middle. Finally, chronological order obscures the connections between items that are not sequential. Something that happened five months ago might be relevant to something that happened last week, but if you have to account for everything that happened in between, your readers may have a hard time seeing the connection.

The shorter the report, the less compelling these objections become. Hundred-page reports are almost never organized chronologically; ten-page reports occasionally are; one- or two-page reports often are. Moreover, you may find that though chronological order is not the right choice for an entire report, it is perfect for a part of it: The historical background of an event, project, area, or idea might be effectively presented in chronological order; the procedures section of an experiment is almost certain to be chronological.

Spatial Order. Spatial order takes the reader from top to bottom, left to right, east to west, outside to inside, and so on. It is an extremely useful order for describing many kinds of mechanisms (see Chapter 10)—both visible ones like a clock, an eyeball, or a gas turbine, and invisible ones like a corporate hierarchy. Because technical writers spend a good deal of time describing objects, places, and mechanisms, you will be using spatial order often.

Like chronological order, spatial order is easy for writers to achieve and easy for readers to follow. But like chronological order, it is dull, it hides important items that fall in the middle, and it obscures relationships between items that don't happen to be next to each other. It does coordinate well with photographs and diagrams, but quite often this artwork gives readers all the spatial orientation they really need, freeing the prose for a different emphasis.

Logical Orders of Presentation

The great advantage of natural order is that it is easy to use. Once you have decided to begin with the upper left-hand corner and to proceed clockwise—or to begin with September 1967 and work up to the present—the material tends to organize itself naturally; everything is sequentially related. Logical order, on the other hand, requires you to create relationships where no natural sequence exists; it requires you to impose an order on your material. It is, therefore, more difficult to achieve than natural order, but it is often the only appropriate way of getting your ideas across.

The particular logical order you use will, of course, depend on the kind of information you are presenting, the reason you are presenting it, and the specific needs of the people who will be reading it.

Order of Decreasing Importance. The great advantage of order of decreasing importance is that it gives busy readers the chance to see your best points right away. Thus, in a report recommending a particular equipment purchase, you would present your most compelling reasons first; by the time readers have gotten to your eleventh reason, they may have already decided to buy.

Order of decreasing importance works especially well in reports whose main points or ideas are parallel:

Six ways of improving your FM reception
The effects of chemotherapy on short-term memory
Seven causes of heat loss in the home
Concentrations of trace elements in the atmosphere

Because the core of a great many reports is devoted to discussing lists of points or arguments or problems, you will use this order repeatedly.

It does not, however, work equally well for all reports. A typically complex report of safety violations in your laboratory, for instance, might have many topics to cover: federal regulations governing laboratory safety, how well your laboratory meets these regulations, how you went about your investigation, suggestions to correct safety violations in your laboratory, and the cost of making the corrections. The most important of these to both you and your readers might be the suggestions for bringing the laboratory up to code, but you can't start there because readers first need to know at least something about the safety code itself. Nevertheless, when you get to the list of code violations and the parallel list of correction proposals, *they* should be in order of decreasing importance.

Order of Increasing Importance. The reverse of the order we have just examined, order of increasing importance begins with the least important point and builds slowly and inexorably to a strong climax. Seldom the major organizing sequence in a technical document, this order is nevertheless useful for clearing the table of minor points before you get down to the more serious ones. Thus you devote the first page and a half to the six minor reasons for doing such and such, and then concentrate on the single major reason for the next sixteen pages. Order of increasing importance will ensure that your readers don't skip over the minor reasons (a problem with decreasing order, as we've seen), and it will save your report from anticlimax. Of course, if the minor points can't be handled and dismissed speedily, you will probably be better off with decreasing order. And, once again, if the points aren't parallel, neither increasing nor decreasing order will be very useful.

Other Hierarchical Orders. If you have a list of four chemicals, seven causes, nine viruses, six good reasons, or whatever, and none seems more important than another, you should consider putting them into an order based on some other hierarchy. Are some of these items more *familiar* to your readers than others? Then try ranking them from most familiar to least familiar (or vice versa). Are some of them more *difficult* to understand? Then try ranking them by complexity. Are some more *controversial* than others? Then arrange them in an order of controversiality. You can build a hierarchy from any criterion— cost, durability, ability to withstand corrosion, and so on—and then order the specific items from most to least or least to most.

In persuasive reports, order of increasing controversiality is often your best bet. You begin with those points least likely to upset a potentially hostile audience, bringing in the more controversial points after you have established your credibility.

General-to-Specific Order. As its name implies, general-to-specific order presents general premises or criteria first and then presents the specific support or specific applications of those general statements. In one sense, virtually all reports are organized in either general-to-specific order or its inverse, specific-to-general order, for the thesis statement can be considered the generalization, and the body of the report can be considered the specific evidence that supports it. If the thesis comes first, as it usually does, the report is general-to-specific. If the report saves the thesis statement until the end, after it has presented all the specific data, the overall structure of the report can be called specific-to-general.

Our concern here, however, is with methods of organizing the body of the report only. In this more limited sense, general-to-specific order establishes general premises or criteria and then presents the specific applications of those general statements. Specifications, which are minute and exact descriptions of all the components of a particular object, are almost always written in general-to-specific order, starting with legal definitions of all the terms, then presenting all the legal standards and requirements, and ending with long lists of the specific nuts and bolts (see Chapter 7). Similarly, reports that assess whether a particular piece of equipment or a particular process or even a particular concept measures up to some preexisting standard are often developed this way. An outline from the body of such a report might look like this:

Standards
  Cost
  Efficiency
  Maintenance
  Marketability
Brand X Gizmo
  Cost
  Efficiency
  Maintenance
  Marketability

The disadvantage of this type of organization is that if you have lots of general standards, or if they require pages of explanation and clarification, the distance between criterion and application may be so great that your readers will have a hard time keeping the former in mind as they study the latter.

Notice, too, that the items under "Standards" and "Brand X Gizmo" are not organized in general-to-specific order: "Cost," obviously, is no more general than "efficiency." Here they are arranged alphabetically, an order that is ideal for reference works—telephone books and handbooks, for instance— where readers know the items in advance and simply want access to them. The items in our example, however, might more profitably be arranged in order of importance. In other reports, the subpoints might be arranged spatially or chronologically. Thus, unless the report has only one criterion and one application—and no subpoints—general-to-specific order needs to be used in conjunction with some other order.

Specific-to-General Order. Like order of increasing importance, which it very closely resembles, specific-to-general order is sometimes useful in persuasive reports. If, for instance, you are trying to convince superiors at your firm to establish new and more stringent safety regulations, you might begin with some pointed examples of accidents or injuries that have occurred and then lead into the conclusion that these injuries could have been avoided if new safety regulations were in effect. (The examples themselves you would arrange in order of decreasing importance.)

Comparative Order. Comparative order is actually a combination of order of importance and general-to-specific order, but it is used so frequently in technical writing that it warrants its own heading. A simple comparison of any two things involves three components: the two things being compared and the criterion by which you are comparing them. Organizing such a simple comparision is easy and straightforward: You begin by identifying the items being compared, then announce the criterion, and finally make the actual comparison:

Items:	Synthetic gypsy moth pheromone vs. natural gypsy moth pheromone
Criterion:	Cost-effectiveness
Comparison:	Cost-effectiveness of synthetic pheromone
	Cost-effectiveness of natural pheromone
Conclusion:	Synthetic pheromone is more cost-effective.

Unfortunately, comparisons are rarely this simple. Most involve more than one criterion, and many involve more than two items. Suppose, for instance, that you are comparing two carburetors and that you are considering these five criteria: fuel efficiency, manufacturing costs, reliability, normal performance, and peak-load performance. You could follow the pattern of a simple comparison and organize your report like this:

I. Items being compared: Lone-Star carburetor vs. Three-Star carburetor
II. Criteria:
   A. fuel efficiency
   B. costs
   C. reliability
   D. normal performance
   E. peak-load performance
III. Lone-Star carburetor
   A. fuel efficiency
   B. costs
   C. reliability
   D. normal performance
   E. peak-load performance
IV. Three-Star carburetor
   A. fuel efficiency
   B. costs
   C. reliability
   D. normal performance
   E. peak-load performance
V. Conclusion: Lone-Star is better than Three-Star.

The trouble with this organization is the same as the trouble we noted with general-to-specific orders—the specific points being compared are too far away from each other. The reliability of Lone-Star may be on page 6 and the reliability of Three-Star may be on page 10. A more useful approach would be to organize criterion by criterion:

> I. Identify Lone-Star and Three-Star
> II. Fuel efficiency
>> A. Lone-Star
>> B. Three-Star
> III. Costs
>> A. Lone-Star
>> B. Three-Star
> IV. Reliability
>> A. Lone-Star
>> B. Three-Star . . .
> VII. Summary of advantages of Lone-Star over Three-Star
> VIII. Conclusion: Lone-Star is better than Three-Star.

This revised version is much tighter and easier to follow. Each point of comparison is complete in itself and isolated from the others, each is organized in general-to-specific order (from criterion to application), and the whole is organized in order of decreasing importance (fuel efficiency to peak-load performance).

The Matrix Problem

The comparison of carburetors that we just examined exemplifies an organizational problem that often faces technical writers. Sometimes called "the matrix problem," it comes up whenever you have two obvious bases for organizing your report—in this case the two models of carburetor and the five criteria. This one was fairly easy to solve because there were only two carburetors being examined; but suppose there had been nine or ten carburetors, each with its own advantages and disadvantages. Such a report could be organized in a number of ways. You might stay with the criterion-by-criterion pattern that we just discussed:

> Fuel efficiency
>> A. Lone-Star
>> B. Three-Star
>> C. Carburetor No. 3
>> D. Carburetor No. 4 . . .
> Costs
>> A. Lone-Star
>> B. Three-Star
>> C. Carburetor No. 3 . . .

But this pattern, though clear and parallel, has one large disadvantage. If Carburetor No. 5 is the most fuel-efficient, Carburetor No. 3 the cheapest, and Carburetor No. 7 the most reliable, the report should be arranged so that these facts are emphasized. As it is here, Carburetor No. 5 will always be discussed fifth under each criterion, and Carburetor No. 7 will always be discussed seventh. One alternative would be to arrange the carburetors by performance under each criterion—that is, by order of decreasing importance within each category:

Fuel efficiency
    Carburetor No. 5
    Carburetor No. 7
    Lone-Star
    Carburetor No. 6 . . .
Costs
    Carburetor No. 3
    Lone-Star
    Carburetor No. 8 . . .
Reliability
    Carburetor No. 7
    Lone-Star
    Carburetor No. 6
    Carburetor No. 2 . . .

Another alternative would be to make your comparison carburetor-by-car-buretor, from best to worst, with the criteria under each one also arranged by importance:

Lone-Star
    Fuel efficiency
    Costs
    Reliability
    Normal performance
    Peak-load performance
Carburetor No. 7 . . .
Three-Star . . .

You could make all these versions less unwieldy by adding appropriate tables and graphs (see Chapter 11, "Graphics"). If, however, your goal is to convince your readers that Lone-Star is better than the other eight carburetors, and if your readers are really more interested in your conclusions than in all the point-by-point comparisons, a simpler pattern is possible:

Criterion 1
  Lone-Star
  all others summarized
Criterion 2
  Lone-Star
  all others summarized
Criterion 3 . . .

This strategy emphasizes the point that you want to get across (that Lone-Star is the best choice), and it reduces wear and tear on your readers.

The best order for your particular piece of writing may be any one of the natural or logical orders we have discussed, or it may be a combination of them. The longer and more complex your subject matter, the more likely you are to need to mix and match orders. Later in the book, as we discuss specific types of technical writing, we will suggest which mixes are most appropriate.

Organizing for Multiple Audiences

In this chapter, we have been discussing strategies for organizing *anything* you write—though we have used the word "report" as shorthand. There is, however, one organizational strategy that applies almost exclusively to reports: organizing by audiences.

As you will recall, a single report may be addressed to many different groups of readers at the same time, each of whom has different needs and backgrounds, and many of whom won't read the report from cover to cover. Organizing a report for a multiple audience like this is tricky, because what might be most important to one group might be totally irrelevant or incomprehensible—or both—to another. If you are faced with this problem, you might try dividing the body of your report into separate sections, each aimed at the specific concerns of a different audience. Each of these sections can be organized in any of the ways we have discussed. And to provide the overall context of the report, you can add summaries and general overviews.

Suppose your report's purpose is to recommend that your company install a new, more efficient water-recycling system. Having analyzed your goals and audiences thoroughly, you know that the financial department will be interested in the projected costs of the switch. The production department will want to know how the new equipment will affect productivity, both during the installation period and after it is in operation. The company's design engineers need details on the technical specifications of the new system and how it can best be integrated with existing equipment. Other audiences are interested in the installation procedures, the maintenance schedules, and so on. If you devote a section to each of these topics, each of your audiences can focus exclusively on that part of your report most relevant to its needs.

Exercise 2.1

Suggest the most appropriate order or orders of presentation for each of the following:

1. A memo to your colleagues suggesting improvements in the design of a product you are all working on.
2. A review of scholarly articles on the effects of violent movies on children.
3. A laboratory report on the effects of high doses of sugar on rat metabolism.
4. A set of instructions on how to build a "live-catch" mousetrap.
5. A report to potential customers explaining how your company's product compares with its three major competitors.
6. A report to management explaining how your company's product compares with its three major competitors.
7. A report to your research director on why your project should be allowed to continue despite your recent lack of success.
8. A report describing a new piece of laboratory equipment and explaining why your company should purchase it.
9. A report analyzing whether the chemical defoliant your research team has developed meets all federal health and safety standards.

▶ Outlining

On your organizing work sheet, you should now have, in addition to your thesis statement, a list of main points with their supporting evidence, all arranged in a suitable order and number-coded to your note cards. What you have, in other words, is that most feared and despised of writers' tools—an *outline*. It probably isn't decked out in Roman numerals, capital and small letters, and the rest, all evenly indented and neatly aligned; but it is an outline nevertheless—a *scratch outline*, a set of codified notes that you can understand and follow. And this is probably enough to get you writing.

The best thing about a scratch outline is that it is easy to rearrange. If you see new and better ways of organizing your material as you begin to write—or if you discover that point number 3 is really just an aspect of point number 7—you can make the adjustment easily (for instance, by drawing a circle around number 3 with an arrow down to number 7).

Some writers, however, prefer a more rigorous and formal outline—and occasionally superiors will require one. A formal outline is shown in Figure 2.2. This is a *topical outline*; each of the entries is a phrase or clause focusing on one of the report's topics or subtopics. Note that the entries at each level are written in parallel form: *I.A.* and *I.B.* are both clauses, *II.A.* and *II.B.* are both noun phrases, and so on.

If the entries are written in sentence form, the outline is called a *sentence outline*. Occasionally, both sentences and phrases are used in a single outline. Here again, the entries at each level must be parallel. If *I.* is a phrase, then *II.*,

FACTORS GOVERNING THE EFFECTIVENESS OF WIRE FISH TRAPS

```
I.    Use of Wire Fish Traps
      A.  Where wire fish traps are used
          1.  In the Caribbean
          2.  In the South Atlantic
      B.  How wire fish traps are used
          1.  Baited or unbaited
          2.  Individual traps or trap lines
II.   Types of Wire Fish Traps
      A.  Features common to all types
      B.  Traditional designs
          1.  Chevron-shaped
          2.  Z-shaped
          3.  S-shaped
      C.  Experimental designs
          1.  D-shaped
          2.  O-shaped
          3.  "Stackable" traps
III.  Variables Affecting Success of Wire Fish Traps
      A.  Fishing techniques
          1.  Trap placement
              a.  location of traps in relation to local
                  fish habitat
                  i.   in habitat
                  ii.  near habitat
              b.  distance between traps
          2.  Trap immersion (length of time trap is submerged)
              a.  effect on variety of fish caught
              b.  effect on quantity of fish caught
      B.  Trap design
          1.  Size of trap
          2.  Construction of trap
              a.  mesh size
              b.  entrance shape
              c.  frame composition
      C.  Meteorological phenomena
          1.  Tidal rhythms/phases of the moon
          2.  Seasonal variations
          3.  Sea conditions
              a.  calm seas:  small catches
              b.  rough seas:  larger catches
      D.  Fish behavior
          1.  Attraction of conspecifics into traps
          2.  Predatory behavior
              a.  prey seeks refuge in traps
              b.  predator seeks prey in trap
```

FIGURE 2.2   A Formal Outline.

III., and all the other Roman numeral headings must also be phrases; if *II.A.* is a sentence, then *II.B.* and *II.C.* must be sentences as well. Each of the major divisions in the outline will become a section heading, each of the divisions at the next level will become a subhead, and so on. Even the relatively minor supporting details have a place in the outline—as subdivisions.

If you can work an outline through, you will be able to write the report itself with much less trouble. In truth, however, though virtually all good writers use some kind of outline, few people spend the time and effort it takes to write a formal one, because it takes nearly as much time to write a sentence outline as it takes to write a first draft. The outline is not an end in itself; rather, it is an aid to the writer, neither more nor less. Whatever type of outline helps you the most is the one you should use.

Review Questions

1. What is the difference between primary research and secondary research?
2. What is the difference between an index and an abstract?
3. What is meant by the phrase "searching the literature"?
4. When should you take verbatim notes from your sources? When should you summarize?
5. What are some ways to determine the reliability of a journal you wish to use? Of an individual author?
6. What is the difference among facts, inferences, and opinions?
7. What is the difference between a thesis statement and a statement of content?
8. What are the main steps in the process of grouping your data?
9. What are natural orders of presentation? How do they differ from logical orders of presentation?
10. What are the advantages of chronological order? What are the disadvantages?
11. In what situations would chronological order be appropriate? Spatial order?
12. Which logical order of presentation is most frequently used? Why?
13. For what types of writing is comparative order most useful?
14. For what types of writing is order of decreasing importance most useful?
15. What is the "matrix problem"? What are your options when you encounter this problem?
16. What does "organizing by audiences" mean?
17. What is the difference between a scratch outline and a formal outline?

Assignments

1. Prepare a working bibliography of at least ten sources on either the health effects of dioxin or the current status of bubble memory in computer

technology. Note which indexes and abstracts you used. Without reading the sources themselves, identify three to five of your bibliography entries that seem likely to prove most useful, and explain why you think so.

2. Choose any article from a technical journal in your field published in 1978. Prepare a bibliography of sources since 1978 that bear importantly on the article's content. Indicate what indexes and abstracts you used in your search.

3. Here is a matrix problem. Your project was to evaluate twelve proposed sites for a new bridge, basing your evaluation on nine predetermined criteria. You are supposed to recommend one site in your report and to justify the recommendation to your company's management. Three sites, numbers 4, 7, and 11, were eliminated early because each failed one of the major criteria: The land for one is unavailable; the second is geologically unstable; the third would threaten an important wildlife breeding area. The other nine sites are acceptable, but your analysis shows that site number 5 is the best. Design a useful matrix for this situation. Make sure you distinguish between the sites that are unacceptable and the ones that are acceptable but inferior to site number 5.

4. Write an outline for a position paper on the following topic: Should technical writing be a required course for all science and engineering majors? Your audience is the dean who will ultimately make the decision. Decide what your position on this topic is; then draft a thesis statement. Next, generate a list of points to support your thesis, and indicate the type of evidence you would need to support each point. Finally, decide on a suitable order for your points and subpoints.

3 Writing Effective Drafts

Turning your outline into a report actually involves two steps: writing a first draft and then revising that draft as often as necessary to make the finished product clear, concise, and correct. In this chapter we focus on ways to generate the sentences and paragraphs of a first draft. In the next, we discuss how to polish both content and style.

Obviously, the more care you take in writing the first draft, the less revising it will need. But how you divide your time between writing and revising will depend both on the length and complexity of the document you are producing and on your own experience and ability. As you become more adept at writing, you will learn to do a good deal of revising in your head, and especially on short documents like letters and memos, your first draft will probably look much like your last. However, until you gain experience, don't try to write and polish at the same time. A first draft should be serviceable, but it need not be perfect. The more polished your first draft is, in fact, the less likely you may be to change it even if you discover a flaw in your organization or an error in your logic. Moreover, trying to write and revise at the same time may cause you to freeze up altogether. Your main goal in writing a first draft should be to get your ideas down on paper in reasonably good shape, and worrying excessively about grammar and style often interferes.

 ## Getting Started

Even writers who understand that their first draft need not be perfect often have a difficult time getting started. The following suggestions should help you over this first hurdle.

Don't Start at the Beginning

You don't need to write your report in the order in which it will be read. Usually, in fact, the best place to start writing is the body. As any experienced writer knows, the introduction is almost invariably the hardest part to write. It must contain not only the thesis statement and the statement of purpose, but also a context-setting statement of the problem being discussed. Establishing the context for your readers is important, of course, but it is often easier to do after the rest of the report is drafted—when you can read over exactly what you have said and thus see more clearly the kind and amount of background material your readers need.

Even within the body, you should start with the *easiest* section, not necessarily the first. Once you have completed a couple of sections, your confidence will grow, and you won't be as likely to freeze when you get to the more difficult ones.

Don't Agonize over a Word, a Sentence, or an Idea

If you don't like a particular word or sentence but can't think of a better way of expressing yourself, make a note of your displeasure and then go on. By the same token, if you find you are having trouble getting an idea to gel, write an approximation of what you want to say; then come back to it later. Especially at the start, the most important thing is to write something that you can work with; revising, remember, comes later.

Don't Get Lost in Your Notes

As you begin to write, you will probably discover that for many of your points you have more evidence than you need. Squirreled away on your note cards may be seven different examples, six expert opinions, and a two-page quotation, all in support of a simple subpoint. Don't feel that you need to include everything. Pick the strongest examples and let the rest go; combine the expert opinions, instead of treating them individually; condense the quotation to a sentence or two if you really need it at all. When you revise your draft, you will probably do even more cutting and condensing, but you should start here.

You may occasionally find, moreover, that you don't have quite the right evidence for a particular point. Don't try to force the evidence you do have to suit your point. Instead, put that section of your report aside, and go on to another. You can find the exact evidence you need later.

As you write, keep track of anything in your report that will need a footnote—anything that you have obtained from an outside source. Don't bother with the footnotes themselves yet, but do *indicate* where each footnote will be needed. You might, for instance, put an asterisk (*) in the text and then write the name of the source of that particular piece of information in the margin.

Writing Introductions

Because the introduction provides the framework that makes everything that follows accessible, the effectiveness of a document hinges to a large extent on the effectiveness of its introduction. Small wonder, then, that so many writers find the introduction to be the hardest section of a draft to write. Postponing the introduction until you have completed the rest of the draft (and thus know what you've actually said in it) should help, but, even so, orienting your readers to what lies ahead can be a pesky affair.

Many novice writers are especially troubled by the belief that their introduction must be enticing—that it must spark readers' curiosity and make them want to read further. This is certainly true of most business and popular writing, but it is less often true of technical writing. Technical readers, as you recall from Chapter 1, read because they must—the information interests them because it's job-related. Thus, you usually won't have to coax them into reading.

What technical readers need instead of enticement is *information*. They need to know what you are writing about and why you are writing about it, and they need enough background information so that they can understand what's to come. Thus, how much information you put in your introduction depends on the complexity and purpose of your document and on the needs of your particular audience. A brief letter to a colleague already familiar with your topic may require only a sentence or two of introduction, whereas a three-hundred-page recommendation report that will be read by dozens of people will probably require several pages.

Most of the information you will need to draft your introduction will already be on your note cards or in your outline; the rest will certainly be in your head. We discuss below the common features of the technical introduction.

Statement of Purpose

The statement of purpose announces the topic and situates your document within that topic:

> This report presents our findings on the salinity levels of water samples taken from the Chesapeake Bay above and below the mouth of the Potomac in June 1982.

> Here, as you requested, are my recommendations for improving the flow rate of your experimental apparatus.

> This report details the progress made to date on our research into digital image processing techniques that might be applicable to our automated assembly line.

Each of these one-sentence statements of purpose tells readers why the report is being written. Often, however, readers won't be able to grasp the full significance of the document without further information. An expanded statement of purpose might include (1) a description of the specific technical problem being addressed and how it relates to any larger practical problems—the background of the problem; (2) a breakdown of the major subproblems or points covered in the document; and (3) an indication of how readers may be able to *use* the information presented in the document.

The following expanded statement of purpose covers all of these points:

The background of the problem—why we use chlorine to begin with.

Power plants require large volumes of cooling water to carry away waste heat. Bacterial and algal slimes will attach themselves to the walls of the piping and decrease the heat transfer across the condensers if fouling is not controlled. The most economical method currently used to control fouling is the addition of chlorine to the cooling water.

The practical problem—chlorine toxicity.

However, chlorine presents problems when its toxic effects carry over into receiving waters. Researchers have established both acute and chronic toxicity thresholds for aquatic organisms, often at levels below those set by the Environmental Protection Agency for the amount of chlorine allowed in the discharge of cooling water. Several states have set stricter levels.

The technical problem—where the chlorine winds up— and the various subproblems.

This study was conducted to determine the nature, levels, and persistence of chlorinated compounds in the discharges of five power plants in Northern California. In-plant studies were conducted to understand the demands for chlorine in the cooling water system and to determine the levels reaching the receiving waters. Decay studies were conducted at the outfalls. Water samples were taken at the discharge pipes and in the receiving waters to determine how long chlorine persisted in the water and how far it spread from the plants.

Significance of the project—who can use it for what.

This report adds significantly to our understanding of residual chlorine behavior in coastal waters. From this work, power plants may be able to restructure their disinfection programs and reduce the amount of chlorine reaching the environment as well as reducing costs. Regulatory programs may be able to use the results to improve effluent guidelines and monitoring programs.

Background

Depending on what you are writing, you may need to talk about the history of the project, the characteristics or history of the organization you work for, the theory or theories that formed the basis of your work, the findings of previous research on the topic, and so on. If this information can be covered briefly, it belongs in the introduction. If your readers (or *some* of your readers) need detailed background, you should put it in a separate section or in an appendix and summarize it in the introduction. For instance, most proposals and research reports require extensive discussion of prior investigations of the subject, both those conducted by the writer and those conducted by other researchers. This discussion usually follows the introduction and is given its own heading, "Review of Literature." The introduction itself should include enough background to provide a framework for the information to come—but no more than that.

Scope and Limitations

A good statement of purpose should give your readers a fairly accurate sense of the scope of the report—what it covers and what it excludes. If necessary, however, by all means take the extra space you need to describe your scope more fully. In particular, try to be explicit about what the report *doesn't* cover that readers might otherwise expect to find (and tell them where to find it or why it is not included). Other limitations of the project—assumptions, methodological problems, and so on—also deserve a prominent place in the introduction.

Other Introductory Material

Virtually all technical documents include the purpose, the scope, and some background in the introduction. Depending on the topic and the intended audience, an introduction might also include any or all of the following: an overview of the procedures used to generate the information discussed in the body; definitions of key terms; a description of the qualifications and duties of the personnel involved in the project that culminated in the document you are currently drafting; lists of findings, conclusions, and recommendations; and a plan of development that explains how the rest of the document will be organized. Later in this book, as we take up each of the various types of technical documents, we will point out the special features of their introductions.

Considerations for Drafting an Introduction

There is no rigid rule governing the order of the elements in an introduction, just as there is no rule governing which elements must be used in any particular introduction. Generally, however, the less familiar your readers are with the subject of your document, the more background they will need *before* they come to your statement of purpose.

Not surprisingly, the first few sentences of the introduction are usually harder to write than anything else in the draft. Inexperienced writers tend to make their opening sentences much too general and all-encompassing, presenting histories and theories that are not directly relevant to the subject of the document. Most reports about aviation don't need to begin with a mention of the Wright brothers, nor do most reports on rocketry need to begin with definitions of gravity, acceleration, and trajectory. To avoid this all-too-common trap, we suggest that after you've drafted your introduction and appended it to the body of your document, you try reading the whole thing without these opening sentences. If it reads smoothly, you can probably omit them permanently.

Weaving the various elements of your introduction into a smooth-flowing series of paragraphs may also prove difficult. If it does, consider dividing the introduction into subsections with headings and subheadings. (Formal report introductions are often formatted this way; letter and memo introductions rarely are, but, by the same token, these are usually short and focused enough that they require only a paragraph or two.)

 # Writing Paragraphs

Whether you are working on the introduction or some other section of your draft, you need to present your ideas in readable increments, developing each idea so that your readers can absorb it. The basic unit of development of a technical document is the paragraph. You have, of course, been writing paragraphs since grammar school, so we will dispense with the basics and focus instead on the specific requirements of *technical* paragraphs.

Deductive Strategy in Paragraph Design

A technical paragraph is much like a technical document in miniature: It contains a unifying idea—a *topic sentence*—and any data, definitions, examples, discussion, or other evidence needed to explain, support, or qualify that idea. Like a technical document, it can be organized in any of the patterns we discussed in the last chapter—chronological, spatial, comparative, and so on. And although the topic sentence can go anywhere in the paragraph, it usually precedes the supporting evidence, just as the thesis statement of a report usually precedes the body.

Putting the topic sentence first like this is called *writing deductively*, and it is a basic principle of technical communication. If you start with your main idea and then supply your supporting information, you maximize clarity; your readers will know from the beginning where you are taking them, and they will find it easier to follow your train of thought. Writing deductively, moreover, allows you to see more clearly what substantiation your topic sentence needs as you draft your paragraph, and it reduces the likelihood of your including data or examples that *don't* relate directly to the topic of that particular paragraph.

The importance of the deductive strategy is worth emphasizing because this strategy is directly opposite to how most scientists and engineers *gather* the information they must write about, and thus many apprentice technical writers find it counterintuitive. Because they are used to working *inductively*, accumulating individual pieces of data through experimentation, observation, and so on, and then basing their conclusions on what they have found, they tend to write the way they investigate. But investigating and writing about an investigation are separate processes with separate goals; putting the "message" before its evidence gives readers what they need where they want it.

Paragraph Length

Because some topic sentences naturally need more developing than others, there can be no hard-and-fast rule governing the correct length of a paragraph, but a paragraph of 100 to 150 words is usually about right for most technical reports, 50 to 100 words for letters and memos. Paying attention to the length of your paragraphs as you draft them will save you time in the long run, for those that deviate radically from the norm will probably need to be revised. Excessively short paragraphs tend to be choppy and incomplete, whereas excessively long ones tend to be disunified and hard to understand.

The major cause of too-short paragraphs is inadequate development. If you find that you are consistently writing paragraphs of only three or four lines, examine them carefully to see if you have provided enough examples, definitions, explanations, and other evidence to satisfy your *readers'* need for complete information. Then develop the paragraph accordingly. If the point you are making is so obvious that it truly doesn't need development, it probably doesn't warrant its own paragraph, either. If this is the case, try to find a way of incorporating it into an adjoining paragraph as an aspect of that paragraph's main point.

Too-short paragraphs also result from a failure to keep a main point and its supporting information together. If you have a generalization supported by three brief examples, write one paragraph, not four. Similarly, don't separate cause from effect, data from conclusion, or statement from explanation.

Paragraphs that are too long, on the other hand, usually result when writers fail to identify specific, discrete points and subpoints in their writing and, instead, indiscriminately group *all* related points together. To test whether a paragraph you have written has more than one main point, try to summarize the paragraph in a single sentence. If you need more than one sentence to cover everything in the paragraph, you probably need more than one paragraph.

 # Writing Sentences

As we have seen, you can control a paragraph's flow of information through its structure and length. The same is true of each sentence in the paragraph. The proper structure and length for any given sentence can be determined by the complexity and importance of the information it contains. In anything you write, some of your information will inevitably be more difficult—more technical, more complex—than the rest, and the sentences that convey this information must make allowances for its difficulty. Similarly, some information will be relatively more important than the rest, and the sentences that convey this information must be designed to emphasize its importance. In this section we will show you how to control the flow of information through the sentences you write.

Sentence Complexity

When we talk of a sentence's complexity, we must consider not only its length, the raw word count, but also its grammar. Simple sentences consist of only a subject and a verb (and usually an object or a verb complement). But sentences can be almost infinitely embellished with dependent clauses, modifying phrases, coordinate constructions, and so on. Though sentence length and grammatical complexity are usually related, they need not be: A series of coordinate constructions (essentially a string of simple sentences joined by "ands" or "buts") may be longer than a grammatically more difficult compound-complex sentence (a sentence with at least one dependent clause and at least two independent clauses).

Much has been written about the ideal level of complexity of a technical sentence, and although there are some minor discrepancies from study to study, virtually everyone agrees that the length and grammatical complexity of a sentence should be determined by the audience and the subject matter. Not surprisingly, the less educated your audience is, the simpler and shorter your sentences should be. Less educated readers are usually comfortable with sentences in the five-to-fifteen-word range. Managers and professionals, who will probably be your most frequent readers, should have no trouble with technical sentences of twenty words or so, but even they may have trouble with sentences that run much longer (more than thirty words), especially if the material is difficult or the grammar is complicated.

Regardless of audience there should be an inverse relationship between the complexity of the subject matter and the complexity of the sentence that expresses it. The denser or more technical the information you are discussing, the less demanding your sentence structure should be.

Sentence Design: Parallelism and Subordination

Though short sentences are easy to read individually, a technical document composed exclusively of short sentences is bound to be choppy and dull. Worse, short sentences tend to give equal emphasis to everything in them, and thus they make it hard for readers to recognize the points *you* find most important. Thus, for most of the things you write, you will need to mix simple sentence structures with those designed to emphasize what you want emphasized. The fundamental principles that control sentence emphasis are parallelism and subordination.

Parallelism. The principle of parallelism is a simple one: Ideas that are equal in importance deserve to be given equal (preferably identical) grammatical form. In the following sentence, the parallel structure gives equal emphasis to all three companies:

> IBM, Victor, and Digital are now vigorously promoting their 16-bit personal computers.

Removing the parallel structure changes the emphasis considerably:

> Along with IBM, Victor and Digital are now vigorously promoting their 16-bit personal computers.

The most basic way to create parallel structures is to reiterate words or parts of speech. Nouns parallel other nouns; verbs parallel other verbs. Sentences are parallel if (like this last one) they contain the same words in the same places, or similar words in the same places, or logically related words in the same places. Structure repetition is central to parallelism, and compound subjects and predicates are just the tip of the iceberg.

One of the most common—and most effective—ways to create parallelism is by stringing together two or more nouns or adjectives:

> Land-grant colleges and universities have a century-old mandate to excel in teaching, research, and public service.

> This interior flat latex wall paint is durable, washable, stain-resistant, and completely lead-free.

Strings of verbs, because they represent actions, are even more commonplace in scientific and technical writing than are strings of nouns or adjectives:

> The procedure involved capturing feral Eastern cottontail rabbits in a semirural area, tagging them, releasing them, and later recapturing them.

Phrases as well as individual words may be strung together in parallel arrangement:

> This project requires six 1′ × 6′ pine planks that are freshly cut, knot-free, and unwarped.

Her primary responsibilities include drafting specifications, preparing budget estimates, designing prototypes, and supervising prototype construction and testing.

To get maximum yield from a vegetable garden, you must be willing to thin it, weed it, mulch it, water it, and fertilize it conscientiously.

Clauses, too, may be arrayed in parallel structures:

Professor Ehrlichman has inquired about the status of his proposal, which I approved on March 19 and which, according to my records, I forwarded to you on March 22.

The patients were examined when they entered the clinic, when they had been on the diet for two weeks, and when they had reached their ideal weight.

The toxicity rose to an unacceptable level in mid-August, and it has continued to rise.

Degrees of Parallelism. You achieve parallelism by creating compound subjects, verbs, or objects, or by composing forceful modifier strings. For additional emphasis, you may showcase that parallelism by repeating the words and phrases that introduce the strings. The following sentence is perfectly parallel:

You must dedicate yourselves to keeping such bias out of your teaching and research.

These versions, however, are more elegant because they *stress* the parallelism by drawing it out:

You must dedicate yourselves to keeping such bias out of your teaching and your research.

You must dedicate yourselves to keeping such bias out of your teaching and out of your research.

You must dedicate yourselves to keeping such bias out of your teaching and to keeping it out of your research as well.

Generally speaking, the more you repeat yourself verbatim in a parallel structure, the more formal it will be.

Parallel constructions, as you can see, are communicated to readers through the grammatical forms, sounds, and meanings of words. They may also be announced in advance typographically—with a dash in informal writing and, in more formal writing, with a colon:

> Air-purifying facepieces fall into three types: the full-face mask, which covers the entire face including the eyes; the half-mask, which covers only the nose and mouth and fits under the chin; and the quarter-mask, which covers only the nose and mouth and fits on the chin.

Note the "double parallelism" here. The three types of masks are presented in parallel, and each is followed by a "which" clause that defines the characteristics of each type. Although commas are usually used to set off the elements of a parallel construction, in cases like this, where commas are required *within* each parallel element, semicolons are used *between* elements.

Sometimes, though not always, lists of parallel ideas are themselves set off typographically and are indented and numbered so that the equal emphasis being placed on them is unmistakable:

> Canister-type gas masks may be subdivided into three categories:
>
> 1. Front- or back-mounted masks, in which the canister is hung in a harness on the wearer's chest or back and connected to the full-facepiece by a hose
> 2. Chin-style masks, in which the canister is mounted directly onto the full-facepiece
> 3. Escape masks, designed for emergency escape, in which the canister is attached to a half-mask facepiece or to a mouthpiece only

Putting Parallel Structures in Order. When you align words, phrases, and clauses with one another, you want to be sure to choose an order for them that is less random than the order in which they occurred to you. In the following sentence, for example, the order is chronological:

> Science is not a solid, indestructible body of immutable truth: it keeps changing, shifting, revising, discovering that it was wrong and then heaving itself explosively apart to redesign everything.

Spatial order, specific-to-general order, order of increasing or decreasing importance, simple-to-complex order, familiar-to-unfamiliar order, and even alphabetical order can all be appropriate for parallel structures. However, it is particularly effective to use parallelism when you want to build momentum in a list, highlight a progression, and emphasize the last entry, as in the following sentence:

❙ This innovation will benefit the company, the industry, and the world.

Using Parallelism for Effect. Parallelism adds balance and emphasis to technical writing. It also adds style. If you do not take the time to structure your language so that it conveys the structure of your thoughts, your readers may well perceive your style as clumsy and amateurish. If, on the other hand, you use parallelism conscientiously, you will find that your writing style will begin to attract compliments. The following paragraph, taken from the 1980 Phi Beta Kappa oration at Harvard University, forcefully illustrates the rhetorical power of formal parallelism. As is appropriate in a highly ceremonial address to a scholarly audience, it uses words, phrases, and clauses in parallel structures embedded within parallel structures. The effect is one of stateliness and extraordinary lucidity.

The author introduces his first three parallel clauses with a colon. They are equal in importance and symmetrical.

Two examples of tiny and fragile organisms follow the colon here. Note that the author has embedded two parallel "which" clauses within the parallel descriptions of chloroplasts and mitochondria. In the next sentence, parallel infinitives are immediately followed by parallel adjectives, as the sentence builds to a climax.

Four successive participial phrases

One major question needing to be examined by scientists is the general *attitude* of nature. A century ago there was a consensus about this: nature was "red in tooth and claw," evolution was a record of open warfare among competing species, the fittest were the strongest aggressors, and so forth. Now it begins to look different. The tiniest and most fragile of organisms dominate the life of earth: the chloroplasts inside the cells of plants, which turn solar energy into food and supply the oxygen for breathing, appear to be the descendants of ancient blue-green algae, living now as permanent lodgers within the cells of "higher" forms; the mitochondria of all nucleated cells, which serve as engines for all the functions of life, are the progeny of bacteria that took to living as cells inside cells long ago. The urge to form partnerships, to link up in collaborative arrangements, is perhaps the oldest, strongest, and most fundamental force in nature.

There are no solitary, free-living creatures; every form of life is dependent on other forms. The great successes in evolution, the mutants who have, so to speak, made it, have done so by fitting in with, and sustaining, the rest of life. Up to now we [humans] might be counted among the brilliant successes, but flashy and perhaps unstable. We should go warily into the future, looking for ways to be end the passage, echoing one another solemnly until the central idea of the address reverberates in the minds of its readers.

more useful, listening more carefully for the signals, watching our step, and having an eye out for partners.

A Note About Errors

There is no such thing as semisymmetry; things either balance or they don't. Once you start to set ideas up in parallel, you must follow through or you will be guilty of faulty parallelism, which is really nothing more than incomplete parallel structure. The following sentence is typical:

> The procedure involves inoculating the samples with test serum, incubating them for 48 hours, and, finally, an analysis of bacteria content will be performed.

We discuss faulty parallelism in the next chapter, pages 120–124.

Subordination. The principle of subordination is a corollary to the principle of parallelism: Ideas that are *not* equal in importance should *not* be given identical grammatical form. Thus, you begin the process of crafting a sentence by determining whether the ideas you intend to include in it are equally important or whether some are more important than others. Then you design your sentence so that the most important idea, if there is one, is in the main clause. Which of the following sentences would you prefer to see in a recommendation letter?

> Although he occasionally makes mistakes, he is—in general—a most meticulous worker.

> Although he is—in general—a most meticulous worker, he occasionally makes mistakes.

Put the two clauses in parallel structure, giving equal emphasis to each of the two ideas, and the comment is neither particularly damning nor particularly complimentary:

> He occasionally makes mistakes, but he is—in general—a most meticulous worker.

By choosing to place two or more ideas in parallel structure or to subordinate one or more of them to others, you communicate to your readers what you consider the relative importance of those ideas to be.

Inexperienced writers tend to favor parallel constructions over subordinate ones, even when the ideas they are presenting are not truly equivalent. Typical is the following sentence, which sounds amateurish because it makes clearly tangential information parallel with important information:

> The pressure plate is made of stainless steel, is coated with three layers of rust-resistant paint, and is designed to hold the electronic components securely in place.

Because what the pressure plate does is what's most important here, subordination, not parallelism, is appropriate.

Degrees of Subordination. Having established that you want to use subordination, you must next decide how much subordination is necessary. You may subordinate an idea by placing it in a subordinate (dependent) clause, or you may express it as a phrase or even as a single word. Consider the following sentence, which uses a parallel construction where a subordinate one would be much more effective:

> Professor Makovsky is a microbiologist experimenting with intergeneric fungal crosses, and she believes that the technique we propose has a high probability of success.

If you put the less important of the two ideas in a subordinate clause, the result is this:

> Professor Makovsky, who is a microbiologist experimenting with intergeneric fungal crosses, believes that the technique we propose has a high probability of success.

If you reduce it to a phrase, here is what you get:

> Professor Makovsky, a microbiologist experimenting with intergeneric fungal crosses, believes that the technique we propose has a high probability of success.

In this sentence, a phrase is all you need. Expressing the idea as a clause—dependent or independent—adds words without adding any more information. A slightly different sentence, however, might require a clause:

> Microbiology professor Barbara Makovsky, who has been experimenting with intergeneric fungal crosses for nearly fifteen years, believes that the technique we propose has a high probability of success.

Clauses are more emphatic than phrases, and phrases are more emphatic than single words. That Professor Makovsky is a microbiologist is mildly important; this information is conveyed in a word. That she has been experimenting with intergeneric fungal crosses is somewhat more important; this information is conveyed in a phrase. That she endorses the proposed technique is extremely important; this information is expressed as a clause, the main clause of the sentence, in fact.

The key to subordinating effectively is remembering that *you* are the one who chooses where the emphasis belongs. Given that virtually any group of related ideas can be worked into a single, grammatically correct sentence, the trick is to determine the relative importance of each, and then to assign each to a grammatical structure—independent clause, dependent clause, phrase, or word—that reflects its importance.

Kinds of Subordination. Words, phrases, and clauses are used to subordinate ideas in two ways: as *appositives* or as *modifiers*. Appositives follow nouns or noun clauses and restate them in different words. Modifiers describe, restrict, or qualify nouns, verbs, or other modifiers. The following sentence contains an assortment of each:

> Professor Barbara Makovsky, a microbiologist experimenting with intergeneric fungal crosses, believes that the technique we propose—protoplast fusion—has a high probability of success.

As you can imagine, only the barest, most elementary sentences are without some kind of subordination. (Stripped of all its subordinating elements, the preceding example would read: "Professor Barbara Makovsky believes.")

Appositives emphasize an idea by reiterating and clarifying it:

> Our goal, to have the new system in operation by October, is now unattainable.

In the preceding sentence, the appositive is an infinitive phrase. In the following sentence, the appositive is a subordinate clause:

> Their principal argument, that we had failed to take into account the possibility of an effect from temperature variation, collapses when you consider that temperature has been shown repeatedly to have no demonstrable effect on the size of tomatoes ripening concurrently.

Long and grammatically complex appositives like the preceding one are usually reserved for highly formal occasions.

Although appositives lend an air of elegance to sentence structure, they should be used judiciously. A long appositive that separates the subject from the verb may make the sentence hard to read, especially if the subject matter is highly technical. In cases like this, choose comprehensibility over sophistication, and don't use the appositive. (Long appositives may also jeopardize your subject-verb agreement. See pages 112–113.)

Technical writers use appositives frequently; they use modifiers constantly. Virtually every sentence you write contains modifying words, phrases, or clauses, and most sentences routinely contain several, as you can see from the following examples:

Because the weather has improved, we now hope to have the cables in place by the end of next week.

The plastic nozzle, which detaches readily from the outflow pipe, should be cleaned thoroughly after each experimental run.

By changing the design of the packing crate, we were able to increase the number of units per crate by 25 percent.

You have, of course, been writing modifiers for as long as you have been writing. What we want to remind you of here is that the decision to use a modifying phrase instead of a single word, or a clause instead of a phrase, rests entirely with you; it is based on your appraisal of the importance of the information you are presenting.

A Note About Errors

Technical writing is especially vulnerable to two types of modifier errors—*misplaced modifiers* and *dangling modifiers*. Here is one of each:

You must enter the room where the experiment is taking place quietly.

In monitoring the emissions carefully, air quality was significantly improved.

We discuss misplaced and dangling modifiers on pages 117–120.

Effective Subordinate Clauses. Subordinate clauses are among the most important modifiers used in technical writing; more than half the sentences in a typical report are complex sentences—that is, those that contain at least one independent clause and one subordinate clause. The two most important types of subordinate clauses for technical writers are (1) those introduced by the relative pronouns "that" and "which" and (2) those introduced by subordinating conjunctions like "if," "although," "because," "after," "while," and so forth. Because technical writing relies so heavily on complex sentences, you need to pay special attention to your subordinate clauses as you compose your draft. The following points will help.

1. *Make sure that "which" always has an antecedent.* Because "which" is a pronoun, it must refer unambiguously to a previously mentioned noun (its antecedent). In the following example, "which" has no antecedent:

> To date, solar energy utilization has been shown to produce no damaging effects to the environment, which helps explain its increasing popularity.

To improve the clarity of this sentence, you can determine what the antecedent of "which" should be, and add it to the sentence. Better yet, recast the sentence to avoid "which" altogether:

> Its increasing popularity can be partly explained by the fact that to date solar energy utilization has been found to produce no damaging effects to the environment.

2. *Insert "that" in those sentences where it is needed to prevent misreading.* Not every "that" clause actually begins with "that." If "that" can be left out without injury to the meaning of the sentence, it often is. In the following sentence, "that" is understood:

> The division director feels [that] we will be able to achieve the necessary staff reduction simply through attrition, with no layoffs.

In some sentences, however, "that" cannot safely be left out:

> We discovered the problem was not in the new software but in the firmware for the system.

This type of sentence often escapes revision because it is clear to the person who created it. Most readers, however, will take in the words "We discovered the problem" as a unit. To prevent that misreading, you should include "that."

3. *Preserve the distinction between "that" and "which."* "That" and "which" are not supposed to be interchangeable. Under traditional rules of usage, "that" is used only in restrictive clauses, those you cannot set off with commas:

> The experiments that we ran last week yielded surprisingly positive results.

"Which," by contrast, is used in nonrestrictive clauses, those that *can* be set off from the rest of the sentence with commas or can even be omitted with no damage to the meaning of the sentence as a whole:

> The experiments, which took the better part of a year to complete, yielded surprisingly positive results.

Not all editors and publishers of technical documents enforce this particular rule. Most journals, in fact, now accept "which" as correct in a restrictive as well as a nonrestrictive clause. But the distinction is worth preserving—especially if your boss is of the old school.

    4. Choose the most precise subordinating conjunction. A subordinating conjunction shows the logical relationship between the idea it introduces and the rest of the sentence. In technical writing, that relationship is most often *temporal*, *causal*, or *conditional*:

Temporal	after, before, since, while, when, whenever, until
Causal	because, as
Conditional	if, although, though, unless, whether, whereas

The differences in meaning among these words are often very subtle, so you must think carefully before you select one. You aren't likely, we realize, to confuse "before" with "after," but you might say "when" when you really meant "whenever." Note the difference between these two sentences:

> When the red light comes on, cut the power by 50 percent.

> Whenever the red light comes on, cut the power by 50 percent.

The difference is subtle, but important. The word "whenever" suggests to readers that the red light can be expected to come on—perhaps even that it comes on regularly, whereas the word "when" offers no such assurance.

    The words "while" and "since" are both ambiguous. Purists restrict "while" to mean "during the time that" and use "although" or "whereas" for other senses; they use "since" to mean "after the time that" and use "because" causally. Be sure when you use "while" or "since" that the context you have supplied makes your meaning clear.

Exercise 3.1

Rewrite the following sentences to eliminate any problems in their subordinate clauses.

1. Influenza spreads rapidly in colleges and universities because many young people in the susceptible age group live closely together in dormitories, which creates a dense population.

2. The rabbits which were fed algal meal gained weight more quickly than the control rabbits.
3. One acre of marsh can support from three to eight million snails, which is brought down to zero in snow goose eat-outs.
4. In the early stages of development, many fish were subject to high rates of mortality, which reduced the number of fish which survived to reproduce.
5. When I opened the supply cabinet, I found the laboratory assistant had not reordered methyl chloride.

Exercise 3.2

The following paragraph sounds choppy and immature because it uses far too many simple sentences. Use subordination to make it flow more smoothly.

The protective apparatus of the eye consists of the outermost coat of the eyeball called the sclera. The sclera is a white, opaque, fibrous coat. The anterior portion, the cornea, is transparent and lies over the colored part of the eye. (The colored part of the eye is called the iris.) The cornea is kept lubricated by secretions from the lacrimal glands. These secretions are more commonly called tears. The lacrimal glands are not part of the eyeball per se, are located at the upper outer margin of each orbit, and are comparable in size and shape to a small almond. However, their secretions are essential. Without them the cornea would be damaged by particulate matter and foreign microorganisms. This would make normal vision impossible. The eyelashes also give some protection against the entrance of foreign substances into the eyes. At the base of the lashes, too, are small glands that secrete lubricating fluid. The eyelids are folds of skin above and below the eyeball. Eyelids are lined with a protective mucous membrane.

Making Your Draft Coherent

A piece of writing is considered coherent when every sentence in it leads seamlessly and inexorably to the next one, every paragraph builds on the one that precedes it, and every movement into a new section or chapter is easy for readers to follow. Writers achieve coherence by providing their work with adequate and effective transitions. Transitions, which may be as brief as a single word or as long as a paragraph, ensure a smooth flow of ideas by keeping readers apprised of where they are, reminding them where they have been, and allowing them to anticipate where they are headed. Since transitions allow readers to see the logic and organization of your work, every sentence, paragraph, and section deserves *some* kind of transition to the adjacent ones.

Transitions Within and Between Paragraphs

There are three ways to create a smooth transition from sentence to sentence within a paragraph or between the end of one paragraph and the beginning of the next one: by substituting pronouns for nouns, by repeating words and sentence structures, and by using transitional words and phrases.

Pronoun Substitution. Most writers (even inexperienced ones) use pronouns—"he," "she," "it," "they," "their," "each," "one," "these," and so on—as a matter of course. The most natural of transitional devices, pronouns provide coherence by "standing in" for a previously mentioned noun. The main value of pronouns is to unclog sentences that would become slow and repetitious if the nouns themselves were to be constantly repeated. Thus, the "liquid metal-cooled fast breeder reactor" in your first sentence can become "it" in your second—providing, of course, that the "it" can refer to no other noun.

Demonstrative pronouns and adjectives—"this," "that," "these," "those"— are also useful as transitions:

> The heavy elements uranium and thorium contain enormous amounts of stored energy in their atomic nuclei. *This* energy is releasable by any process that causes the atomic nuclei of *these* elements to divide.

In the preceding example, the demonstratives are being used as adjectives modifying "energy" and "elements." They can also function as pronouns:

> There are four halogens. Of *these*, only iodine does not undergo halogenation.

> *This* is the first in a series of six experiments.

A Note About Errors

Pronouns must be *exact* substitutes for the nouns they replace, and they must refer to their antecedents unambiguously. The following passage contains two pronoun reference errors:

> The following steps explain to the beginner how to load a roll of film into their Okati III camera. It does not describe its mode selections.

We discuss pronoun errors in the next chapter, pages 115–116.

Word and Structure Repetition. Pronouns are most useful for building coherence between and within sentences, much less so in extended discussions. As readers move farther and farther away from the original noun, they are bound to have some difficulty remembering what the pronoun is standing in for. Thus, a second type of transitional device becomes necessary: repetition.

Word Repetition. Especially when you are introducing new terms or new concepts, repeating them often keeps your readers zeroed in on what each one means. Consider this sentence:

> Phosphates from faulty septic systems can increase the productivity of algae; in 30 percent of the cases studied, the nutrients resulted in significant phytoplankton blooms.

This sentence is fine for biologists, but not for lay readers, who may not know, may not remember, or just may not stop to think that phosphates are one kind of nutrient and algae are one kind of phytoplankton. For these readers, repetition will dramatically increase the coherence of the sentence:

> Phosphates from faulty septic systems can increase the productivity of algae; in 30 percent of the cases studied, the phosphates resulted in significant algal blooms.

Unlike many other kinds of writing, technical writing avoids verbal variety for its own sake. The right word bears repeating. But if you are worried about a term's becoming monotonous, try to find a balance between word repetition and word variety—including pronouns. Notice the balance between key term repetition and pronoun substitution in this excerpt from an expanded definition:

> Immunity differs from natural resistance in that (1) it is not inherent in the body but is acquired during life and (2) it is specific for a single type of microorganism or substance or for closely related groups of microorganisms. It may be active or passive.
> Active immunity results when antibodies are produced by the cells or tissues of the host as a result of contact with microorganisms or their products. This type of immunity is of relatively long duration, ranging from a period of a few months to many years, depending on the antigen and the host response. Active immunity may be acquired by natural means while an individual is recovering from an infectious disease such as chicken pox, mumps, measles, poliomyelitis, or typhoid fever, or it may be conferred by inapparent or subclinical cases of some diseases. Active immunity may also be acquired artificially by injection of an antigen into the body of the host, as in vaccination for smallpox.

The repetition of the key terms "immunity" and "active immunity" makes for smooth transition through the two paragraphs. Note, too, that the second sentence of the second paragraph couples a demonstrative with a version of the key term: "This type of immunity" replaces "active immunity." Such variations keep the repeated term from *sounding* repetitious—without impeding the coherence.

      Structure Repetition: Parallelism. Another good way to create transition in your writing is to repeat sentence structures—that is, to make them parallel. We have already discussed parallelism *within* a sentence; it is equally useful for creating transition *between* sentences. The passage on active immunity illustrates both uses. In the first paragraph, the two ways in which immunity differs from natural resistance are presented in parallel construction:

(1) it   is not   inherent in the body . . .
(2) it   is   specific for a single type . . .

And in the second paragraph, the author uses parallel construction both within a sentence and between sentences:

Active immunity   may   be   acquired by . . .
or it   may   be   conferred by . . .
Active immunity   may also be   acquired artificially by . . .

This kind of structure repetition provides transition by signaling to the reader that the *ideas* presented in these parallel constructions are, like the grammatical elements that convey them, essentially equivalent.

Exercise 3.3

Improve the coherence of the following passages.

1. Kwashiorkor has been described as the most serious of all the diseases afflicting the human population. In parts of the tropical world, four-fifths of the children are either killed or crippled for life by protein deficiency. Moreover, protein malnutrition inhibits full development of the brain.
2. Air sampling for contaminants in an industrial setting is useful, because it usually consists of areas where a specific and predictable type of contaminant is likely to be found. It has serious limitations in a laboratory setting, however, where high concentrations of any one compound are rarely found. The type of compound in use may also change from week to week.
3. Flash has two important uses in underwater photography. You can use it as a primary light source when there isn't enough daylight available to take pictures. Flash also brightens colors that appear dull underwater.

Transitional Words and Phrases. Pronouns, word repetition, and parallel construction do the major job of building coherence. Their great advantage is that they are integral parts of the sentences that use them. Often, however, another type of transitional device is needed: transitional words and phrases. Consider this passage:

> In the past decades the greatest priority in U.S. energy research has been accorded to the development of economical fission power. The energy crisis and the continuing controversy concerning the environmental hazards associated with fission energy have led to a dramatic increase in the levels of funding devoted to solar energy. A thorough understanding of solar's potential as an alternative to fission is some years off.

This paragraph is not so disorganized as it seems. Here it is again, this time with transitional words reinstated:

> In the past decades the greatest priority in U.S. energy research has been accorded to the development of economical fission power. *Recently, however*, the energy crisis and the continuing controversy concerning the environmental hazards associated with fission energy have led to a dramatic increase in the levels of funding devoted to solar energy. *Unfortunately*, a thorough understanding of solar's potential as an alternative to fission is *still* some years off.

Suddenly the passage makes perfect sense: "Recently" links up with "In the past decades" to signal a passage of time; "however" signals the reader that the idea of this sentence will be in contrast to the idea of the last one; "unfortunately" and "still" signal a second turn in the line of argument.

Common Transitional Words and Phrases. The value of transitional words and phrases is that they show readers the *nature of the relationship* between two ideas or two sentences. Here is a list of some commonly used transitional words and phrases, preceded by the relationships they express.

Result	therefore, thus, as a result, hence, consequently
Contrast	however, nevertheless, on the other hand, still
Likeness	similarly, likewise
Alternative	instead
Addition	furthermore, moreover, in addition, also
Example	for instance, for example
Emphasis	indeed, of course, in fact
Repetition	that is, in other words, as mentioned earlier
Conclusion	in conclusion, to summarize, in short
Concession	granted, admittedly
Time	recently, now, afterward, later, meanwhile, first, next, then, simultaneously, finally
Place	here, elsewhere, below, adjacent to, to the left

Using Transitional Words and Phrases Effectively. Although transitional words and phrases are indispensable, not every sentence needs one. A series of short sentences, each beginning with one or another of the words or phrases listed earlier, might be superficially coherent, but there are usually more effective ways of achieving coherence. For instance, instead of stringing together several short sentences with words like "moreover," "furthermore," and "also," try combining the sentences by means of parallelism. Similarly, before you link pairs of short sentences together with a transitional word or phrase, see if you can combine them with a subordinate construction:

The volcano had been dormant for years. Therefore, no one was prepared for its horrifying eruption.

Because the volcano had been dormant for years, no one was prepared for its horrifying eruption.

Subordinating conjunctions, like the transitional words and phrases we have been discussing, demonstrate coherence by announcing the relationship between two ideas; in fact, because they make one idea *grammatically* dependent on another, the coherence signals they send are especially strong. In a well-designed paragraph, sentences containing subordinate constructions should thus be mingled with sentences containing transitional words and phrases.

Although transitional words and phrases (the grammatical term for these is *conjunctive adverbs*) and subordinating conjunctions have similar functions, they are grammatically different, and they cannot be interchanged. A clause that begins with a subordinating conjunction cannot stand alone; it is *subordinate*. However, clauses that begin with conjunctive adverbs *can* stand alone. Compare the following:

However, a glucose tolerance test indicated that the patient was hypoglycemic.

Although a glucose tolerance test indicated that the patient was hypoglycemic.

Both "however" and "although" signal a contrast. But only the first example is grammatically correct; the second is a sentence fragment, a dependent clause that requires an independent clause to complete it. If you are unsure whether the word you want to use to begin a clause is a subordinating conjunction or a conjunctive adverb, there is a simple test you can perform. See if you can move the word to another place in the sentence; if you can, the word is an adverb and the clause it introduces can stand alone:

A glucose tolerance test, however, indicated that the patient was hypoglycemic.

| A glucose tolerance test indicated, however, that the patient was hypo-glycemic.

| A glucose tolerance test indicated that the patient was hypoglycemic, however.

Some of these versions of the sentence are more felicitous than others, but all are correct. If you try the same test with "although," on the other hand, you will see that it cannot be moved without turning the sentence into nonsense:

| A glucose tolerance test, although, indicated that the patient was hypoglycemic.

If the word cannot be moved, it is a subordinating conjunction, and the clause it begins *must* be attached to an independent clause. If it can be moved, it is a conjunctive adverb, and the clause it begins can stand independently. (Remember, however, that if you put this independent clause together with another independent clause, you must separate them with a semicolon, a colon, or an "and," "but," or other conjunction, *never* with a comma. See pages 111–112 for a discussion of comma splices.)

Exercise 3.4

Improve the coherence of this passage.

Using microbial pathogens and other natural predators may be the best method of controlling gypsy moths. It is relatively harmless to other forms of life. It leaves no toxic residues. The natural control method is more expensive than the more popular chemical method of control in the short run. In the long run, it is more economical because it is essentially permanent. Insects can develop resistance to any chemical pesticide, but resistance to microbial pathogens is probably very slow. There has been no reported instance of resistance to a microbial pathogen introduced for biological control. Predators and parasites should not be casually released into the environment. Before any natural control method is used, the gypsy moth population must be carefully monitored.

Transitions Between Major Ideas

If the document you are drafting is short and highly focused, pursuing a single idea through several paragraphs, the devices of transition we have discussed so far will probably suffice to make it coherent. But most technical writing is so complicated that it can't be presented in a single, unbroken series of paragraphs. Instead, it must be broken up into more manageable groups of paragraphs—into sections and subsections—each devoted to one of the document's major ideas. Two types of transitions help move readers from one major idea to the next: (1) transitional sentences or paragraphs, and (2) headings.

Transitional Sentences or Paragraphs. When a paragraph brings a major idea to completion, the next often begins with a sentence-length transition that provides coherence by briefly summarizing what has just been covered and briefly nodding toward what is to come. The most obvious sentence-length transition is one like this:

> Having discussed pronouns, word repetition, and parallelism, we will now discuss transitional words and phrases.

Compare this to the transitional passage that actually appears on page 89:

> Pronouns, word repetition, and parallel construction do the major job of building coherence. Their great advantage is that they are integral parts of the sentences that use them. Often, however, another type of transitional device is needed: transitional words and phrases.

This revision, though a little longer, is less mechanical-sounding; it glances back and glances forward without actually saying "in the last section" or "we will now discuss." Moreover, it puts emphasis on the *ideas* rather than on the mere fact that these ideas have been or are going to be expressed.

A transition between major sections might run several sentences or even paragraphs, depending on the length and complexity of the sections it is bridging. Whatever its length, its purpose is the same—to show your readers how the information you have just given them is relevant to the information you are about to give them.

Headings. The movement from one discrete section of a technical document to another is often marked by a heading as well as by a transitional sentence or paragraph. Traditional headings like "Introduction," "Results," and "Discussion" tell readers the general nature of the material they are about to read, but they don't tell them much about the material itself. Technical writers, therefore, often add subheadings and sub-subheadings to provide more specific information.

Although no universally accepted format for headings exists, all formats are designed to distinguish major topics from subtopics and sub-subtopics. If the organization you work for has a standardized format for headings, follow it. If it doesn't, you can safely use the format in Figure 3.1.

```
                  CHAPTER OR SECTION HEADING

      The major heading is centered on the page, typed in
all capitals, and underlined.  A line of space should be
left both above it and below it.

MAJOR TOPIC HEADING

      The major topic heading is typed flush against the
left margin.  It should be underlined, and all the letters
should be capitalized.  Again, a line of space should be
left both above it and below it.

Subtopic Heading

      Like the major topic heading, this one is typed flush
left.  But only the first letter of each word is
capitalized.  It is underlined and set off with a line of
space above and below.

      Sub-subtopic Heading.  This heading is indented; only
the first letter of each word is capitalized.  Like the
other headings, it is underlined.  Unlike them, however, it
is followed by a period to set it off from the text, which
begins on the same line.
```

FIGURE 3.1   A Standardized Format for Headings.

Headings are frequently used in conjunction with a numbering system. Numbering systems further clarify the relationships among the various sections and subsections, and they greatly simplify the task of locating—or *re-locating*—sections of particular interest to your readers. In addition to the familiar Roman numeral outline system, one system common in technical documents is the decimal numbering system:

1. CHAPTER OR SECTION HEADING

   1.1 MAJOR TOPIC HEADING
      1.1.1 Subtopic Heading
      1.1.2 Subtopic Heading
         1.1.2.1 Sub-subtopic Heading.
         1.1.2.2 Sub-subtopic Heading.
   1.2 MAJOR TOPIC HEADING

2. CHAPTER OR SECTION HEADING
(Chapter or section headings frequently start on a fresh page.)

Format considerations aside, the two most important points about headings are that they should be informative and that they should be parallel.

Making Headings Informative. If your document is organized properly, every section and every subsection will have a specific point to make. Use the key terms of that point in your heading. Suppose, for instance, that you are writing a report on fire safety and prevention at your company. If part of your report discusses methods of speeding up response time to fires on company grounds, you could entitle that section simply "Response Time." But there are more informative alternatives:

❙ Improving Response Time to On-Site Fires

Or, perhaps,

❙ How Response Time to On-site Fires Can Be Improved

The point is that although many headings are only a word or two, you should not feel constrained to stay within this limit. If you need a longer phrase or a clause to make your heading informative, use it.

Keeping Headings Parallel. The reason for having different ranks of headings—some centered in all capitals, some flush left, some indented—is to allow your readers to see at a glance how the different parts of your report fit together. The headings help readers distinguish the major points from the minor ones, and they show readers which points are equivalent. Because headings of the same rank represent sections or subsections that are essentially equivalent (all major topics, all supporting points, and so on), they should be grammatically as well as typographically parallel:

2.1 Designing the System
2.2 Constructing a Prototype
2.3 Testing the Prototype

If the last of these headings were changed to "Tests Performed" or "How It Was Tested," the point that designing, constructing, and testing were equivalent phases of the project would be obscured.

Headings of *different* rank, however, need not be parallel. Moreover, the subheadings under one major heading need not be parallel to the subheadings under another major heading:

How the Problem Originated
   Neglect by Local Officials
   Neglect by the Company

How the Problem Can Be Solved
    Petitioning Government
        Attend Town Meetings
        Write to Congress
    Seeking Legal Redress
        Injunctions and Court Orders
        Lawsuits

Headings are one of the first and one of the last things you write when you draft a technical document. They make their first appearance—usually in very rough form—when you work up your outline. But until you complete your draft—until you know precisely what you have said in your document—you cannot write finished versions of your headings.

Review Questions

1. Why isn't the introduction necessarily the best place to begin writing a rough draft?
2. Define each of the components of a technical introduction.
3. Why should most technical paragraphs be written deductively rather than inductively?
4. What is the ideal length of a technical sentence?
5. When should you make grammatical structures parallel? When should you subordinate one structure to another?
6. How can you make a parallel structure even more emphatic?
7. Discuss the options you have as a writer once you have decided to subordinate one idea to another.
8. What are the two principal kinds of subordination?
9. What is a demonstrative pronoun? A demonstrative adjective?
10. How can you prevent word repetition from *sounding* repetitive?
11. How do transitional words and phrases differ from pronoun substitution and word and structure repetition in the way they provide coherence?
12. What is the difference between a subordinating conjunction and a conjunctive adverb? Why are they easily confused?
13. Why is it important to keep headings parallel?

Assignments

1. Obtain a copy of a technical report. (The government documents section of your library is a good place to look.) Examine the document's introduction, its paragraph design, its headings. Then, focusing on a section of three or four paragraphs, analyze the sentence structure. Identify parallel and

subordinate structures, transitional words and phrases, and transitional sentences. Report your findings to your instructor, suggesting specific revisions where appropriate.

2. In Assignment Number 4 at the end of Chapter 2 (page 66) we asked you to write an outline for a position paper addressed to the dean on the topic of requiring technical writing for all engineers and science majors. Now draft this paper.

3. You work for a company that designs computer games. As part of a research project, your supervisor has asked you to make field observations of college students as they play these games. Your supervisor is particularly interested in any interactions between players and watchers and between or among players themselves. Go to a student center and observe these interactions, and then draft a report of your findings.

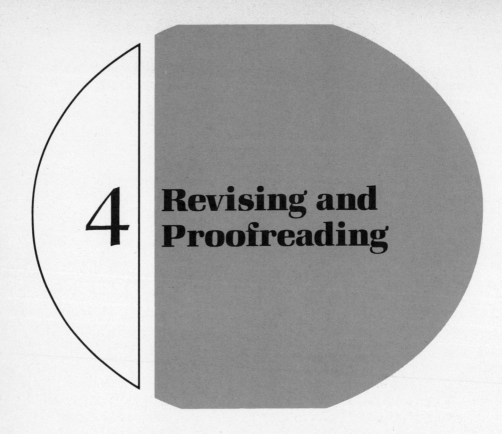

4 Revising and Proofreading

Because the key to effective writing is choosing the best alternative, in a sense this entire book is about revising. If your first draft settled for some less-than-best alternative, "choosing" the best becomes "substituting" the best— that is, revising. For reasons of speed and efficiency, you obviously would like to choose the best alternative the first time. Much of this book aims at helping you achieve that goal, at improving the quality of your first drafts. But trying to write and edit simultaneously, as we pointed out in the last chapter, can easily paralyze an inexperienced writer; at some point you will have to say "that will do for now," resolve to come back to it later, and move on. Even professionals find much to improve in a draft that looked fine when they wrote it several days earlier. At every level of expertise, then, revising remains an essential skill.

The Psychology of Revising

How much time and energy a writer invests in the revision process de-pends on many factors—the importance of the document, the imminence of the deadline, the care and planning that went into the first draft, and so on. On the whole, experienced writers devote more effort to revision than inex-perienced ones do. But the relationship isn't linear. Beginning writers hardly revise at all; their relief at having finished the draft typically generates a glow that blinds them to its flaws. But as a writer begins to learn about writing, the flaws become more visible, and revision becomes more arduous. Then, as the writer learns more about writing, the draft improves and needs less revision.

There are two keys to effective revision: abandoning your emotional commitment to what you already wrote, and switching from the writer's perspective to the reader's.

Abandoning Your Emotional Commitment

It is one of the strange paradoxes of human nature that the more uncertain we are about a job as we do it, the more we tend to resist change once it is done. Every beginning writer has experienced this paradox firsthand. As you write, you complain that the draft is terrible, yet two days later you are prepared to defend every word. This is a natural tendency, but to succeed at revision, you must remember what you didn't like about the draft when you were writing it.

Some of the suggestions we made in the last chapter should help here. If you have put notes in your draft to indicate the ideas you didn't express as well as you had hoped, the sentences that felt awkward to you, the phrases and words that weren't quite right, you will find it a bit easier to detach yourself from what you have written. Similarly, you should begin revising before you recopy your draft. A neat, legible draft ought to be easier to revise than a sloppy one, but sometimes it turns out to be harder—psychologically, at least. It looks too finished—too permanent. (For the same reason, if you are writing your drafts on a word processor as more and more professionals and students are doing, you should resist the temptation to run the spelling check or to format the draft before revising.) All of this, of course, is a matter of attitude: To revise successfully, you must be able to view your drafts as trial runs, not as finished products.

Switching to the Reader's Perspective

Writers necessarily and appropriately think like writers, focusing chiefly on the information that they intend to communicate. The problem comes when writers try to become their own editors. As you sit down to revise your own first paragraph, you already know what is in your final paragraph; in fact, you know things that didn't end up in the draft at all—what you meant to say, what it means, and why it is important. This makes it extremely difficult for you to determine whether or not you have organized and worded your draft effectively.

Your readers will have the opposite problem. They won't know anything except what you tell them. Most readers read fairly quickly and trustingly, for overall meaning rather than for precise detail. If the words are strung together acceptably, readers will rarely notice that a piece of writing doesn't make sense. If words or ideas are missing, they will supply them unconsciously (and they won't always plug in the right ones). If the wording is unintelligible or the organization inconsistent, readers usually assume that the problem lies with them. Rather than admit they don't understand, they forge ahead. The stark fact is that readers can easily think they understand a piece of writing without really understanding it. The true test of understanding comes when they try to *do* something with it—follow a set of instructions, summarize a research procedure, restate an argument in their own words. Many first drafts that look perfectly sound on superficial reading would fail such a stress test.

The only person who can successfully revise a piece of writing, then, is the writer (who knows what it should say), consciously trying to think like the reader (who doesn't). This change from the writer's perspective to the reader's requires an effort of will. Start by recalling who your reader is. As you begin revising, keep asking yourself this question: Could my reader explain what I have written to someone else in his or her own words? The first few times you revise your work, in fact, get the help of an actual reader. Ask him or her to study your draft and explain it back to you. As you listen, notice what your reader leaves out or garbles, but notice also what your reader quotes verbatim from your draft. *Both* are problem areas. After you have done this a few times, you will begin to get the knack of thinking like a reader yourself.

 ## Steps in the Revision Process

Different writers prefer different methods for revising, but all agree that you need more than one pass through the draft. We recommend the following procedure.

Step One: Let It Age

Unless your deadline is imminent, wait a while after finishing your draft before you start revising it. Give yourself some time to forget exactly what you *meant* to say, so that when you begin to revise, you can focus on what you *did* say. Even a few hours will help give you the distance you need to see where the problems are; a space of a few days is ideal.

Step Two: Get an Overview of What You Wrote

Start by reading the entire draft through for overall impressions. Don't change anything yet. Just make notes to yourself: This paragraph seems out of place; that one is boring; this section should be shorter; that one needs more detail.

Step Three: Check the Facts

Now go through the draft again slowly, concentrating on the facts. As you look at each piece of information in your draft, ask yourself if it is accurate; check all figures, formulas, names, dates, and quotations against your notes or—better still—against the original source. This is the time to think about the amount of information you have included and the danger of information overload. First drafts often contain a lot of detail that the reader doesn't really need. Remind yourself sternly of your goals, recall your audience profile, and look for ways to save the reader's time by demoting certain information to an appendix or leaving it out altogether. Finally, stay alert for missing facts, information the reader *does* need that you forgot to include.

Step Four: Examine the Contexts of Your Facts

Having completed Step Three, you know why the facts in your draft are there. But does your reader know? The purpose of Step Four is to make sure that the draft provides adequate context, that it explains the meaning and importance of the facts it includes. Context is information, too—and because the writer knows it so well, it is the sort of information most likely to be left out.

To revise for context, read the entire draft again, asking yourself continually whether readers will understand at each stage where they are, where they are going, and why. Locate the actual words in the text that provide the necessary orientation. If a particular section is not self-explanatory or if it will be clear only after the reader has finished a later section, either add a few sentences of explanation or move the section to a more appropriate part of the draft.

Step Five: Test the Macrostructure

Ignoring all previous outlines, now outline your draft. Is the structure a logical and appropriate one for your goals and audiences? If it isn't, adjust the outline until it is; then cut and paste the draft to match your new outline. In this step you are working for unity and coherence: You want a structure that starts somewhere, ends somewhere, and covers the ground in between in an order that makes sense.

If you make changes in the macrostructure, be sure to make the corresponding changes in your headings and transitional sentences and paragraphs. These devices of coherence are often overlooked in revision.

Step Six: Test the Microstructure

Once the overall structure is coherent, test each part individually. One by one, check your sentences to make sure that each is firmly hooked to the one before and the one after. Within sentences, look to see where you have subordinated ideas and where you have set them up in parallel. Are the main ideas in the main clauses? Are the parallel ideas equal in importance? Also look for items that seem out of place or out of order; if you find any, move them or supply transitions.

Step Seven: Get It into Print

Now is the time to type your draft (or if you are using a word processor, to get a fresh printout) with the revisions you have made so far. If you are still harboring an emotional attachment to what you wrote, seeing it in print will help sever the ties. Your original draft, no matter how crossed out and scratched up, will still look and sound familiar to you. A clean, typed copy will seem less familiar, and you will be able to respond to it more as your readers will.

Step Eight: Prune for Conciseness

You already cut the excess information out of your draft in Step Three; now cut the flab, the extra words and phrases that take up space but contribute little or no meaning. Concise writing is easier to understand and more effective than flabby writing. Finding flab isn't difficult, but it is time-consuming. You must subject every phrase you have written to a battery of questions: Is it precise? Is it relevant? Is it necessary? Is it direct? A "no" to any of these questions indicates the presence of flab. Later in this chapter we will introduce you to the most common types of flab and how to remove them.

Step Nine: Check for Clarity

Occasionally, conciseness is carried too far, and some meat gets cut along with the flab. Thus, once you have scissored out all the unnecessary words, make sure that what is left will be clear to the reader. If necessary, restore the missing information.

At the same time, and remembering your reader's level of expertise, go through your draft for purposeless jargon, undefined terms, and the like. Make every concept, every term, as precise and concrete as it can be. (Chapter 12, "Technical Report Style," will discuss the empty, abstract language that is the principal enemy of clarity in technical writing.) Now—not before—struggle to find the perfect word.

Step Ten: Check for Errors

At last we come to what most beginners think revision is all about—checking for errors of spelling, punctuation, and grammar. For this last step, you will need a dictionary and a grammar book. If you don't remember whether "receive" is spelled "receive" or "recieve," look it up. If you don't remember whether the comma goes inside or outside the quotation mark, look it up. If a sentence just doesn't feel right to you but you can't put your finger on what is wrong, change it to something you *know* is right. ("The equipment has laid unused for six months." Or is it "lain unused"? If you can't decide, play it safe with "remained unused," or—better still—change it to "No one has used the equipment for six months.")

The problem with looking for errors, of course, is that you have to *recognize* them as errors before you can fix them. What this means, in effect, is that you usually have to make them once so that your teacher can identify them for you. To help resolve this problem, we have compiled a list of errors that appear frequently in technical writing; we describe these errors (and how to avoid them) in a separate section later in this chapter. Throughout the book, moreover, we will be calling your attention to particular errors that plague particular kinds of technical writing so that you can be on the alert for them when you revise.

▶ Revising for Conciseness

Almost all first drafts are flabby, bloated out with words that contribute no information. Clipping out a word here and a phrase there takes time, but it will improve your finished product immeasurably. Flabby writing slows down the communication process, forcing readers to extract your meaning from the swelled sentences that contain it. Because concise writing allows your ideas to stand out clearly and precisely, it is both more efficient and more effective.

Consider the following sentence:

> It should be noted in regard to the catalytic converter installed as part of the demonstration project that it seems that this unit is not functioning quite as well or as efficiently as anticipated.

Now we will analyze the sentence phrase by phrase. "It should be noted" is irrelevant; the reader needs the information, not the fact that the information should be noted. "In regard to" doesn't mean anything. "It seems that" is vacillating and indirect. "This unit" is repetitious; the reader knows which unit you are talking about. "Quite" is superfluous. "As well" is both imprecise and repetitious; "as efficiently" carries the same meaning more precisely. Here is what is left when the flab is eliminated:

> The catalytic converter installed as part of the demonstration project is not functioning as efficiently as anticipated.

This concise sentence runs seventeen words, as opposed to thirty-four for the flabby original.

While you are learning to search for flab in your own work, set yourself an arbitrary goal: Try to eliminate at least 20 percent of the words on every page. In time, you will learn to do some of this revising in your head, and your first drafts will become less flabby. To help you learn to revise for conciseness, we have compiled a list of stylistic traps that usually lead to an accumulation of flab.

- Flabby compounds ("along the lines of")
- Redundancies ("round in shape")
- Modifier strings ("difficult, challenging, and complex")
- Indirect constructions ("there is")
- Unnecessary hedging ("it seems that")
- Throat clearers ("needless to say")

Flabby Compounds

Prepositions and conjunctions provide your sentences with their essential connective tissue, but you should hold them to a single word. If you expand them into whole phrases, you are simply adding words without adding meaning. Keep an eye out as well for swollen adjectives and adverbs.

Flabby Compound	*Concise Synonym*
along the lines of	like
a majority of	most
a number of	many
as a consequence of	because
as of now	now
as to	about
at the present time	now
due to the fact that	because
for the purpose of	for, to
for the reason that	because
from the point of view that	as, as if
in accordance with	by
inasmuch as	because
in favor of	for
in order to	to
in terms of	in, about
in the case of	concerning
in the event that	if
in the nature of	like
in the near future	soon
in the region of	about
in view of the fact that	because
on the basis of	by
prior to	before
through the use of	by, with
with a view to	to
with reference to	about
with the exception of	except
with the result that	so that

Redundancies

In writing, *redundancy* is a synonym for useless repetition. "Round in shape" is typical. Since "round" is a particular kind of shape (objects cannot be round in color), the last two words add nothing to the phrase. The same goes for "yellow in color," "attach together," "rise to a higher level," "brief in duration," and so on. Sometimes the repetition is less obvious. "At the present time" is redundant because the present can only be a time; "at present" is therefore all you need.

Redundancies are harder to find when they are spread throughout the sentence, but they are redundancies nonetheless. Consider this sentence:

> The photographer simultaneously snapped the picture while adjusting the lights at the same time.

The idea of simultaneity is expressed here three times. Once will do.

Modifier Strings

In first drafts, modifiers tend to multiply as writers struggle to find the right one. Here is a typical first draft sentence:

> The innovation is significant, important to the company, and potentially very profitable.

There is nothing wrong with this sentence in a first draft. Although it is flabby, it makes a definite point; however, attentive revising is needed to bring that point out concisely. "Significant" means the same thing as "important," but "to the company" narrows the claim usefully. And "potentially very profitable" goes to the heart of the matter, telling *why* the innovation is significant/important to the company. If you suspect your readers may miss the importance of profit, your final draft should read as follows: "The innovation is important to the company because it is potentially very profitable." Otherwise, keep it short: "The innovation is potentially very profitable to the company."

Whenever you see two or more adjectives connected by "and," suspect flab. To test your suspicion, start by asking yourself whether there is any relationship between your adjectives that deserves to be made explicit (important because profitable). If not, see which adjectives you can cut on the grounds that other adjectives in your string include their meaning already ("important" means "significant"). If that doesn't work either, try to come up with a new adjective that can replace several in your string. There are two ways to do this. You can think of a more general word that encompasses the meaning of the ones you have now. Or you can think of a more specific word that focuses on the precise thought you meant to communicate. Try the second approach first; you don't want to replace flabbiness with vagueness. If none of these strategies yields a better sentence, stick with the original. But nine times in ten one of them will.

Adjectives are by far the main offenders, but beware of adverb strings too: "We reviewed each step of the process slowly, carefully, and cautiously." "Carefully" alone will probably do, though, depending on your meaning, "meticulously" might do better. Watch out as well for flabby strings of nouns and verbs: "Next we considered and analyzed the findings and results." "Analyzed the results" is sufficient.

Indirect Constructions

As a rule, the best structure for a sentence is the most straightforward. Indirect constructions waste space and confuse the reader. Yet many technical people, addicted to the roundabout approach, begin many of their sentences with "there is" or "it is." Consider the following sentence:

> There is one point in the proposal's favor, which is that it has a low budget.

Depending on what you wish to stress, you can pare this down in any of several ways:

> One point in the proposal's favor is its low budget.

> The proposal's low budget is one point in its favor.

> The proposal has one point in its favor: a low budget.

All these revisions are shorter and less twisted than the original.

Unnecessary Hedging

To their credit, scientists and engineers are wary of stating conclusions that sound more certain than the evidence justifies. Sometimes, however, this professional caution gets out of control and leads to unnecessary hedging:

> It seems that the data indicate . . .

> This method may be feasible if the circumstances remain unchanged.

> The project should cost in the area of $2 million depending on the adequacy of our approximate calculations.

At its most extreme, such intellectual squeamishness borders on the ridiculous:

> Because symptomatic autonomic neuropathy in diabetics carries a poor prognosis with an increased risk of sudden, unexplained death and cardiorespiratory arrests, its early detection would seem desirable.

We are not suggesting that you treat every tentative conclusion as if it were cast in concrete. But most technical readers are familiar with the central assumption underlying the scientific method—that conclusions are probabilistic, based on existing data, and true only until someone disproves them. Instead of "it seems that the data indicate," you can safely write that "the data indicate" (but not that "the data prove conclusively").

Strangely enough, habitual hedgers often succumb to the opposite habit as well: They overuse intensifiers. Nothing is essential, expensive, or encouraging to these writers; everything is absolutely essential, prohibitively expensive, or tremendously encouraging. (They even qualify the unqualifiable, creating phrases like "very unique.") Apart from being flabby, these extra words actually sap the strength from the statements they are supposed to be fortifying. "I am certain" *sounds* certain. "I am terribly certain" sounds less so.

Throat Clearers

Concise sentences usually get off to a fast start. When a sentence begins with anything other than its main point, look twice for flab. The following phrases are typical slow starters:

It may be said that . . .
Bear in mind that . . .
It should be pointed out that . . .
You may remember that . . .
It seems to me that . . .
I can assure you that . . .
In this context we should consider that . . .

These phrases add nothing to what is coming except to notify the reader that *something* is coming—and they delay its arrival. Some technical writers string these throat clearers together like a chain of empty boxcars: "In this connection we would do well to note that what should be pointed out here is that the hypothesis was confirmed."

The most offensive throat clearer is probably "needless to say." If it is needless to say, don't say it. If you need to say it, don't tell readers you don't.

Exercise 4.1

Each of the sentences that follow is flabby in one or more of the ways we have described. Identify the source of the flab; then rewrite each sentence to eliminate it.

1. With regard to the modifications made to the settling tank, it should go without saying that the odor level is still completely unacceptable.
2. Please be so kind as to handle the samples extremely carefully and gently, as they are fragile and easily broken.
3. In the event of unforeseen complications, you may freely use the telephone number given above in order to reach us quickly.
4. There is an important drawback to the first method, which is that the equipment that it uses costs appreciably more than the equipment required for the second method.
5. The enclosed catalog materials should be useful, helpful, and valuable to you in deciding which Rototiller best suits your needs at the present time.
6. It was the reluctant conclusion of the committee that the project could not be funded this year.
7. Inasmuch as the desalinization project is now nearing its completion, we are expecting to receive your initial payment on the contract within a period of six weeks.
8. Malnutrition among the elderly was again the subject of yet another research proposal last year.
9. The work of the Department of Agriculture is not limited to farming alone but also encompasses forestry and fisheries as well.
10. It is a mistake to suppose that efficiency and efficiency alone was the sole criterion used to judge the designs for the new power plant.
11. We were surprised at how apparently successful the final experiment seemed to be.
12. The initial chromatographs came out rather blurry in our judgment, quite probably because the new equipment had been calibrated extremely carelessly. The results of the second trial, after recalibration, seem to suggest that the equipment is capable of achieving totally acceptable results, at least under the conditions of the study, if the calibration is very precise.

Exercise 4.2

Revise the following paragraph to make it more concise, more concrete, and clearer.

It is no doubt true that there is no single analytical determination that has had a more profound influence on the design, development, creation, and effective operation of water treatment facilities than the accurate and precise measurement of water turbidity. Turbidity is commonly thought of by most people as the relative clarity of liquids. High turbidity is brought about by the presence in water of algae, rust, bacteria, or particulate matter such as clay or mud in suspension. As well as being entirely unacceptable aesthetically, water that is murky and high in turbidity can in some circumstances be regarded as a potential hazard to human health. This is because certain kinds of turbidity interfere with the processes of filtration and disinfection in wastewater treatment. When one takes into consideration the incidence throughout history of diseases caused by water-borne pathogens, it becomes abundantly clear that turbidity is one of the most vitally important parameters in the process of preparing drinking water for human consumption.

Avoiding Common Errors

Sifting through your draft for grammatical errors before you are sure that each of your sentences belongs where it is and says exactly what you want it to say would be an inefficient way to revise—you certainly don't want to spend time making an irrelevant, inaccurate, or flabby sentence correct. Thus, revising for correctness is the last thing you do before preparing your final draft.

Grammatical correctness is important for two reasons. First, even if some conventions of grammar seem merely arbitrary, having no great influence on a reader's understanding, most do, in fact, clarify the language and help prevent ambiguity. Second, although readers aren't likely to think more of you or your ideas just because you *haven't* made any errors, they are very likely to think less of you and your ideas if you *have.*

Before you can correct a grammatical error, of course, you must be able to recognize it. In this section we review the errors most frequently committed in technical writing:

- Sentence fragments
- Comma splices
- Faulty subject/verb agreement
- Faulty predication
- Faulty pronoun reference
- Misplaced modifiers
- Dangling modifiers
- Faulty parallelism

This review should enable you to spot these errors as you revise or—better still—to avoid them in the first place.

Sentence Fragments

Sentence fragments, though they are capitalized and punctuated like real sentences, either lack a subject or a verb, or else fail to express a complete thought. In technical writing most sentence fragments are subordinate clauses or verbal phrases.

Subordinate clauses contain a subject and a verb, but they cannot stand as complete units of thought; they are dependent for their meaning on an independent clause.

A permanent repository for spent nuclear fuel is not expected to be ready until near the end of the century. Even though the temporary storage pools of twenty-eight reactors will be completely filled between 1986 and 1990.

With sentence fragments like this one, repunctuation is all that is required:

A permanent repository for spent nuclear fuel is not expected to be ready until near the end of the century, even though the temporary storage pools of twenty-eight reactors will be completely filled between 1986 and 1990.

Because verbals look much like legitimate verbs, verbal phrases—especially participial phrases—are frequently mistaken for complete sentences:

Our understanding being that the electrophoretic separator would be installed within three weeks.

To correct fragments like this, you have to rewrite the sentence, usually changing the verbal into a verb:

Our understanding was that the electrophoretic separator would be installed within three weeks.

Comma Splices

A comma splice occurs when two independent clauses are separated only by a comma:

The hermit crab has few predators, it is generally free to feed along the beach or in seabeds.

You can correct comma splices in a number of ways.

(1) Make two sentences:

The hermit crab has few predators. It is generally free to feed along the beach or in seabeds. (Although these sentences are now correct, the fact that the two ideas they present are closely related has been lost.)

(2) Replace the comma with a semicolon or with a semicolon and a conjunctive adverb:

> The hermit crab has few predators; it is generally free to feed along the beach or in seabeds.

> The hermit crab has few predators; thus, it is generally free to feed along the beach or in seabeds.

(3) Leave the comma, but follow it with a conjunction:

> The hermit crab has few predators, so it is generally free to feed along the beach or in seabeds.

(4) Make one of the clauses subordinate to the other:

> Because the hermit crab has few predators, it is generally free to feed along the beach or in seabeds.

Technical writers frequently write comma splices when they use conjunctive adverbs as if they were conjunctions:

> The sebaceous glands are highly active in youth, however, their output decreases with age.

The clauses on either side of "however" are both independent, and conjunctive adverbs cannot be used to separate two independent clauses. Replacing the comma before "however" with a semicolon is the easiest solution, but all the options we have just discussed are available.

Faulty Subject/Verb Agreement

A sentence's verb must agree in number with its subject—a singular subject takes a singular verb, and a plural subject takes a plural verb. Agreement problems rarely occur in simple sentences, in which the verb usually follows the subject directly. But as subject and verb move farther apart, agreement is often harder to see:

> The effect of these changes on production costs are unknown.

Despite all the intervening plural nouns, the subject of this sentence is still "effect," so the verb should be "is."

Subject/verb agreement often goes awry when the subject of the sentence is one of these singular pronouns: "each," "one," "either," "none," "neither," and "everyone."

❚ Each of the options have been considered.

❚ Neither of the extender brackets show signs of metal fatigue.

The preceding verbs should be changed to "has been considered" and "shows," respectively.

When two or more subjects are joined by "and," the verb should be plural. The following example is correct:

❚ The weight of the rock sample and the specific location within the test site from which it came suggest recent meteoric activity.

However, when the subjects are joined by "or" or "nor," the verb agrees with the subject closest to it. Both of the following are correct:

❚ Neither the power switch nor the indicator lights were functioning.

❚ Neither the indicator lights nor the power switch was functioning.

Phrases that begin with expressions like "as well as," "along with," "together with," or "in addition to" do not affect the number of the verb, even if they directly follow the subject; thus, the following sentence is correct:

❚ Magnesium, as well as carbon and iron, was found in trace amounts.

Faulty Predication

When a sentence's predicate (its verb, along with any objects, modifiers, and complements that accompany the verb) is not logically related to its subject, the sentence is said to have faulty predication. Like faulty subject/verb agreement, faulty predication tends to occur in long, complex sentences in which the subject and verb are far apart. The most common error of predication arises in those sentences whose subject cannot perform the action of their verb:

> The success of the hydroponics program that we proposed to the Department of Agriculture last year will succeed only if local farmers can be persuaded to participate.

It is "the hydroponics program," not "success," that will succeed; thus, the sentence can be corrected in two ways:

> The success of the hydroponics program that we proposed to the Department of Agriculture last year will come about only if local farmers can be persuaded to participate. (This is correct, but a bit wordy.)

> The hydroponics program that we proposed to the Department of Agriculture last year will succeed only if local farmers can be persuaded to participate.

Sentences with linking verbs are especially vulnerable to faulty predication. A linking verb is like an equals sign; the words on either side of it must balance exactly. Here is a sentence that doesn't balance:

> The effect of the drug is a stimulant to the hypothalamus.

The "effect" is not a stimulant; it is the *drug* that is the stimulant. However, the effect of the drug is *to stimulate*. Thus, three revisions are possible:

> The drug is a stimulant to the hypothalamus.

> The effect of the drug is to stimulate the hypothalamus.

> The drug stimulates the hypothalamus.

The version you choose will depend on the emphasis you want.

Exercise 4.3

Rewrite each of the following sentences to eliminate any sentence fragments, comma splices, faulty predication, or faulty subject/verb agreement.

1. Conversion to resistant varieties, though usually effective with other vascular wilts, are not satisfactory in the case of oak wilt.

2. Logic design, the study of computer architecture, including the schematic representation of computer elements, the technical terminology common to computer machinery, and the structure of computer components.
3. Several infectious diseases are regularly accompanied by malabsorption, however, the mechanisms triggering this condition are unknown.
4. A requirement that all personnel learn emergency evacuation procedures was recently required by the company's safety committee.
5. The power-producing capabilities that this generator demonstrated in the recently completed pilot study makes it an unusually attractive prospect.
6. The tests indicated that neither the tires nor the landing gear were responsible for the accident.
7. Members of the Southshore Fishermen's Association were angered, however, when they learned that the south end of the bay, as well as the entire river, were put off limits to commercial fishing boats.
8. The purpose of this procedure, which has been used effectively in similar experiments with other grasses, will demonstrate the sex characteristics and reproductive requirements of *Buchloe dactyloides.*

Faulty Pronoun Reference

Faulty pronoun reference can take two forms: *faulty pronoun/antecedent agreement* and *ambiguous pronoun reference.* The former occurs when a pronoun differs in person or number from its antecedent; the latter, when a pronoun could refer to more than one antecedent. Consider this passage from a set of instructions on capturing live rattlesnakes:

> Once the rattlesnake is inside the sack and the sack's mouth is securely knotted, never place your hands or fingers on the sack beneath the knot. They are capable of biting through cloth.

Because the intended antecedent of "they" is "rattlesnake," the pronoun should be singular. Making this change, however, would only create a new problem:

> Once the rattlesnake is inside the sack and the sack's mouth is securely knotted, never place your hands or fingers on the sack beneath the knot. It is capable of biting through cloth.

The agreement problem is gone, replaced by an ambiguous reference problem. Because the singular noun "knot" directly precedes the pronoun, while the intended antecedent "rattlesnake" is at the other end of the sentence, the "it" seems to refer to "knot." Here the best solution is to change the pronoun "it" to "a rattlesnake":

> Once the rattlesnake is inside the sack and the sack's mouth is securely knotted, never place your hands or fingers on the sack beneath the knot. A rattlesnake is capable of biting through cloth.

Among the most common agreement problems are those involving indefinite pronouns like "each," "everyone," "something," "neither," and "nobody." These are all singular, so pronouns referring to them must also be singular. ("Everyone had their own opinion on the subject" is wrong; replace "their" with "his," "her," or "his or her," depending on the situation.)

Another common pronoun agreement error involves nouns like "group," "company," and "association." Most writers today use a singular verb with these nouns: "The committee was vocal in their opposition." Because the verb is singular, the pronoun must be singular as well: "The committee was vocal in its opposition."

Ambiguous reference problems, on the other hand, are most often caused by the misuse of demonstrative pronouns. Hasty writers who wish to refer to the whole idea of a preceding sentence or paragraph often begin their new sentence with a vague demonstrative: "This is because ..."; "This shows us that ..."; "This leads to ..." This is fine, so long as there is no possibility that readers will think that the "this" refers to some specific antecedent. Unfortunately, that possibility too often exists, as you can see from the following example:

> The process has two distinct steps, the first step being the removal of nondegradables by filtration, and the second step—by far the more complex of the two—being the oxidation by "hungry" microorganisms of the remaining organic matter. This will be described in detail in section 4.2.

Will the whole process be described in section 4.2? Or will only the complex second step be described there? In this case, you can restore clarity simply by turning the demonstrative pronoun into an adjective:

| This two-step process will be described in detail in section 4.2.

or

| This complex step will be described in detail in section 4.2.

As you revise your own work, watch out for sentences beginning with "this." In the few cases where the "this" is absolutely unambiguous, you can safely leave it in. More often than not, however, you will need to revise.

Exercise 4.4

Eliminate the pronoun errors in the following sentences.

1. Although the typical engine runs well in moderate temperatures, they often stall in extreme cold.
2. Students majoring in English should know how to write well; after all, his or her grades are based entirely on the papers he or she hands in.
3. Each pin is encased in a spring that is held in place by a washer. This enables the pin to return to its original position after each compression.
4. Once you have separated the supernatant from the liquid, boil it for three minutes.
5. This investigation revealed that the laboratory was poorly supervised and that the technicians were seriously undertrained. This was probably the major contributing factor in the failure of the experiment. This illustrates the need for an increase in training funds.
6. The company has decided to move their manufacturing plant to Moline.

Misplaced Modifiers

Misplaced modifiers result when a word, phrase, or clause is not adjacent to the word it is intended to modify and therefore gives the impression that it modifies something else, usually the word it *is* sitting next to. Sometimes the result is comic; other times, it is merely ambiguous:

The research project is progressing more rapidly than anyone had anticipated under the supervision of Dr. Mildred Latham.

Next, locate the switch on the console that is made of plastic.

In the first example, the phrase, "under the supervision of Dr. Mildred Latham," fits grammatically in several places, but using it after "anticipated" fosters an unfortunate misreading. In the second example, the clause "that is made of plastic" modifies "console." If the console—*not* the switch—is made of plastic, then the sentence is correct; however, if only the *switch* is plastic, the modifier is misplaced. To correct a misplaced modifier, rearrange or rewrite the sentence so that the modifier is next to the word it modifies:

The research project, under the supervision of Dr. Mildred Latham, is progressing more rapidly than anyone had anticipated.

Next, locate on the console the switch that is made of plastic.

Next, locate the plastic switch on the console.

Although most misplaced modifiers are phrases or clauses, the single word "only" is a big offender. Notice the differences in meaning among these sentences:

I *Only* she reported her findings at the conference. (No one else did.)

I She *only* reported her findings at the conference. (She did not analyze or discuss them.)

I She reported *only* her findings at the conference. (Nothing else.)

I She reported her findings *only* at the conference. (Nowhere else.)

When you revise, move your modifiers around until they are adjacent to the words they are intended to modify.

Dangling Modifiers

Dangling modifiers differ from misplaced modifiers in that, when you try to move them next to the word they modify, you discover that they don't actually modify *any* of the words in the sentence.

I While eating, working, and sleeping, body cells are reacting in repeated cycles.

I Of what little water is available in these areas, it is too salty to be used on plants.

I By tightening the set screw, the swing arm pivots counterclockwise.

There are two ways to correct a dangling modifier. Leaving the main clause as it stands, you can convert the dangling modifier to a subordinate clause:

I While you are eating, working, and sleeping, your body cells are reacting in repeated cycles.

Or, leaving the modifier as it stands, you can add to the main clause the word that the dangler is supposed to modify:

> Of what little water is available in these areas, most is too salty to be used on plants.

All dangling modifiers will respond to one or the other of these revisions; some will respond equally well to both:

> When you tighten the set screw, the swing arm pivots counterclockwise.

> By tightening the set screw, you cause the swing arm to pivot counter-clockwise.

Dangling modifiers are probably the most frequent error in technical writing. They usually appear in sentences written in the passive voice where the modifying phrase refers to a word that would be in the sentence if the sentence were in the *active* voice:

> Alerted to the problem, the ball bearings were replaced.

Like other danglers, this one can be corrected in two ways:

> Once we had been alerted to the problem, the ball bearings were re-placed.

(Notice that if you transform the modifying phrase into a subordinate clause, it must be in the passive voice.)

> Alerted to the problem, we replaced the ball bearings.

(Notice that if you add the word that the modifying clause is supposed to modify, the main clause will shift into the active voice.)

As you revise, watch out for introductory phrases containing *-ing, -ed,* or *-en* words. If they are followed by a main clause in the passive voice, they probably dangle. (We discuss the use and abuse of the passive voice on pages 268–272.)

Now that we have warned you away from dangling modifiers, we would like to acquaint you with the one construction that is allowed to dangle: the absolute construction. Absolute constructions are verbal or prepositional phrases that traditionally stand by themselves and are accepted as modifying the entire clause to which they are adjacent. Here is a common example:

> Taking everything into consideration, the procedure was not overcautious.

Verbal phrases built around "considering" ("Considering what might have gone wrong ...") and "assuming" ("Assuming that angle *A* is a right angle ...") are often absolute constructions. The word "hopefully" is frequently used as an absolute in conversation, though it is not yet fully accepted as such in formal writing.

Exercise 4.5

Repair any misplaced or dangling modifiers in the following sentences.

1. When put on a restricted diet of 500–700 milligrams of salt per day, Dr. Cutler observed that the rats did not become hypertensive.
2. After drowning in cold water, the rescuer should immediately resuscitate the victim.
3. In order to understand the process of taking a photograph, the terms "aperture" and "shutter speed" must be defined.
4. By irrigating only a small part of the soil surface next to the crop itself, weeds cannot grow in the dry areas between rows.
5. By increasing the amount of chlorine, algal growth was contained.
6. In timber-producing forests, artificial pruning is a technique used to speed removal of dead branches.
7. Put the cap bearing the treatment number you have just prepared on top of each bottle, with the number facing up.
8. Using a high-powered telescope, the planet's third moon can be clearly seen.

Faulty Parallelism

You set ideas up in parallel by placing them in similar (or identical) grammatical structures. Faulty parallelism can take two forms: *errors of form*, in which the ideas are parallel but the grammatical structures conveying them are not, and *errors of substance*, in which the grammatical structures are parallel but the ideas they convey are not.

Errors of Form. The first entry in a parallel construction establishes the grammatical form that all subsequent entries must follow. When you begin with a noun, for example, the rules of parallelism forbid your switching over to prepositional phrases in the middle. The following sentence makes an unwarranted shift from phrase to clause:

> Her primary responsibilities included drafting specifications, preparing budget estimates, designing prototypes, and she also supervised prototype construction and testing.

The last of these responsibilities was probably tacked onto the sentence as an afterthought. To make this sentence parallel, you need only convert "she also supervised" to "supervising." You can do it by ear; "-ing" balances "-ing."

Sentences containing correlative conjunctions ("either ... or," "neither ... nor," "both ... and," and so on) often go out of parallel:

> The problem is either in the outflow valve or it is in the main pump.

The grammatical elements following both "halves" of the correlative conjunction must be the same. All the following are correct:

> Either the problem is in the outflow valve, or it is in the main pump. (two independent clauses)

> The problem is either in the outflow valve or in the main pump. (two prepositional phrases)

> The problem is in either the outflow valve or the main pump. (two noun phrases)

The sentence that follows contains a much more sophisticated error:

> This report presents work done between October and December toward construction of a dairy operation that is productive, economically and environmentally sound.

In this sentence, two parallel structures have been fused together. Assuming that its author meant for "economically" to modify "sound," the sentence could be recast this way:

> This report presents work done between October and December toward construction of a dairy operation that is [both] productive and economically and environmentally sound.

When you have a parallel structure within another parallel structure, you need both "ands"; "productive" must be linked to "sound," and "economically" must be linked to "environmentally." Here are two alternate revisions that make the chain of "ands" less obtrusive:

> This report presents work done between October and December toward construction of a dairy operation that is productive and also sound both economically and environmentally.

> This report presents work done between October and December toward construction of a dairy operation that is productive, economically feasible, and environmentally sound.

The first of these preserves the double parallel structure. The second abandons it in favor of a single parallel structure with three equal elements.

Arranging parallel structures around coordinating conjunctions can cause problems in another way as well. Consider the following sentence:

> To avoid accidental inhaling and staining of furniture, draperies, and rugs, use these sprays outside.

Syntactically, "of furniture, draperies, and rugs" can apply to "inhaling" as well as to "staining." The solution here is to supply "inhaling" with its own modifying phrase:

> To avoid accidental inhaling of vapors and staining of furniture, draperies, and rugs, use these sprays outside.

As you review your draft for errors of form, make sure that the modifiers accompanying one element in a parallel construction don't inadvertently modify other elements as well.

Errors of Substance. You also violate the principle of parallelism when you set up in parallel two or more ideas that are not, in fact, equal. Here is a simple example:

> This issue is of vital importance to biologists, to botanists, and to geneticists.

Not only do these categories overlap significantly, but they are at very different levels of generality. Geneticists constitute a far narrower group than biologists do. You could diagram the three groups as in Figure 4.1a. For true parallelism,

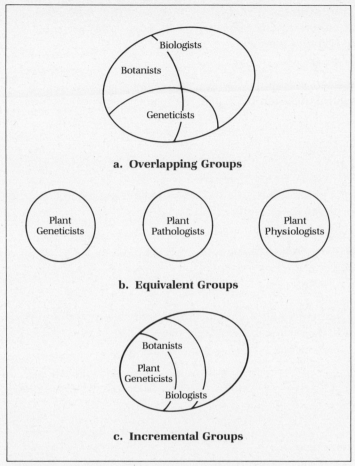

a. **Overlapping Groups**

b. **Equivalent Groups**

c. **Incremental Groups**

FIGURE 4.1 **Three Types of Groupings. Only the second and third lend themselves to parallel constructions.**

you want groups, objects, or concepts that are equivalent, as in Figure 4.1*b*, or incremental, as in Figure 4.1*c*.

Errors of substance are particularly likely to arise, if—in all other respects—the sentence containing the error is perfectly parallel. They are somewhat easier to avoid, however, if the sentence contains an external operator:

> Trickle irrigation has numerous *advantages*: it saves water, labor, and fertilizer; it promotes higher yields; it controls weeds, pests, and diseases; it prevents soil erosion; and it shortens the growing season, producing earlier crops.

Here, the word "advantages" tells us that everything in the list to follow is an advantage. If something that is not an advantage creeps in, the sentence will be out of parallel. An external operator may precede a list (as "advantages" does) or follow it; either way, it regulates what things may be listed. The following sentence is out of parallel:

> Biochemistry and the rising cost of medical school were two very important factors in my decision to change my major to medical technology.

"Biochemistry" is not a "factor" all by itself. Possibly the writer meant to say, "My mediocre performance in biochemistry."

Of all the errors of substance, the most serious is using a parallel structure where a subordinate structure is called for:

> The new computer will go on line tomorrow and should cut production time by 30 percent.

Depending on the audience and the context, this sentence might be revised in a number of ways. Following are two likely possibilities:

> The new computer, which will go on line tomorrow, should cut production time by 30 percent.

> The new computer, which should cut production time by 30 percent, will go on line tomorrow.

If the ideas you are expressing are not truly equivalent, setting them up in parallel will send an erroneous signal to your readers. (For a detailed discussion of when to use subordination instead of parallelism, see pages 80–85 of the preceding chapter.)

Exercise 4.6

The following sentences contain errors of form or substance. Revise them to make them parallel.

1. After you tune the Toyota, check the oil, the transmission fluid, tires, belts, hoses, and the water level in the radiator.
2. Energy from wood waste can be obtained in four ways: by direct combustion, through bioconversion, by reducing it chemically, and through pyrolysis.
3. The need for air, maximum depth and height, and temperature change are all limitations faced by every deep-sea diver.

4. You replace the dirty filter with a fresh one; this is then screwed down hand-tight.
5. What I want is to work in product development at a research firm, to do applied research at a university, or that I might be accepted into a graduate program in plant pathology.
6. The corporate sales structure allows each individual salesperson to gain a top financial and prestigious position in the company.
7. Their conclusions were neither based on sufficient data, nor were they presented clearly.
8. According to some nutritionists, certain common food constituents generate the symptoms of hyperactivity and may be divided into three main categories: (1) salicylatelike compounds, (2) low molecular weight food additives, and (3) artificial colors and flavors.
9. Legislation in the wealthier industrialized countries has been effective in reducing smoke and smog pollution, keeping airborne effluent from factory chimneys under control, and in a general improvement of the quality of the atmosphere.
10. Applicants exceptionally qualified and who have been strongly recommended by the Admissions Committee will be invited for interviews.

▶ Proofreading

Why take the time to make your fully revised, neatly recopied document free from mechanical errors? In truth, with a modest amount of extra effort, a reader can read past your mechanical problems and figure out what you are trying to say. But there are contrary truths.

Many readers in the "real world" will not invest the extra effort needed to read a piece of writing that is substandard. They will simply read something else instead, and your message will be lost. Even if the reader does struggle through your writing, he or she will understand it less fully if it is mechanically inept. Research has shown that this is true even when the reader does not consciously notice the mechanical errors.

Aside from understanding less, the reader of a technical document will be less impressed by its reasoning and less inclined to follow its recommendations if it is riddled with mechanical errors. Once again, this is true even if the reader isn't consciously aware of the errors. Readers who *do* notice mechanical errors typically think less of the writer as a result.

Identifying Mechanical Errors

You should proofread your document for three types of errors: grammatical errors, typographical errors, and spelling errors. Of course, you have already revised your draft for grammar. This time through, you need to look not only for errors that you missed then, but also for new errors that you made as you recopied the draft. Punctuation errors are especially likely to slip through to your final draft, but keep an eye out for all the errors we discussed in the last section.

Spelling also needs to be checked as carefully during proofreading as during revision. As writers revise, they are usually quite diligent in checking the spelling of all the technical words they have used. Seldom, however, do they pay equal attention to their nontechnical words. As a result, these are the words most frequently misspelled in technical documents. Most of the spelling errors, in fact, come simply from making a careless choice between homonyms (words that sound alike) and near-homonyms. Following is a list of frequently confused homonyms; you would do well to familiarize yourself with these words now, so that you can be especially attentive to them when you proofread:

accept	except	
affect	effect	
allude	elude	
ante-	anti-	
apprise	appraise	
ascent	assent	
aught	ought	
bear	bare	
buy	by	bye
capital	capitol	
censor	sensor	
coarse	course	
complement	compliment	
council	counsel	
desert	dessert	
disburse	disperse	
discreet	discrete	
do	due	
elicit	illicit	
faze	phase	
fir	fur	
forth	fourth	
hail	hale	
hear	here	
hole	whole	
incite	insight	
its	it's	
lead	led	
liable	libel	
loose	lose	
oral	aural	
overdo	overdue	
pair	pare	

peace	piece	
peak	peek	
peal	peel	
pedal	peddle	
persecute	prosecute	
personal	personnel	
perspective	prospective	
plain	plane	
precede	proceed	
prescribe	proscribe	
principal	principle	
rational	rationale	
reality	realty	
role	roll	
sight	site	cite
stationary	stationery	
steal	steel	
their	there	they're
thorough	though	through
to	too	two
undo	undue	
vain	vane	vein
waive	wave	
who's	whose	
your	you're	

If you aren't absolutely certain of the meaning of a word in this list, look it up now and mark it in the text. No amount of careful proofreading will help you if you are simply ignorant of the difference between "its" and "it's" or "affect" and "effect."

Typographical errors, unlike grammatical and spelling errors, are caused purely by carelessness, never by ignorance, so time and patience are all you need to catch them. Here are the mistakes you should look for:

- Missing letters, for example, "th" for "the"
- Extra letters, for example, "ther" for "the"
- Transposed letters, for example, "hte" for "the"
- Wrong letters, for example, "thr" for "the"
- Missing words, for example, "on right" for "on the right"
- Extra words, for example, "on the the right" for "on the right"
- Transposed words, for example, "the on right" for "on the right"
- Missing spaces, for example, "onthe" for "on the"
- Missing punctuation, especially parentheses and quotation marks.

A word of caution to writers who use word processors: Many software packages are available to check for typing and spelling errors, and some even check grammar. By all means, use them. But use them *in addition to* proofreading—

not as a substitute for it. Your word processor will have no trouble finding errors like "afferct" for "affect" or "theg" for "then," but it will not be able to detect context-dependent errors like "affect" for "effect" or "them" for "then"; only you can catch these.

Proofreading Techniques

It is a truism, of course, that careless errors can be caught by careful proofreading. But it is hard to see errors in your own work because you know what is *supposed* to be there. Here are some of the tricks that professionals use to catch errors that would otherwise escape their notice.

1. Proofread out loud—better yet, get someone to read your report out loud to you.
2. Look at every word. If you find yourself skimming, use an index card to cover up the lines below the one you are reading.
3. Use a dictionary whenever a word doesn't look quite right.
4. Proofread several times. The first time through, read out loud for missing or repeated words and phrases. The second time, use an index card to check every word. The third time, watch especially for grammatical problems. The fourth time, read against your rough draft to make sure you haven't dropped a whole paragraph or page.
5. Try to proofread several days after you have finished revising.
6. At least once in your proofreading, don't think about the meaning of the words. Reading backwards is a good way to focus on words instead of ideas.
7. When you can, ask a friend to proofread your work after you have proofread it.
8. Be especially alert for writing errors you have made in the past.

Proofreading Symbols

The most efficient way to mark the errors you find is simply to cross out the wrong words and write in the correct ones immediately above. In addition, you can make corrections using proofreaders' symbols. Figure 4.2 demonstrates the symbols you are most likely to use; you can find the complete array in any good dictionary.

Producing Perfect Copy

If the document you are proofreading is the actual document that your readers will see—that is, if it isn't going to be typeset or retyped by a professional—it should be more than correct; it should be perfect. A report filled with proofreaders' symbols and cross-outs looks almost as slovenly as one filled with errors. Instead of simply hand-marking corrections, then, you will

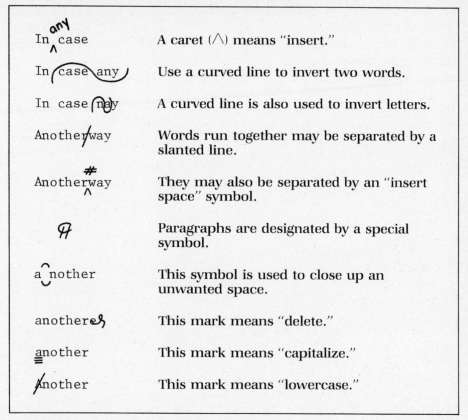

FIGURE 4.2   Common Proofreading Symbols.

want to white out errors and retype the corrections. Nor will you be able to white out more than about two or three errors per page; you will have to retype. However, if you have access to a photocopier that reproduces on bond paper, you can use all the white-out you want; when the original is photocopied, the white-out won't show. The single greatest advantage of word processors over typewriters, of course, is that you can generate perfect copy without extensive retyping; once you have made your corrections, you simply command the machine to print out a clean copy.

Review Questions

1. Why is it difficult for beginning writers to revise?
2. Why should you let a paper age before you revise it?
3. How do you go about checking the contexts of the facts you have cited in your draft?
4. When should you make a clean copy of your draft? Why?

5. What is the best test to use to determine whether your document will be clear to your readers?
6. What are modifier strings? Indirect constructions? Throat clearers?
7. What are the four ways to correct a comma splice?
8. In what kinds of sentences are faulty subject/verb agreement and faulty predication most likely to occur?
9. What is an antecedent? What is an ambiguous pronoun reference?
10. How can misplaced modifiers be corrected? Dangling modifiers?
11. What is faulty parallelism? What is the most serious error of substance?
12. How does proofreading differ from revising?
13. Why is proofreading important?
14. Describe the process of proofreading.

Components of Technical Documents

PART 2

5 | Definitions

Of all the components of technical writing that we discuss in this segment of the book, definitions are undoubtedly the most fundamental. If an idea—*any* idea—is to be communicated effectively, all the terms used to formulate it must be clearly understood by the readers. When readers come across a term they don't know or aren't sure of, they don't usually run to a dictionary or technical encyclopedia to look it up. Instead, they either try to guess the term's meaning from the context in which it is being used, or else simply skip over the sentence altogether and move on to the next one. In the process, they may misread—or miss completely—an important point. Definitions prevent this from happening; they are, in a sense, insurance against misreading.

 Goal and Audience

The purpose of all definitions is to establish common ground between you and your readers. Because much of the writing you will be doing on the job will be directed to managers, technicians, customers, and other people who do not share your technical expertise or the language that goes with it, many of the words and concepts that are the stock-in-trade of your own discipline will need to be defined for them.

When you write to your colleagues or to other experts in your field, on the other hand, you will probably need only an occasional definition—common ground will already exist. But some terms will need to be defined even for them. Scientific and technical knowledge is constantly changing, so definitions are always being refined and updated. The definition of "virus," for instance, is much different today from what it was in 1892, when the term was coined; as new evidence mounts, virologists will continue to revise their definition. Moreover, new products, new technologies, and even whole new areas of study are constantly coming into being, bringing with them new words and concepts unfamiliar to laypeople and experts alike.

How do you know which of the terms you use in a given piece of technical writing need to be defined? Locating your term on the following grid should help you decide:

134

What to Define

	Readers know the concept	Readers do not know the concept
Readers know the term		
Readers do not know the term		

Obviously, if your readers already know the term and the concept, no definition is necessary, and if they don't know either the term or the concept, then you must define it. The other two categories are trickier.

Reader Knows Term/Does Not Know Concept

Quite often your readers will be familiar with a term but not with its meaning, at least not with its meaning the way you plan to use it. Suppose you were writing about the efficiency of a particular make and model of air conditioner. Most laypeople know that something is efficient if it does the job well without costing too much time, money, or effort. But they don't know that the efficiency of a machine is the ratio of energy delivered to energy supplied. If you want to tell laypeople about the efficiency of this air conditioner, therefore, you should define "efficiency." An audience of mechanical engineers wouldn't need the definition, but an audience of lawyers almost certainly would.

Vast numbers of words have specialized meanings in one technical field that differ from their popular meanings or from their meanings in other technical fields. Special care is needed in defining terms like these because your readers are apt to think that they know what you mean even though they don't. To most readers, for instance, "velocity" and "speed" are synonyms; an attentive technical writer will take pains to point out that "velocity" is speed *in a specific direction*.

Also in this category are those terms that depend on a particular context for their meaning because they are inherently vague or abstract. Legal documents like insurance policies and technical specifications, for example, usually begin with a series of definitions that stipulate meanings for terms like "purchaser," "benefits," "property," "defects," "reconstruction," and so on. These terms have no exact meaning outside of the document in which they appear.

Reader Knows Concept/Does Not Know Term

The remaining category is also problematic. Your readers may know a term's meaning, but not the term itself. People who go fishing undoubtedly know that fish can be grown in small ponds for later stocking—but they may not know that the process is called "aquaculture." Readers who have taken a pill to relieve the pain and fever caused by a black eye may not realize that they have taken an "analgesic" and "antipyretic" drug for their "circumorbital hematoma." Whenever you are tempted to use technical terminology for a familiar concept, *force* yourself to consider whether you really need the term. More often than not, you don't; you can write about fish hatcheries and black eyes. But if the only alternative is an awkward or roundabout phrase, or if you plan to spend many pages on the concept, or if one of your goals is to teach readers the term, then use it—and, of course, define it.

 ## Methods of Defining

Not every term that needs a definition needs the same kind of definition. The best definition for a given term may be as short as a single word or as long as several pages, and it may be very simple or extremely complex and technical. What goes into a definition depends on what your readers already know about the term itself; what they know about your field in general; and, given the context in which the term appears, what you think they need to know to understand your point. Take the word "protozoa," for example. If you were writing a short article for a general audience on the animal and plant life of a forest pond, you could probably define "protozoa" as "single-celled animals" and leave it at that. In a microbiology textbook aimed at undergraduates, however, you would probably devote several pages—perhaps even a whole chapter—to your definition of protozoa, describing their size, shape, and physical characteristics; classifying them on the basis of some specific criterion; and providing examples of each type. And if you were writing your doctoral dissertation on a specific protozoan, you might begin with a highly technical definition distinguishing your protozoan from others.

There are five basic types of definitions: informal, formal, operational, stipulative, and expanded. Depending on why you are writing the definition and for whom you are writing it, one of these will be the best one to establish common ground between you and your readers.

Informal Definitions

An informal definition provides a quick clue to the meaning of a term. Usually consisting of a one-word synonym or a short phrase, it is most useful when you want to clarify a term without interrupting the flow of your writing for a more elaborate definition:

Although a small segment of the rock sample was retiform (composed of crossing lines), the rest was completely smooth.

If the solution remains turbid (cloudy), it should be discarded.

> This fierce-looking animal is, surprisingly, a herbivore: it feeds only on plants.

In each of these examples, an informal definition is all the reader needs, because the meaning of the unfamiliar term is uncomplicated and familiar. (But as we said in the last section, before you use technical terms with familiar meanings, you must decide whether your readers really need the terms at all: "If the solution remains cloudy . . ." might suffice.)

Informal definitions can also be useful for explaining a difficult, unfamiliar term that is not central to the point you are making. Because informal definitions are so short, they can't be completely thorough, but often an approximate definition is all your readers really need.

> Impurities can then be removed by electrophoresis, a separation technique involving electrical current.

> The experimenters then sterilized their surgical instruments in an autoclave, which is essentially a sophisticated pressure cooker.

Formal Definitions

The formal definition, sometimes called the sentence definition, is more complete and rigorous than the informal definition. This is the classic method of defining a term, based on a system of logical analysis developed by Aristotle. The formal definition consists of three parts: the term to be defined, the class of objects or ideas to which it belongs, and the characteristics that distinguish it from the other members of its class. It takes the form of a simple formula:

Term =	*Class* +	*Distinguishing Characteristics*
overburden	soil, rock, or other materials	that overlie mineral deposits and are removed in surface mining
photovoltaic conversion	the transformation	of sunlight directly into electricity by use of a solar cell
distemper	a contagious viral disease	primarily affecting dogs and characterized by fever, shivering, convulsions, respiratory failure, and paralysis
caliper	a measuring device	with two adjustable arms, used to find thickness, diameter, and distance

Selecting the Appropriate Class. In any formal definition, you should choose the narrowest class that you are certain is familiar to your audience. Stay away from words like "instrument," "device," and so on when formulating your class—and if you can't, at least add some qualifying words: "a measuring device," "a navigational instrument." And don't begin by calling the term you are defining a "term" or a "word"—"Electron microscopy is a term used to . . ." This merely puts "electron microscopy" in a class that includes every word in existence.

Selecting Distinguishing Characteristics. Finding the best distinguishing characteristics for the term you are defining requires thinking hard about what makes your term different from all the other members of its class. What makes it unique may be how it looks, what it is made of, where it is found, what it is used for, how it works, and so on. Whatever specific traits you select should penetrate to the essence of the term you are defining. Shape and composition, for instance, are certainly part of the essence of a brick; a conical paper brick would be a contradiction. Shape is also an essential characteristic of calipers, because all of them must have two arms; but composition isn't—they can be made of a variety of materials. Like many mechanical devices, calipers are unique because of what they do.

Choosing the distinguishing characteristics for a formal definition is difficult because you have to be sure neither to exclude too much nor to include too much. Defining "bird" as a warm-blooded animal that flies, for example, leaves penguins, roadrunners, and ostriches out and lets bats in. Biologists therefore rely instead on feathers, hard-shelled eggs, and other such characteristics to make the definition airtight.

To check if you are excluding too much, try to think of an example of the term you are defining that doesn't fit your definition. Then broaden the definition, either by changing the distinguishing characteristics or by adding a qualifier like "usually" to cover the exceptions. Our definition of "distemper," for instance, says that it is a disease "*primarily* affecting dogs"; that qualification is necessary because wolves, cats, raccoons, and weasels are also susceptible to distemper.

Similarly, to check if your distinguishing characteristics are too broad, try turning the definition around: "A warm-blooded animal that flies is a ____." If you can fill that blank in with a term other than "bird," your definition is too broad.

Exercise 5.1

Assume that each of the following definitions was written for the typical American college student. Decide whether these are adequate formal definitions. Is the class too narrow or too broad? Do the distinguishing characteristics miss the essence of the term? Does the definition include or exclude too much?

1. An elevator is a mechanical device for conveying people or goods from one place to another.
2. Hepatitis is an inflammation of the liver that is common among college students.
3. A college is an educational institution that awards a bachelor's degree in science or the humanities to students who have passed its curriculum.
4. A typewriter is a machine for writing that produces characters similar to printers' types.
5. A frog is a member of the Ranidae family that is tailless and can leap.
6. Tobacco is a substance that is cured, aged, and then smoked, usually in a pipe or a paper wrapper.
7. A motorcycle is propelled by a gasoline-powered engine.

Writing Effective Formal Definitions. Even if the class is narrow and the distinguishing characteristics are relevant, a formal definition can be hard to understand. Here are four suggestions for writing clear and effective definitions.

Don't Overload Your Definitions. As definition writers think up more and more distinguishing characteristics, the tendency is merely to string them all together with a series of ands and affix them to the end of the definition. But this not only produces awkward sentences; it also obscures major distinguishing characteristics among the minor ones. Choose your characteristics carefully, and subordinate the less important ones.

Avoid Circular Definitions. A definition should not merely repeat the term it is defining. Defining "cost-benefit analysis" as an analysis of costs and benefits cheats your readers of information. The only time this *isn't* cheating is when the term to be defined includes a familiar word along with one or more unfamiliar ones. There is nothing wrong with defining a "sedimentation tank" as a particular kind of tank, or with defining a "pontoon bridge" as a particular kind of bridge.

Use Familiar Words. As much as possible, limit your definition to words the reader already knows, so you won't need additional definitions to explain the first one. As an entertaining bad example, remember Samuel Johnson's definition of "network": "anything reticulated or decussated at equal distances with interstices between the intersections." If you can't define your terms without making use of other equally unfamiliar terms, then you probably shouldn't be writing a formal definition at all. Perhaps all that your readers need is a less precise informal definition; if not, you will need to expand your formal definition into as many sentences or paragraphs as are required to make the term clear.

Keep Your Definitions Neutral. The purpose of a definition is to inform, not to influence; therefore, don't judge the term you are defining. If you are defining "diesel engine," it won't do to say it is "a much better engine than the gasoline engine"—this is an opinion, not a definition.

Operational Definitions

An operational definition is useful for explaining terms that involve processes. Here, for instance, is an operational definition of the term "Marsh Arsenic Test" taken from a technical encyclopedia:

> MARSH ARSENIC TEST. A test for the presence of arsenic, performed by treating the solution to be tested with zinc and hydrochloric acid, whereby the arsenic is reduced to its hydride AsH_3. This hydride and the liberated hydrogen are passed through a long tube that is heated to decompose the arsine, so that a deposit of metallic arsenic forms in the cool portion of the tube, close to its end.[1]

Like a formal definition, the operational definition begins with the class to which the term belongs—in this case, "a test for the presence of arsenic." But then it goes on to distinguish the term by explaining how the test is performed. Thus, an operational definition is also a type of *process description*—these we discuss in detail in Chapter 9. In social science, similarly, operational definitions often specify the measurement procedures to be used.

Stipulative Definitions

You are under no obligation to define every meaning of a term; the meaning you plan to use is enough. To keep your readers unconfused, however, it is usually a good idea to *stipulate* that you are assigning the term only one specific meaning. This is especially true when your readers are likely to know the meanings that you *don't* intend; stipulating your meaning will keep them from thinking that you have gotten the term wrong. Usually, all you need to add to your definition is a phrase like "in meteorology, this term means" or "chemical engineers define this as" to keep the record straight.

If you plan to define a term differently from the way your field has usually defined it, or if the term itself is inherently vague, then you need to stipulate your meaning very explicitly. For example, in announcing one of its grant programs, the Department of Energy provided a list of who was eligible to apply. Included in this list were "individuals," "state and local agencies," "Native American tribes," and "small businesses." All of these are self-explanatory, except for the last—how small is a small business? To make its meaning entirely clear, the Department of Energy provided the following stipulative definition:

> For the purpose of this solicitation, a small business is a concern, including its affiliates, which is organized for profit, is independently owned and operated, is not dominant in the field of operation in which it is submitting a proposal to DOE, and has 100 employees or less.[2]

Expanded Definitions

Although formal definitions are more thorough than informal ones, they may still be inadequate for certain readers. Our definition of "distemper," for instance, might be enough for most readers, but veterinary students would need much more detail to feel that they really understood the term. Virtually all formal definitions, in fact, *can* be made more detailed and complete; whether they *should* be, of course, depends on your goal and audience.

An expanded definition provides additional information that cannot be comfortably squeezed into a formal definition. This supplementary information may require a sentence or two, or a whole chapter, or a whole book.

Methods of Expanding a Definition. Any of the following techniques—or any combination of them—may be used to expand a definition.

Exemplification. Probably the most useful addition is an example or a series of examples to make your definition more concrete:

> The order Marsupiala is characterized by the absence of a placenta and by the presence of a pouch on the female's abdomen, which contains the teats and provides protection for the young, some of which are born in a nearly embryonic state. Included in this group are the kangaroos, wallabies, wombats, bandicoots, pouched moles, and opossums.

Examples are especially useful when the term you are defining is abstract or general, or when your readers are not experts. A definition of "corrosion" aimed at laypeople, for instance, should probably include some easily recognizable examples—a rusted water pipe, tarnished silver, and so on.

Comparison and Contrast. Showing the similarities or differences between your term and some other term is often a useful technique, especially if the two terms are similar enough to be confused:

> A harbor is any body of water that provides ships with a shelter against rough seas, high winds, or other damaging natural phenomena. A harbor is not necessarily a port, though the terms are often used interchangeably. A port, in fact, is a particular *type* of harbor—specifically, one that provides facilities for moving passengers or cargo between ship and dry land.

This technique is also good if your readers are already thoroughly familiar with one of the two terms. If your readers have mastered the four-stroke engine, defining "two-stroke" engine for them won't be difficult. ("Like the four-stroke engine, the two-stroke . . ." "The two-stroke, on the other hand, does not . . .")

Description. A definition of an object can be expanded with physical description—what it looks like, what its parts are, and so on. An expanded definition of "cumulus cloud," for instance, might include the following description:

> Cumulus clouds are fluffy, white masses that bunch up on top of each other in the middle or low altitudes. They have flat bases and rounded outlines.

Illustration. A photograph of a cumulus cloud would certainly make the task of describing one easier. Some definitions, in fact, would be virtually impossible to follow without an accompanying illustration. The definition presented in Figure 5.1 provides a good example.

Classification. The technique of grouping items into logical categories and subcategories based on shared attributes is called classification (see Chapter 6). If the term you are defining represents a group of items—that is, if a single term encompasses a number of variant types—you might consider expanding your definition with a classification. The definition of circular curves (Figure 5.1) is a classification. Another is presented as follows:

> Distillation is the process of evaporating and recondensing liquids at controlled temperatures and pressures. There are three basic types of distillation: destructive distillation, fractional distillation, and rectification distillation. Destructive distillation is . . .

In this example, each type of distillation would next be defined, examples provided, and similarities and differences explained.

Etymology. Often, explaining a word's origin will help to clarify its meaning:

> "Phototropism" is derived from the Greek words "phot-" meaning light and "trope" meaning turn. Phototropism is plant growth in the direction of a light source.

> An amoeba is a rhizopod protozoan characterized by an irregular and ever-changing shape. In fact, its name comes from the Greek word "amoibe" meaning change.

Etymologies are especially useful early in an expanded definition, before you have gone into too much detail. They are not useful, of course, for words

CIRCULAR CURVES. From a mathematical standpoint, a circular curve is an arc having a constant radius; but it is used in civil engineering as a general heading to cover simple, compound and reversed curves.

A circular arc joining two tangents (straight lines) is called a simple curve (Fig. 1). Large radius simple curves are used in highways to provide a means of gradually changing the direction of the center line of a roadway. Simple curves connected by tangents were formerly used on railroads but they have been superseded by the combination of spiral and simple curves. This practice is also followed in modern highways built for high speeds. In Fig. 1, Point A is called the point of curvature (P.C.) and point B the point of tangency (P.T.). Point C is known as the point of intersection (P.I.) of the tangents. In highway practice the length of the curve is generally represented by the length of the circular arc but in railroad practice it is given in terms of chord lengths.

Figure 1

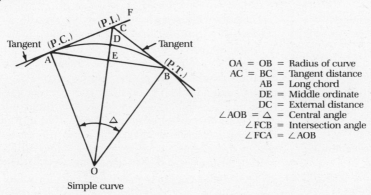

OA = OB = Radius of curve
AC = BC = Tangent distance
AB = Long chord
DE = Middle ordinate
DC = External distance
∠AOB = Δ = Central angle
∠FCB = Intersection angle
∠FCA = ∠AOB

Simple curve

A curve made up of two or more simple curves, each having a common tangent point at their junction and lying on the same side of the tangent, is called a compound curve. Compound curves have an advantage over simple curves since they may be easily adapted to the natural topography of a particular location.

A curve made up of two simple curves, having a common point of tangency at their junction and lying on opposite sides of the common tangent is called a reversed curve (Fig. 2). This type of curve is advantageous for use in connection with railroad crossovers and spur tracks but should never be employed for main lines. Reversed curves are used in highway location when the alignment requires an abrupt reversal in direction.

Figure 2

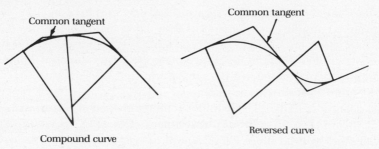

FIGURE 5.1   An Illustrated Definition.

(From *Van Nostrand Scientific Encyclopedia*, Sixth Edition, Douglas M. Considine, ed. Copyright © 1983 by Van Nostrand Reinhold Co., Inc. All rights reserved.)

whose current meanings are far from their origins. "Herpes," for instance, is derived from the Greek "herpein" (to creep)—a connection too tenuous to be of much use.

Analogy. An analogy is a comparison between two things that one doesn't ordinarily think of as being similar. Coming up with an analogy that isn't farfetched or strained is difficult, but a good one can be very effective, especially if you are addressing a general audience. The following definition ends with a useful analogy. It also makes use of several of the other techniques we have just discussed:

Formal definition	Carbohydrates are organic compounds composed of carbon, hydrogen, and oxygen. They are a major source of energy in the human diet, providing about 60% of our total intake of calories. Formed primarily by green plants, some carbohydrates, like wood and cellulose, cannot be digested by humans. Most, however, are easily digested.
Classification	Digestible carbohydrates are divided into two categories: simple and complex. Simple carbohydrates are sugars—including table sugar (sucrose) and fruit sugar (fructose). Complex carbohydrates are starches—potatoes, corn, and rice, for instance, are mostly starch. Both provide "fuel" from which we get the energy to live, but there's a difference in the way our bodies burn that fuel. To understand this difference, think of the two types of carbohydrates as two unlit campfires, one of softwood and one of hardwood. When you light them, the softwood fire will give off as much heat as the hardwood fire, but it will burn out faster, and the fuel will need to be replaced before the hardwood has stopped burning. Simple carbohydrates are the softwood in our digestive systems, and complex carbohydrates are the hardwood.
Informal definitions	
Examples	
Analogy	

The various techniques we have discussed in this section by no means constitute a complete list. Expanded definitions may include discussions of cause and effect, background history, negation (explanation of what a term is *not*), geographical location, necessary materials or equipment—anything that will help readers understand the concept in the way you plan to use it. But bear in mind that all this is *additional information*; it cannot replace your formal definition. Expanded definitions, therefore, usually begin with a formal definition and then go on from there. Note that we said "usually"; sometimes the definition will be more effective and clearer if it begins with a series of examples or with the term's etymology or with a startling contrast and *then* presents the formal definition. Building up to the formal definition is especially effective when you are writing for general readers, who need to begin with terms and concepts firmly rooted in their reality.

Writing Effective Expanded Definitions. Like any good technical piece, your expanded definition should be carefully organized. It should not be a random accumulation of information somehow pertaining to your term. Instead, all the information contained in it should be relevant, logical, and coherent. Consider the following definition of "spermaceti":

> Spermaceti, comparable to beeswax, is a waxy substance obtained from certain whales, especially the sperm whale. It was once widely employed for candles and provides the legal standard of illumination: one spermaceti candle = one foot candle. Spermaceti is used today in many cosmetics, as a stiffener in cold creams and a glossing agent in lipsticks. Spermaceti, unique in nature, is difficult to substitute for and is the main cause of sperm whale exploitation. The seed oil of the jojoba desert shrub, however, is almost identical in chemical composition and is a completely adequate replacement.

This definition is neither relevant nor coherent. It focuses, first off, on how spermaceti is used, not on what spermaceti *is*. Thus, much of the information presented in the definition doesn't really help to clarify the term. The point that spermaceti was used for candles or that it is still used in cosmetics might be worth making, but not until the meaning of the term itself is clear. For the same reason, the analogy to beeswax needs to be developed; readers will get a better sense of what spermaceti is if they are told *how* it resembles beeswax— is it like beeswax in composition or in method of production or in use?

Moreover, the definition throws together so many pieces of tangential information that it doesn't flow smoothly and logically from point to point. Even the writer seems to have gotten lost among the facts: If spermaceti is "comparable" to beeswax and "almost identical in chemical composition" to jojoba oil, how can it be "unique in nature" and "difficult" to replace?

As you can see from the spermaceti example, once you have decided which of the techniques for expanding your definition you actually need, you must spend some time weaving them together so that they read smoothly. If your definition is longer than a paragraph, make sure that each paragraph is unified and that every sentence follows logically from the one before it. Transitions are especially useful in improving the flow.

Exercise 5.2

The following paragraph from a student's definition of corrosion is choppy and hard to follow. Revise it. Feel free to add, delete, or transpose as you see fit.

> The corrosion of iron involves oxygen. If there is no oxygen present, the process of rusting will not take place. The oxygen mixes with moisture in the air and attacks the iron. This forms hydrated iron oxide, commonly known as rust. Other factors promote rusting. Salt acts as a catalyst (helps the reaction occur faster) to corrode the metal quicker. If the metal is under stress, corrosion will occur faster. Without oxygen neither of these would cause corrosion.

Placing Definitions in a Report

Because definitions rarely exist as separate, self-contained documents, you must decide where to place them in a report. There are three choices: in the text itself, in footnotes, or in a glossary at the beginning or end of the report.

Definitions in the Text

The advantage of placing your definitions right in the text is that it gives readers the meaning of the term when they most need it—at the precise moment they encounter the term. The main disadvantage is that placing them in the text may interrupt the flow of your writing. If you have to stop halfway through a point you are developing to define a couple of your terms, your readers may lose the thread. Working informal definitions and short formal definitions smoothly into your text isn't difficult—you can interpolate them into a sentence with commas or parentheses, or append them to the end with a colon or dash. Expanded definitions are harder to work in, but the effort is usually worth it. If the term is important enough to merit an expanded definition, the expanded definition is important enough to merit inclusion in the text.

Where in the text should you place expanded definitions? If the term is essential to the whole report, the definition belongs in the introduction. If the term is essential to only a section of the report, the definition belongs at the beginning of that section. If the term is relevant to only a paragraph or two, it probably doesn't merit an expanded definition—try to cut it down.

Definitions in Footnotes

If you have defined some terms because you aren't sure whether your readers are familiar with them—or because you know that the report is going to several groups of readers, some of whom will and some of whom won't know the terms—consider putting the definitions in footnotes at the bottom of the page. If readers need a definition, they can glance down and get it; if they don't, they can read on, uninterrupted. However, if a single page has more than two or three footnotes, or if the footnotes are so long that they eat into text space, a glossary may be the answer.

Definitions in a Glossary

A glossary is an alphabetized list of words with their meanings, placed either before or after the text of the report itself. As glossaries are in their own separate section, they obviously won't interrupt the flow of your writing. Yet readers often find them a bit unwieldy—readers have to stop, mark their place, flip through the report until they find the glossary, locate the term in the list, then turn back to the text and, doubtless, reread the sentence containing the term. Thus, glossaries shouldn't be used in place of in-text definitions. They can, however, be used effectively *in conjunction with* in-text definitions, especially when the report is long. The first time you use a term, you should define it in the text. Later, when you use the term again, those readers who don't remember its meaning can find it in the glossary rather than having to thumb back through your report looking for your original definition.

Review Questions

1. Describe the difference between definitions written for readers who know the term but not its exact meaning and those written for readers who know the meaning but not the term itself.
2. Under what circumstances might you be better off *omitting* a term rather than defining it?
3. When are informal definitions most useful? Formal definitions?
4. On what basis should you select the *class* for a given term? The *distinguishing characteristics*?
5. Characterize the language most appropriate for defining a complex, technical term.
6. What is an operational definition?
7. Under what circumstances might you need to expand a definition?
8. Describe the most common methods of expanding a definition.
9. How can a word's etymology help clarify its meaning?
10. What are the characteristics of a well-written expanded definition?
11. What are the advantages and disadvantages of placing definitions in the text of your report? In footnotes? In a glossary?

Assignments

1. Write formal definitions of three important terms in your field. Then write informal definitions of the terms, and stipulate an audience and an occasion for which each informal definition would be adequate.
2. Look up an expanded definition of an important term in your field in a standard technical reference book. Rewrite the definition for a different audience, specifying what audience you have in mind.
3. Choose one important term in your field, and specify in detail three different situations that would call for expanded definitions of the term. (If the term were *osmosis*, for example, one of the definitions might be for a repair manual for a filtering apparatus relying on osmosis.) Write the three definitions.
4. As the writer of a freshman-orientation packet, draw up a comprehensive list of academic terms that freshmen might need to know during the first few weeks of school. (Included on the list might be such terms as "drop/

add period," "Dean of Academic Affairs," "credit hours," and so on.) Indicate *how* you would define each term—by formal, informal, or expanded definition. Then define five of the terms.

Notes

1. "Marsh Arsenic Test," *Van Nostrand's Scientific Encyclopedia*, 4th ed. (New York: Van Nostrand Reinhold Co., 1968).
2. "Program Announcement and Grant Application for Northeast Region," Appropriate Technology Small Grant Program, U.S. Department of Energy, 1980.

6 Analysis: Partition and Classification

Analysis, like definition, is an indispensable component not only of technical documents but also of the technical writing process itself. When you sift through and sort information, grouping together related facts and issues and separating them from other facts and issues to which they are not related, you are *analyzing*. What we have been calling "organization" is, in essence, data analysis, and it is at the center of all scientific and technical thinking. It is the first step in bringing order out of chaos.

Types of Analysis: Partition, Logical Division, Classification

In its broadest sense, the term "analysis" encompasses any process exploring the relationship between a whole and its component parts. More strictly defined, however, "analysis" means breaking things down into smaller units. Chemical analysis, for instance, involves the division of compounds and mixtures into their constituent elements. In this strict sense, there are two kinds of analysis: *partition* and *logical division*. When you take a singular object (a hammer, a bicycle, a microprocessor) and divide it into parts, that is *partition*. When you begin with objects in the plural (hammers, bicycles, microprocessors) and divide by types, that is *logical division*. These two types of analysis lead to entirely different subcategories. For instance, if you were asked to partition an egg, you would probably divide it into a shell, a white, and a yolk. Logical division of the plural "eggs," however, would yield subcategories such as Pee Wee, Small, Medium, Large, Extra Large, and Jumbo if you were using size as a criterion, and—if the criterion were quality—Grade AA, Grade A, Grade B, and Grade C. Other criteria for dividing "eggs" logically might be shell color, species, freshness, fertility, or even degree of doneness (for example, hard-boiled, soft-boiled, raw).

You may be aware that the United States Department of Agriculture refers to its system for analyzing eggs by size and grade as a *classification*, not as a logical division. Technically, classification is a grouping process, not a process of division. When you begin with a disorderly jumble of discrete components and assemble them into a whole, that is classification. Classification is nothing more than logical division in reverse, and, for simplicity's sake, the term "classification" is frequently used to encompass both analysis by logical division and analysis by grouping. The distinction between the two is really more theoretical than practical—dividing the concept "eggs" into categories (logical division) versus collecting individual eggs into groups (classification). We will follow custom and use the term "classification" for both processes.

Partition, logical division, and classification are all ways of making sense of the world by finding some reasonable compromise between the undifferentiated whole and the chaotic pile of details. You cannot say very much about a camera without dividing it into parts. You cannot say very much about engines without dividing them into categories. And you cannot say very much about the hundreds of diseases that attack garden tomatoes without grouping them into types. Thinking through an analysis—that is, deciding where to divide and how to group—is the most complex part of writing one. Most of this chapter will be devoted to the thinking process that precedes the writing of an analytical paper. We will begin with partition.

 # Partition

Partition is basically dissection—that is, division into parts. When you analyze a mechanism or a process in order to explain it to readers, you begin by partitioning it. Virtually all partitions in technical writing are components of longer technical documents such as operational definitions, specifications, instructions, process descriptions, and mechanism descriptions. Writing a partition may be as simple as supplying an exploded diagram in a set of assembly instructions, or it may be as detailed and time-consuming as compiling a complete parts list to be attached to the shop manual for an automobile. The type of partition you would normally include in a set of specifications differs somewhat from the type of partition you would include in a mechanism description. Therefore, before you can learn to create either of those documents, you must learn what the different types of partitions are and how they differ from one another.

Types of Partitions

Partitions are either simple or complex, depending on the number of steps they involve. In a simple partition, you divide an object or mechanism directly into its component parts or constituent elements; in a complex partition, you must divide that object or mechanism first into groups of related parts, then into individual parts. Most of the partitions you will need to write will be complex partitions; however, simple partitions are required occasionally in technical writing, so you should know how to do them too.

Simple Partitions. When you give instructions to novice cooks on how to separate eggs for a soufflé, you advise them to throw away the shell, which they don't need, and to place the white and the yolk in separate containers to be added separately to the soufflé. This partition is a simple one because no further subdivision is necessary: You could divide egg yolks and egg whites into *their* components—and you might do so for a class in poultry science—but you would not do so in a recipe. A simple partition is nothing more than a list of parts. There may be thousands of parts in the list, but as long as they are simply listed without being grouped, the partition is a simple one.

Simple partitions appear most often in manuals and in specifications. When you write them, you have only two tasks—preparing a *complete* list of parts or constituent elements and choosing an order in which to arrange the entries. The technical writer who prepares a label for a jar of mayonnaise need only determine from factory specifications what went into the mayonnaise and then select an order in which to list the ingredients on the label. Even that task is made easy by FDA regulations, which require that ingredients be listed on food labels in order of volume.

A slightly more elaborate list than a list of ingredients might include some description of each of the component parts. A routine blood chemistry analysis, for example, lists the important biochemical constituents of blood and gives the quantity of each found in the test sample. The only thing you really have to worry about when preparing a simple partition is an inadvertent omission.

Complex Partitions. Complex partitions are necessary when the *relationships among parts* are important. In a simple parts list, each part is a discrete unit, but in a complex partition, the object or mechanism being partitioned is first divided into groups of related parts; then *within those groups* the parts are arranged in some type of logical order. There are two types of complex partitions used in technical writing—partition by attribute and partition by function.

Partition by Attribute. When you partition an object or an abstract entity by attribute, you divide it into parts in order to deal with the *parts* effectively. When hardware stores divide thermos bottles into metal, glass, rubber, and plastic parts or into breakable and unbreakable parts, their object is to separate from one another those parts that must be treated differently. They separate replacement glass liners from replacement stoppers and lids because the plastic parts sell much more slowly than the glass parts do. An automotive parts store catalog divides a car into its parts generically: All the filters are grouped together, as are all the belts, all the hoses, and all the bolts. In a partition by attribute, grouping together those parts with similar properties and differentiating among them are far more important than showing how the parts fit together to form a whole. Screws go with other screws, not with the plates to which they are attached.

Partition by Function. In a partition by function—more commonly called a functional analysis—you divide an object or system into groups of related parts, not on the basis of a common attribute but on the basis of a common *purpose.* Toggle pins, legs, and washers go together with whatever it is they anchor, support, or seal. In a functional analysis you are not particularly concerned with the parts themselves; instead, you are interested in how they work together in the mechanism as a whole—and so are your readers, who may be seeking some understanding of the theoretical principles underlying the operation of a wastewater treatment plant or who may have to assemble a bicycle from a crate of parts. In the crate, those parts may be packed by attribute—wheels together, chains together, and all the bolts in one plastic bag—but in the directions the bicycle had better be divided functionally, or its new owner will never be able to put it together successfully.

You will write partitions by function more often than any other type. They form the backbone of instructions, process descriptions, and mechanism descriptions—three of the most common types of technical writing. Most technical readers want more than a list of parts; they want to know how those parts fit together within a process or mechanism. For those readers, only a functional analysis will suffice.

Steps in the Partition Process

Once you have decided what type of partition to undertake, systematic partition of a concrete object or an abstract entity has two further steps: (1) dividing and subdividing and (2) organizing and writing the partition.

Dividing and Subdividing. The actual process of partition begins with the systematic disassembly of the concrete object or abstract entity that is the subject of your analysis. If you are working with a concrete object that is fairly small, but not too small, such as a piston pump, a pencil sharpener, or a frog, you can probably take it apart yourself. If yours is a simple partition, decide on the order in which you plan to list the parts in the final document; then line them up in that order as you remove them to save yourself the trouble of sorting them later.

If yours is a partition by attribute, you must begin by selecting the attribute or attributes that you plan to use. As you remove the parts, you are going to have to assign them to groups on the basis of their size, weight, constituent material, perishability, durability, flammability, price, or whatever. The list of attributes you can use is virtually limitless, so choose an attribute or property that is relevant to the needs of your readers. If they must decide which components to ship air freight and which to ship by cheaper surface mail, choose "weight" as a criterion. If they must decide which components are to be stored in what facility, choose "perishability."

If yours is a partition by function, this first step will probably be the hardest of all. When you partition by function, you must divide the object or entity that you are analyzing into subsystems, and to do that you must know what its subsystems are. That means you must understand the scientific principles underlying the design of an object and its operation. A cigarette lighter, for example, works by combustion, requiring heat, air, and fuel. If you know this, it is a relatively simple task, when dismantling the lighter, to assign its parts to the fuel delivery system, the air delivery system, or the ignition system. If you do *not* know how a cigarette lighter works, you will still be able to disassemble it into a heap of parts, but you won't be able to organize those parts by function. Therefore, whenever you are obliged to partition an object or a system that is unfamiliar to you, you must begin by acquiring a working knowledge of the principle or principles of operation that define its subsystems.

Understanding the Subsystems. There are two ways in which you can teach yourself how an object or system works: You can find out for yourself by taking the object apart and putting it back together, or you can get someone who already knows how it works to teach you what the subsystems are.

In many of the partitions that technical writers are required to write, actual physical disassembly is not feasible. Enormous, complex mechanisms like nuclear reactors and tiny mechanisms like microprocessors (whose parts can be seen only through an electron microscope) cannot be taken apart easily. Abstract entities such as corporate hierarchies and industrial processes do not permit any sort of literal disassembly at all. And many concrete objects are, for practical purposes, unavailable to a technical writer for actual disassembly—for instance, you couldn't take a human lung apart without access to a dissecting room. In these cases, you have no choice but to disassemble on paper.

Dissecting objects on paper is much less satisfactory than actually disassembling them because it is necessarily secondhand. Because you cannot observe the parts directly, you run a good risk of forgetting some of them when you compile preliminary parts tallies. Worse still, because you are prevented from observing for yourself how parts fit together, you must know before you start what binds those individual parts into coherent subsystems. If you are assigned to analyze a Wankel engine or a pond ecosystem but have no firsthand knowledge of how one works, you will have to learn the theory from someone who knows. This is the approach used by journalists and by technical writers who are not themselves technical experts. They interview as many experts as possible to get a balanced and complete view of the underlying theory; they study the specifications to learn the parts; then they systematically organize the parts according to the theory.

Defining the Subsystems for Readers. Occasionally, when you undertake a partition by function, you may find that the theoretical subsystems and the physical parts do not exactly correspond. In that case, you will need to define the subsystems for your readers and to draw clear lines of demarcation between them—arbitrarily, if necessary. Consider the following excerpt from an encyclopedia article on the human respiratory system:

> The respiratory system in man consists of the nasal cavity, the throat (or pharynx), the voice box (or larynx), the windpipe (or trachea), the bronchi, and the lungs, all of which are involved in the act of breathing.

The Nasal Cavity. The nasal cavity is a space of complex shape, lined with mucous membrane and bounded below by the mouth and above by two air-filled cavities, the frontal and sphenoidal sinuses, and by a third cavity, the cranial cavity, which encloses the brain. A bony and cartilaginous partition, the nasal septum, divides the nasal cavity into right and left sides. Into the cavity's lower wall, three ridges project . . .[1]

The paragraphs that follow this one give, in sequence, the parts that comprise the throat, the voice box, the windpipe, the bronchi, and the lungs.

Notice that this partition of the human respiratory system does not include the diaphragm, which plays an important role in respiration, but does include the pharynx, which is as much a part of the digestive system as it is of the respiratory system. The author of this article made an arbitrary decision as to which organs to assign to the respiratory system. The author began the article, in fact, by acknowledging the role of the diaphragm in respiration and then deliberately excluding it from this particular system. You, too, may have to choose arbitrarily where systems and subsystems in your partitions will begin and end.

When you partition by function, you may wish to assign a multipurpose part to all the systems in which it plays a part or to only one. Either tactic is fine as long as the arrangement you choose helps readers understand how each part fits into the whole. Like the author of the article on respiration, if you are faced with an arbitrary decision about which parts to assign to which system, simply assign them to one or the other, and then explain to your readers what you have done and why you have done it. Define the systems clearly in your introduction, and point out where you have drawn the lines between them. Your readers may disagree with you, but as long as they understand you, your partition has accomplished its purpose.

Organizing and Writing the Partition. Organizing and writing a partition are easy once you have thought it through. Tallying all the parts and assigning them to categories or to subsystems is the hard part. To write the partition, all you need to do is to convert your scratch-paper lists to paragraphs and add whatever descriptive detail, if any, your audience needs to understand the lists. You may elaborate a partition, if you choose, with definitions of the various parts, with descriptions of their physical properties, with their numerical quantities, or with any other details about them that your readers need. Partitions vary considerably in scope and in level of specificity. They all begin, however, in the same way—with an introductory overview.

Introducing a Partition. A partition usually has only a brief introduction of its own, because it is generally part of a much longer document that has its own introduction. Virtually all partitions begin, however, with a short overview of what is being partitioned. This overview may consist of no more than the name of what is being partitioned and a statement of the division criterion (function, price, flammability, and so on). For simple partitions and partitions by attribute, that is all that is needed. Partitions by function often begin with a definition of the mechanism or system as a whole, most frequently an operational definition. The introduction to a partition by function may also include a short description of what the mechanism or system—as a whole—does. That provides a context for later descriptions of what each of the subsystems does.

Partitions of mechanisms or processes often begin with a drawing that itself initiates the partition process. Drawings such as cross sections, cutaway views, and exploded diagrams show both a mechanism as a whole and all its parts simultaneously. Charts and graphs may be used to partition abstract entities. Pie charts (circle graphs) are often used to partition mathematical wholes. Other graphics commonly used in partitions include flowcharts, which are used to partition processes, and organizational charts, which are used to partition abstract entities such as administrative hierarchies. You will find examples of all of these types of charts in Chapter 11, "Graphics."

The introduction to a partition ends with a statement of the categories or subsystems to be used. Announcement of these major subdivisions leads smoothly into the body of the partition, in which each of the categories or subsystems is discussed in detail.

Developing the Body of the Partition. When you divide wholes directly into parts, the verbal text of your analysis may comprise nothing more than a list of those parts, or it may contain lengthy descriptions of each part in relation to all of the others. Either way, the organization of the partition will proceed part by part, so the parts should not be taken up in random order. They are usually ordered chronologically, spatially, or in order of importance. Industrial formulas begin with lists of ingredients in the order in which they will be used. The exhaustive parts lists that make up much of a set of construction specifications (see Chapter 7) are arranged in general-to-specific order and in order of decreasing importance. The partition of the human respiratory system that we quoted earlier is arranged in operating order, which also happens to be a spatial order moving from the outside of the body to the inside. And reference documents such as inventories are usually arranged numerically or alphabetically, so that readers can locate individual entries easily.

When you divide wholes into categories or subsystems, a more complex presentation is necessary. First, you must state the basis of the partition ("For insurance purposes, we have divided into flammable and nonflammable materials . . ."). If you have stated the basis of the partition in your introduction, you need not repeat it in the body, but you should mention it somewhere so that your readers will be properly oriented. Then you should announce the first set of major subdivisions (if you have not done so in your introduction), providing as much definition and description as you think your readers will need to tell them apart. You may have as few as two major subdivisions and, if you are partitioning by attribute, as many as ten or twelve. If you are partitioning by function, you should have no more than about five or six subsystems. If you have more, you have probably gone directly to *sub*-subsystems. Look to see if any of your subsystems can be grouped together.

Once you have established what your major subdivisions will be, you must choose an order in which to present them. You may choose spatial order, chronological order, order of importance, or any other order that suits your goal and audience. In the following example, an excerpt from Robert Pirsig's now-classic analysis of the motorcycle in *Zen and the Art of Motorcycle Maintenance*, the subsystems are presented in operating order:

If you are partitioning something as complex as a motorcycle, you may have to divide and subdivide repeatedly before you arrive at individual parts.

Pirsig himself points out this analysis is "duller than ditchwater." Purely analytic writing, without description, has a kind of static elegance, but it rarely comes alive.

A motorcycle may be divided for purposes of classical rational analysis by means of its component assemblies and by means of its functions.

If [it is] divided by means of its component assemblies, its most basic division is into a power assembly and a running assembly.

The power assembly may be divided into the engine and the power-delivery system. The engine will be taken up first.

The engine consists of a housing containing a power train, a fuel-air system, an ignition system, a feedback system, and a lubrication system.

The power train consists of cylinders, pistons, connecting rods, a crankshaft, and a flywheel. . . .[2]

There are only two hard-and-fast rules in organizing partitions: List your categories or subsystems before you discuss them, and discuss them in the order in which you have listed them. You yourself must decide when to divide "across" the categories or subsystems and when to divide "down" them. *Dividing across* means that you describe all parallel categories or subsystems in general before describing any one of them in detail. *Dividing down* simply means that you analyze the first category or subsystem completely (as Pirsig does) before moving on to the second.

Assume, for example, that you are dividing a computer for a nontechnical audience. When you say: "A computer consists of an input system (or systems), a processor, a memory, and an output system (or systems)," you are dividing across. When you say: "The first important part of a computer is its input system, which may consist of one or several input devices such as a card reader, a terminal, magnetic tape, or magnetic disks," you are dividing down. It is best to divide across when you are dealing in generalities and down when you are concentrating on particulars. Short definitions of "input system," "processor," "memory," and "output system," or one-sentence summaries of their

what if an item falls into more than one classification? — Just so your consistent purposes, help readers understand how a computer functions *as a whole*. Once you begin describing in detail how any one subsystem functions, however, it is best to go on dividing and finish off that subsystem before moving along to the next.

Concluding a Partition. Partitions do not really need much of a conclusion; the body of an analysis contains the information its readers are seeking. Therefore, all you need to do to conclude a partition is to warn readers in some way that the discussion is over, so they won't think they are missing the last few pages. There are two ways to provide such a warning. One is with a summary overview, or resynthesis, of what you have just analyzed. The second way to end is with a generalization that puts the organizing principles underlying the object or abstract entity into some sort of perspective. Here is one example:

> As this analysis clearly demonstrates, a computer is *not* a mechanical brain. It is part of a brain; the memory and the processor are derived from biological models. But a computer does not *think*. The computer does not yet exist that can write its own programming unassisted or debug itself when its programming fails. Computer technicians have an expression: "Garbage in; garbage out." They—better than anyone—know that the information you get from a computer is precisely as good or as bad as the information you put in.

▶ Classification

As we stressed at the beginning of this chapter, classification is a grouping process, not a process of division. When you classify "aircraft," for example, you are interested in *kinds of aircraft*, not *parts of an aircraft*.

Classification imposes an order on objects, facts, or ideas that are randomly arranged. When you classify, you group together everything that is alike, differentiate what is *in* each group from what is outside it, and then subdivide each group into mutually exclusive categories. For instance, when you classify insects, you must omit spiders. Arachnids look very similar to insects (especially to nonscientists), but their eight legs make them significantly different. Similarly, bicycles must be distinguished from tricycles and motorcycles before they can be broadly divided by the number of gears (1-speed, 3-speed, 5-speed, 10-speed), then perhaps subdivided by wheel diameter or frame design.

Partition and classification have much in common, of course: Both have as their primary goal the creation of meaningful relationships between small and large units. But the differences are crucial. When you partition, you are dealing with a finite number of parts. When you have covered every part of the object or abstract entity you began with, you are finished. Moreover, in partition, the relationships among the parts are predetermined. You do not *create* relationships among components when you partition; you *discover* them. When you classify, on the other hand, you create a system of categories and subcategories into which new entities can logically be fitted. Should a new entity appear that does not fit the categories—a new kind of aircraft— the categories must be revised to incorporate it.

The challenge of classification is coming up with the characteristic or characteristics on which you hinge your categories and subcategories. This is called the *classification criterion*, and settling on one can be a major intellectual task. Because of the search for a sound criterion, the process of thinking through a classification is quite different from the process of thinking through a partition. Your choice of a criterion will depend ultimately on your goal and on the needs of your audience. However, you may arrive at that criterion through either of two approaches—deductive or inductive. The first step in the classification process is deciding which of these two approaches to use.

Types of Classification

There are three types of classification performed by technical writers: deductive classification, limited inductive classification, and formal inductive classification. The first two of these are fairly easy to do; the third poses an enormous intellectual challenge.

Deductive Classification. You are classifying deductively when you decide on the criterion *before* you begin grouping. For example, when you classify stereos according to price and then subclassify them by quality within each price range, you are using a deductive approach. This approach is most appropriate when you are sure of your exact audience and your exact goal— when you know, for instance, that your particular readers want to see automobiles classified by the number of cylinders, oranges by size, or cables by tensile strength. You need not go to the trouble of deriving the criterion if you have been told it in advance or have guessed it, and you need not experiment with various other criteria in search of a better one. What you do need to watch for when you classify deductively is to make sure that your dividing lines between groups are clear—"under $100," not "cheap."

Limited Inductive Classification. You are classifying inductively when you generate a criterion by arranging and rearranging the items to be classified until they fall into mutually exclusive groups. In a limited inductive classification, you have a finite array of items to be classified. Assume, for example, that you work part-time for a tire store and that one of your jobs is to classify the inventory of tires in the warehouse. You might classify them by brand name, by purpose (snow tires, all-weather tires, and so on), by design (bias ply/radial or steel-belted/glass-belted)—it would be up to you to choose a criterion that would help salespeople locate the right tire for a customer as quickly as possible. But you would be under no obligation to devise a system that included every brand name or every conceivable type of tire design. Your system would be complete as long as it included every brand and every design that your particular store carried. You also perform limited inductive classification when you classify the facts and ideas to be incorporated in a written document into a scratch outline (see pp. 63–65). In limited inductive classification, all you need is a place for everything you have, not a place for everything you *might* have.

Formal Inductive Classification. In formal inductive classification, you have an infinite array of items to be classified. You must classify *all* insects, *all* beverages, *all* flowering trees. Scientific classification is inductive; biological taxonomy, for example, has come about through detailed and systematic observation of nature by thousands of scientists over several centuries. Even today, the classification of insects is still being refined as new varieties are discovered and identified, and the classification of viruses is still in its initial stages, limited by our incomplete knowledge of these organisms. In formal inductive classification, you gather together physically or on paper a representative sample of insects, beverages, or flowering trees and then examine them for common denominators. Next, from your list of common denominators, you select the criterion for grouping them that seems most relevant to the needs of your audience.

Formal inductive classification is time-consuming, difficult, and frustrating, chiefly because it must meet three unchanging standards: parallelism, mutual exclusivity, and completeness.

Standards for Formal Classification

The three standards for formal classification—parallelism, mutual exclusivity, and completeness—apply to all kinds of classifications, both deductive and inductive. But, as you will see, they are easy to achieve for deductive classifications and for limited inductive classifications, both of which are self-contained. Achieving them in a formal inductive classification is much more difficult.

Parallelism. Categories are parallel when the same criterion has been applied to arrive at each of them. For instance, the classification air pollution/water pollution/land pollution/industrial pollution is not parallel. Between land pollution and industrial pollution the criterion has changed from the nature of the recipient of the pollution to the nature of the pollution source. Unless you use the same criterion throughout, you are not really *dividing* at all; that is why a classification that is not parallel is flawed. It is permissible to change criteria when you move *down*, but not while you are moving *across*. That is, you may divide pollution into air pollution, water pollution, and land pollution; and then *sub*divide one or all of those groups into industrial, residential, municipal, and so on. But each level of the classification must itself be parallel.

An extreme form of criterion change is *re*dividing rather than *sub*dividing. When you change criteria every time you create a new category, you are not actually dividing the whole into categories at all; instead, you are dissecting it several different times, each time in a different way. Here, for example, is an unsuccessful classification of pencils.

> I. Purpose
>   A. Writing instruments
>   B. Marking instruments
>   C. Drawing instruments
> II. Type of Lead
>   A. Graphite pencils
>   B. Charcoal pencils
>   C. Chalk pencils
>   D. Grease pencils
>   E. Wax pencils
> III. Style
>   A. Retractable
>   B. Nonretractable

For this classification to work, I, II, and III must be the names of subgroups, not the names of criteria. Here is a revised outline that works:

> I. Marking Instruments
>   A. Grease pencils
>   B. Wax pencils
> II. Drawing Instruments
>   A. Wax pencils
>   B. Chalk pencils
>   C. Charcoal pencils
>   D. Graphite pencils
> III. Writing Instruments
>   A. Wax pencils
>   B. Graphite pencils
>   1. Retractable
>   2. Nonretractable

An excellent way to ensure that you do not redivide rather than subdivide at any point is to sketch out your classification scheme in the form of a tree diagram, as in Figure 6.1. A tree diagram helps to ensure not only the parallelism of categories, but also their mutual exclusivity.

Aside from being derived by the same criterion, parallel categories should also be roughly equal in scope or size. The rule of parallelism in classification is based on the principle of balance. Occasionally you will violate this principle

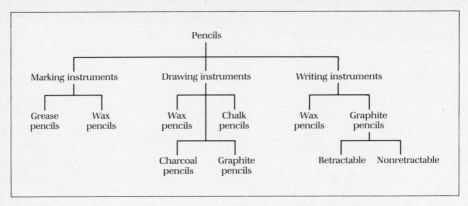

FIGURE 6.1   Classification of Pencils in the Form of a Tree Diagram.

by creating a category that is too narrow, but more often the problem is an overly broad category. Dividing a large group into a subgroup two-thirds its size simply isn't useful.

Mutual Exclusivity. Each category in a classification must be independent of all the others with absolutely *no* overlap. If any entry on your master list—or any potential entry on that list—could fit into more than one category, your categories are not mutually exclusive, and your classification is flawed. To test for mutual exclusivity, consider each entry on your master list to see if it can fit into more than one category. Then consider the categories in pairs to see if you can think of anything that could fit into both of them. Suppose you are classifying "pollution." Smog is chemical pollution, industrial pollution, and air pollution. Volcanic ash is natural pollution, chemical pollution, air pollution, and land pollution. Once you have established that you have an overlap problem, define the offending categories more precisely. Keep refining until no single example goes in more than one place. Then reexamine your final categories carefully to make certain they are still parallel.

One way to guarantee that your categories will fail the test of mutual exclusivity is to choose terms for them that are not themselves mutually exclusive. If you classify pollution as either natural or man-made (or "people-made"), you have categories that *look* mutually exclusive but actually are not. Sewage, for instance, is both natural and man-made.

One caution about mutual exclusivity: The rule is that no individual example may go in more than one place in the scheme, but that does not mean that the same *subcategory* cannot turn up in more than one *category*. On the contrary, parallel subcategories are desirable. In our classification of pencils, the subgroup "wax pencils" appears in three different places on the tree diagram. Each time, however, it is assigned to a different *main* group. The terms "wax marking pencils," "wax drawing pencils," and "wax writing pencils" each appear only once.

Completeness. Mutual exclusivity requires that no item fall into more than one category. Completeness is its opposite, requiring that every item (or potential item) fall into some category. Air pollution/water pollution/land pollution does not constitute a complete classification system for pollution, despite its elegant parallelism, because it has no place in it for noise pollution or for high levels of mercury in canned tuna.

Failures of completeness result from insufficient data collection, from not having enough entries on your master list. Unfortunately, this type of failure is hard to guard against. It will help to keep the entries on your master list as specific as possible and to pick them from the broadest range of possibilities. And make the list as long as you can; an extra item—for example, the "greenhouse effect"—can keep you from missing an entire category—in this case, thermal pollution. Another way to help ensure completeness is to limit the "universe" of your classification. Instead of classifying pollution in general, classify common types of pollution in industrial areas of the eastern United States. And, if you must, rely on the all-purpose category "other" to house odds and ends. As long as the miscellany that winds up there isn't terribly important, no one is likely to mind.

But even though it is easy to be technically complete, genuine intellectual completeness is almost as tough to achieve as absolute mutual exclusivity. No matter how exhaustive a classification you make of available coal gasification processes, as soon as someone invents a new one, your classification may be obsolete. Moreover, completeness is vital because an incomplete classification falsifies reality and may leave readers misinformed. Detectable incompleteness damages your credibility, whereas spurious completeness blinds readers to those missing categories and thus misleads them.

Scientists strive for completeness in their taxonomies by attempting to make all categories dichotomous—that is, by creating pairs of categories that are antithetical and that, taken together, encompass the entire universe. The categories matter/energy, organic/inorganic, natural/synthetic are all dichotomous. They have the considerable advantage of being not only complete but also mutually exclusive and parallel.

Achieving Parallelism, Mutual Exclusivity, and Completeness. The three standards for assessing the value of a classification are not equally important. Failures of completeness are serious; if you leave an option out of your cost-benefit analysis, it calls the entire analysis into question. The cost of mutual exclusivity problems is slightly lower, but still serious; you will not know where to put things, and your readers will not know where to find them. The cost of parallelism failures is lowest; assuming completeness and mutual exclusivity, parallelism adds elegance but is not absolutely necessary.

Organizing a Classification

Organizing a classification seldom poses any special problems *once you have thought it through*. It is common practice to begin a classification with a definition of what is being classified, along with a careful notation of the scope and limits of the analysis (*common* types of pollution in *industrial* areas of the *eastern United States*). Next, readers should be told what criterion has been chosen. If your reasons for selecting a particular criterion are not obvious, you may wish to discuss them before proceeding to the classification proper, perhaps naming some of the other criteria you might have used but chose not to. The introduction to a classification also customarily contains a statement of the purpose of the analysis, along with a brief discussion of the relationship between that purpose and your choice of criterion. The following introduction, from a U.S. Geological Survey circular on reclamation of surface-mined areas, concentrates on the purpose of the analysis and its scope:

The classification is limited to major U.S. methods, and the six to be covered are listed.	The variety of state mining laws and regulations shown in Table 1 covers all the surface mining methods now in use in the United States. . . . The major methods are open pit, quarry pit, area, contour, auger, and dredge mining. A brief description of each of these mining methods is included to define the method, point out basic characteristics, and show the wide range of surface disturbance that can occur. Basic mined-area reclamation techniques and associated constraints are also presented.
Part of its purpose is background, but the classification will discuss surface disturbances and reclamation techniques for each method.	

Here is another, more discursive introduction, which uses a series of definitions to make a highly technical topic accessible.

A Classification of Passive, Linear, One-Port Electronic Components

The classification opens with the most basic definition in the series.	An electronic "component" is any device having a number of input and output terminals. An "active" component is one that requires an external source of power to operate. Often, this external power is used to "amplify" one of the input currents or voltages. A "passive" component requires no external power supply. It therefore cannot provide current "amplification"; all of the current that enters the device must leave—no more, no less.
A passive component is defined by contrasting it with an active component.	
The description of a passive component builds on its definition.	A "one-port" component has only two terminals—one input and one output. If the component is passive, i_{in} must equal i_{out}, which is the current passing *through* the component. An easy way to relate v_{in} to v_{out} is to describe the *voltage difference across* the component as v_{in} minus v_{out}.

Because the last paragraph is highly technical, the author summarizes it here before proceeding.

The last paragraph helps make it clear that the characteristics "passive," "one-port," and "linear" are related to one another. The introduction ends with a list of the major subcategories and a brief mention of the criterion used to group them.

The behavior of a passive one-port component can be thoroughly described by two parameters, the current through it, i, and the voltage drop across it, v. Since i and v are often functions of time, they are often notated $i(t)$ and $v(t)$. Either $i(t)$ or $v(t)$ can be considered the input, but once one of these parameters is known, the other can be determined from the relationship imposed by the device.

In a "linear" passive one-port component, there is a *linear* relationship between $i(t)$ and $v(t)$ over the useful operating life of the device. There are three types of passive linear one-port components—resistors, capacitors, and inductors—each of which has a different function.

As you might predict, the body of this classification proceeds as follows:

I. Resistors
   A. Carbon resistors
   B. Precision resistors
   C. Wire-wound resistors
   D. Noninductive resistors
   E. Variable resistors
II. Capacitors
   A. Disc (parallel-plate) capacitors
   B. Electrolytic capacitors
   C. Variable capacitors
III. Inductors
   A. Air-core inductors
   B. Soft iron-core inductors
   C. Litz-wire inductors
   D. Toroidal core inductors
   E. Variable inductors
   F. Hard ferrite inductors
   G. Artificial inductors

An outline such as this one can suffice as the body of a classification all by itself, as can a detailed tree diagram, but few classifications are *that* spare and unelaborated. Most are developed verbally by definition and description, proceeding category by category, with subcategories wherever necessary. The discussion of each category should emphasize the characteristics that distinguish its entries from entries in all the other categories and should clarify the relationship between those defining characteristics and the purpose of the analysis. Categories are usually arranged in order of importance, often in a particular type of order of importance, such as most familiar to least familiar, cheapest to most expensive, or most versatile to most highly specialized. Subcategories, too, are arrayed in order of importance and discussed in terms of their special characteristics. The following discussion of resistors uses a combination of definition and comparison/contrast to differentiate two types:

Resistors

The definition of "resistor" is presented in terms of criteria already established in the introduction. The discussion of carbon resistors establishes the criteria by which all other types of resistors will be assessed: their constituent material, the type of circuit in which they may be used, their tolerance, their "noise" level, and their frequency range.

Precision resistors are compared and contrasted with carbon resistors in terms of constituent material, type of circuit, tolerance, noise level, and frequency range. Once a basis for comparison has been established, carbon resistors and precision resistors may be evaluated relative to one another.

The "i" vs. "v" relation for a resistor is $Ri(t) = v(t)$ for R△-resistance in ohms. This can also be stated as $Gv(t) = i(t)$ for G△-conductance $= 1/R$.

Carbon resistors are the most common type. Made of a mixture of carbon and clay, they are used only in low-power circuits because the carbon mixture, when exposed to high currents, tends to overheat and undergo molecular change which changes the resistance. The resistance values of carbon resistors cannot be predicted very accurately; the resistors are generally sold with tolerance levels of 20%, 10% or 5%. A carbon resistor marked 1000 ohms $\pm$ 20% might have an actual value anywhere between 800 and 1200 ohms. Carbon resistors are "noisy": Thermal agitation of carbon molecules causes the generation of small random voltages that can interfere with the desired signal voltage, especially at high operating temperatures. These resistors are essentially linear over a wide frequency range from DC to fairly high AC frequencies.

Precision resistors are made of films of resistive metal. They are used in low-power circuits and typically have tolerances of 1/2% or 1%. They are stable: The resistance value of precision resistors remains constant over wide temperature ranges and does not change with age. They are less noisy than carbon resistors, but cannot be used at very high frequencies because the layers of metal can act like capacitor plates (see **Capacitors**). This capacitive effect is negligible at low frequencies, but at high frequencies it can cause great deviations from ideal behavior.

Like most classifications, this one proceeds through all types of resistors before moving on to capacitors and inductors.

Classifications, like partitions, need conclusions so that readers will know when they have reached the end of the discussion. Summary overviews and evaluative commentaries are the most common means of providing readers of classifications with a sense of closure. The author of "A Classification of Passive, Linear, One-Port Electronic Components" saves his statement of scope and limitations till the last:

> Other one-port passive devices certainly exist. But most are unsophisticated, low-frequency devices more apt to be used by electronics amateurs than by design engineers. Resistors, capacitors, and inductors, from the most primitive to the most sophisticated, suffice to meet all of our current needs.

◗ The Value and the Danger of Analysis

You cannot survive as a technical writer without analysis—it will provide the structure for much, perhaps most, of what you write. Analytical thinking is fundamental to both scientific and humanistic learning. The belief that analysis is a way to truth is as old as human knowledge itself.

But though analysis is fundamental, it is far from easy. Experienced taxonomists know that the lines between subgroups are often less clearly defined than we would wish them to be. A fungus, for instance, possesses some of the attributes once thought to be unique to animals and some of the attributes once thought to be unique to plants. As a novice taxonomist, you must realize that *all classification systems are arbitrary*. Martin Goldstein and Inge F. Goldstein, in "Classification as the Starting Point of Science," put it this way:

> We have learned in school that whales and bats are classified together as mammals; whales are not classed with fishes, nor are bats classed with birds. This is done not because there is some absolute objective proof that a whale is more like a bat than it is like a herring—in some ways the whale and the bat resemble each other and in others they do not. It is because biologists feel that the ways whales and bats resemble each other are more important than the ways they differ.[3]

There are no absolute criteria for classifying anything, even in science. Nor are there absolute criteria for partition. Pirsig ends his careful, exhaustive partition of the motorcycle with the following warning:

> [T]here is a knife moving here. A very deadly one; an intellectual scalpel so swift and sharp you sometimes don't see it moving. You get the illusion that all

those parts are just there and are being named as they exist. But they can be named quite differently and organized quite differently depending on how the knife moves.

For example, the feedback mechanism which includes the camshaft and cam chain and tappets and distributor exists only because of an unusual cut of this analytic knife. If you were to go to a motorcycle-parts department and ask them for a feedback assembly they wouldn't know what the hell you were talking about. They don't split it up that way. . . .

It is important to see this knife for what it is and not to be fooled into thinking that motorcycles or anything else are the way they are just because the knife happened to cut it up that way. . . .[4]

Don't fall into the trap of thinking that objects and other mechanisms can be partitioned only one way or that a classification system is inherent in whatever it is you are classifying. You don't *find* the system; you *create* it.

Review Questions

1. What is the difference between partition and classification?
2. What is a partition by attribute? A partition by function?
3. What is the principal value of analysis in technical writing? What is its principal danger?
4. Why is it more difficult to partition by function than to partition by attribute?
5. How should a partition be organized?
6. What is deductive classification? Limited inductive classification? Formal inductive classification?
7. What three attributes must any good classification possess?
8. What is the purpose of classification—particularly of scientific classification?
9. Why are all classifications necessarily comparative?

Assignments

1. Write a 500-to-800-word partition of one of the following: a floor lamp, a tennis racquet, a pair of glasses, an automobile jack, a hammer, an oyster, a ski, a bridge, a solenoid. Specify the audience and type of partition.

2. For each of the following terms, choose two appropriate classification criteria, one for each audience: (a) cereals—for grocery store owners and nutritionists; (b) cameras—for a professional about to go on a shoot and an amateur about to shop for a first camera; (c) pesticides—for a state health agency and a vegetable farmer.

3. Develop a classification scheme for one of the following: knives, cars, beverages, shoes, boats, soils, mulches, cats, jeans. Present your classification

in the form of a tree diagram accompanied by an explanatory text of approximately 500 words. Use at least four orders of subcategories. In your explanatory text, explain the system you are using, specify your audience, justify your choice of division criteria, explain the characteristics that differentiate items in one category from items in another category, discuss representative examples, and point out which categories are particularly troublesome.

Notes

1. "Human Respiratory System," *Encyclopaedia Britannica*, vol. 15 (Chicago: William Benton, 1975), p. 764.
2. Robert M. Pirsig, *Zen and the Art of Motorcycle Maintenance* (New York: Bantam Books, 1974), p. 69 ff.
3. Martin Goldstein and Inge F. Goldstein, "Classification as the Starting Point of Science," in Martin Goldstein and Inge F. Goldstein, *How We Know* (New York: Plenum Press, 1978), pp. 120-122.
4. Pirsig, p. 72.

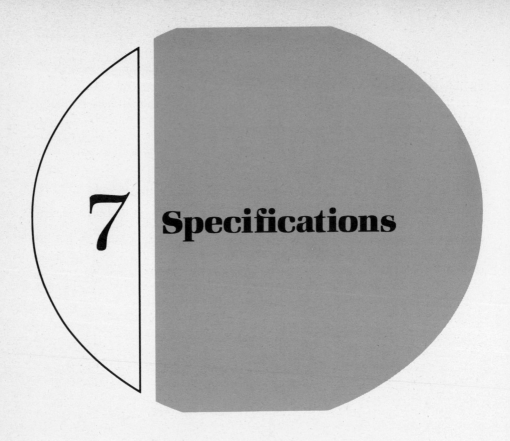

7 Specifications

Specifications are highly concentrated, self-contained, and complete verbal, graphic, and/or symbolic descriptions of all the minute particulars of any physical object or conceptual system. They are written by technical professionals for other technical professionals, normally by people who want something done for people who can do it. Architects write specifications for builders; systems designers write specifications for software engineers; prospective buyers write specifications for manufacturers. The goal is always the same: to ensure that the finished product is *exactly* what is desired.

Here is a familiar example:

> Make thee an ark of gopher wood; rooms shalt thou make in the ark, and thou shalt pitch it within and without with pitch. And this is the fashion that thou shalt make it of: The length of the ark shall be three hundred cubits, the breadth of it fifty cubits, and the height of it thirty cubits. A window shalt thou make to the ark, and in a cubit shalt thou finish it above; and the door of the ark shalt thou set in the side thereof; with lower, second, and third stories shalt thou make it.

Construction specifications are among the oldest, the most widely used, and the most carefully standardized.

 # Kinds of Specifications

Specifications may be open (performance-oriented), closed (prescriptive), or restrictive, depending on what they describe and for whom the description is intended. Open specifications specify what a product should *do* rather than what it should *be*:

> Exterior paint shall be able to resist cracking and peeling down to temperatures of −30° F.

The writer of these specifications is "open" to any brand of paint whose performance meets these preset criteria.

Closed specifications, by contrast, are prescriptive; they identify the precise product desired—often by brand name—and permit no substitutions:

> We want a Pioneer SX 6000 receiver with 2 CS 88-A speakers, a Sansui turntable mode SR 2050C, and an Akai tape deck, the GX 280D.

171

Restrictive specifications fall somewhere in between; they provide a list of alternatives, any of which will be acceptable:

| Get a liquid detergent with a grease-cutting agent in it.

"Liquid" is a prescriptive specification; "with a grease-cutting agent" is a performance specification. The fundamental difference between the two is that a performance specification describes a product in terms of its *results*.

Prescriptive and performance specifications differ primarily in their wording. A specification for a certain color, for instance, is always prescriptive because it identifies an attribute. However, that attribute may or may not be performance-oriented. The specification "white" for a lab coat, though standard, is unrelated to the function of that garment. However, when that specification is applied to clothing destined to be worn in the tropics, "white" becomes an indirect way of saying "shall reflect heat."

Whether you choose to word specifications prescriptively or in terms of performance criteria will depend largely on your goal and audience. If you want to know as exactly as possible what the final product will be and to control its physical attributes as completely as you can, you will probably opt for prescriptive wording. Prescriptive specifications are best for nontechnical readers who lack the background to translate performance criteria into concrete choices. Automobile owner's manuals, for instance, prescribe specific oil viscosities and cold tire pressures. But if you want a creative solution from a knowledgeable reader, performance specifications are more appropriate than prescriptive ones.

The great advantage of prescriptive over performance specifications is that prescriptive specifications are clearer. The specification "Firefighters must weigh at least 190 pounds" may be capricious, unfair, and discriminatory . . . but it is unlikely to be misunderstood. The wisdom of prescriptive specifications may be arguable, but their *wording* is not. With performance specifications, on the other hand, the words *are* the specifications, so using the right words is essential. "Must withstand cold" seems clear enough, but what does "withstand" really mean? "Keep from freezing"? "Keep cold from penetrating through any but the outermost layer"? "Keep from cracking or splitting"? And at what temperature? When specifications turn out to be ambiguous, the courts customarily rule in favor of the reader; the specification writer (or his or her company) must pay the cost of having been ambiguous. Because writing performance specifications presents special problems, we will start with the task of writing prescriptive specifications, one that is frequently assigned to novice technical writers.

 # Writing Prescriptive Specifications

The goal of prescriptive specifications is to create in advance a *complete* description of the physical properties of the desired final product in words, pictures, symbols, or a combination of these. The simplest sort is reference specifications, which send readers to an official source[1] and require that they duplicate the model the source provides:

> All peanut butter sold to the Burberry County public schools shall conform to Federal Specification Z-P-196C.

Specifications are also not difficult to write when they consist of long lists of known requisites with a finite number of possible variations.

> The fan shall be a wall-mounted centrifugal type of fan of all-aluminum construction including aluminum centrifugal wheels and aluminum hardware. The metal in the fan shall not be thinner than sixty-four thousandths (.0064) of an inch. The fan shall be a direct-drive, propeller-type fan mounted on a ball bearing totally enclosed one-sixth horsepower (1/6 H.P.) motor supported by a rigid frame attached to the fan housing. The air inlet shall not be less than ten inches (10″) in diameter.[2]

Omissions are the only real danger inherent in the writing of closed or prescriptive specifications. Consequently, closed specs are almost never drafted from scratch, but instead are transcribed from prewritten models. Technical writers used to have the job of assembling prefabricated subparagraphs into a comprehensive document using the cut-and-paste method, but today nearly all invariable prescriptive specifications have been placed in computer storage. They can be retrieved and assembled by a word processor far more efficiently than by a human being, with almost no risk of an inadvertent omission. What this means is that technical writers now have the job of preparing only those specifications that *do* have to be composed from scratch. Most often, these are restrictive rather than closed specifications, allowing room for some controlled variation in the final product, for example, trade-offs among attributes, special features, individual differences. Frequently, such specifications are labeled "conditions of acceptability," and they are expressed in terms of tolerances.

Most conditions of acceptability are implicitly performance-oriented. They represent the opinions of experts as to the upper and lower limits beyond which a product cannot range and still perform acceptably. Here is an example:

> *Mortar for masonry* shall consist of one part portland cement to not more than three parts of sand; with not more than ten percent of the cement replaced with hydrated lime if required and with just sufficient water for mixing in a mechanical mixer.

Phrased as a performance specification, that same requirement would come out sounding like this:

> *Concrete in foundations* shall have a compressive strength of not less than 2,500 pounds per square inch at 28 days of age.

Prescriptive specifications are standard for routine business transactions, but the trend now in government and industry is toward performance specifications, especially in highly competitive areas such as the computer software market. That is so because even strict adherence to prescriptive specifications is no guarantee of successful results. Nor is it always certain that a product that falls outside the boundaries will fail to perform adequately. The edges of the tolerance range are always set arbitrarily, and prescriptive specifications cannot account for intangible or unexpected variables. Prescriptive specifications are inflexible; they do not allow for individual differences. If the *goal*, the *result*, is what matters and if the method of arriving at that goal is immaterial, then only performance specifications will do.

▶ Writing Performance Specifications

Writing performance specifications is largely a matter of choosing the right words. It is no light matter either, because the words you choose are legally binding. The literal meanings of words, their *denotations*, their dictionary definitions, are the meanings that count in specifications. That is why numbers and symbols are always preferable to words. But performance specifications depend on words because you cannot always express potential in terms of numbers. Performance specifications center on verbs—what the product *does*. They require concrete verbs like "prevents, "kills," "protects," "traps," "repels." Abstract verbs like "discourages" are unacceptable because they are so relative in their meaning. The action "discourages" cannot easily be quantified, unlike the action "kills." Whenever possible, specification writers should use verbs that belong to the appropriate technical jargon; then, at least, most professional readers will agree, more or less, on what the verbs mean. For instance, take the verb "paint," as in "The contractor shall be responsible for painting all interior woodwork." To nontechnical readers, "paint" means "paint." To a painter, however, it means "sand, prime, and cover with two light coats of paint, allowing enough time between applications for the first coat to dry completely." Obviously, the cost in units per hour is a great deal higher according to the second definition.

▶ Organizing Specifications

Both prescriptive and performance specifications must be organized in such a way that any one individual specification can easily be located. Specifications are atomic, after all; they contain dozens, hundreds, often thousands of minute particulars, each one a discrete item. Therefore, the structure of all specifications is always general-to-specific. The most common macrostructure for specifications is this one:

I. General Conditions
II. Special Conditions

III. Technical Division
   A. Technical Section I
      1. General conditions
      2. Special conditions
      3. Scope of the work
      4. Work not included in this section
      5. Standards and requirements
         Article 1. First standard or requirement
         Article 2. Second standard or requirement . . .
   B. Technical Section II . . .

General Conditions

General conditions are very broad specifications that apply to the entire project: its name and nature (1982 Dodge Charger sports coupe; RCA XL-100 solid state color television console); its cost; its delivery date; definitions of terms used in the specifications; scope of the project; and work *not* included (to absolve the vendor of responsibility for whatever is not present in the finished product). These various conditions are arranged from the most general to the most specific.

Special Conditions

Special conditions are specifications that are unusual or perhaps even unique. The specifications for the space shuttle *Columbia*, for instance, identified a particular reentry temperature that the exterior tiles would have to withstand—hardly a usual condition in the construction of aircraft! Special conditions, too, are arranged from the most general to the most specific. Any special conditions that apply only to one part of the project, however, are usually assigned to the technical section that describes that part.

Technical Sections

In the technical sections of a set of specifications, the object or system the specifications describe is broken down into its components—that is, partitioned. Each technical section carries the name of one component or subsystem, and the names given to technical sections vary from industry to industry. In the construction industry, for instance, the technical sections are called "trade sections" and carry the names of jobs (carpentry, masonry, plumbing, and so on). These jobs represent the division of labor among subcontractors and are standard throughout the industry. Engineering specifications (like those for an automobile, for instance) are usually divided according to the components of the system being engineered (for example, engine, engine lubrication, cooling system, battery, alternator, starter, clutch, transmission, and so on). Technical sections may represent subtopics, subsystems, constituent elements, or any other parts of the whole. Luckily, most industries use standard systems, so writers of specifications are normally spared the ordeal of partitioning a wastewater treatment plant down to the last screw.

If possible, the technical sections in a set of specifications should be ordered from the most general to the most specific. In an automobile, the engine and the engine lubrication system are more general than the carburetor, so they customarily head the list of technical sections in an automobile owner's manual. However, many technical sections are equally specific, or parallel: A battery is not inherently more general than an alternator, nor is carpentry more general than masonry. Technical sections that are parallel may be arranged alphabetically, in chronological sequence (that makes the most sense for architectural specifications—masonry precedes carpentry, which precedes plumbing), or in order of increasing or decreasing importance. One commonly used order of importance is from the cheapest to the most expensive. The number of technical sections varies, of course, from project to project. There may be only two or three, or hundreds.

The microstructure of each technical section mirrors the general-to-specific organization of the specifications as a whole. Technical sections begin with the broadest and most all-encompassing points—usually, definitions of terms—and funnel down to particulars. Each technical section is divided into "general conditions," "scope of the work," "work not included in this section," and one or more subsections advancing the actual standards and requirements one by one. These parallel subsections, like the larger technical sections, are usually arranged in chronological order or in order of importance.

"Standards and requirements" subsections are in turn divided into articles, which are divided into numbered paragraphs, which are divided into subparagraphs. For example:

Article 2—Organization

2.1 *Microstructure.* The microstructure of the specifications shall mirror their macrostructure.

> 2.1.1 All technical sections of specifications shall be divided into articles, paragraphs, and subparagraphs.
>
> 2.1.2 All subdivisions within a set of specifications shall be arranged in descending order from the most general to the most specific unless sections are equally specific (that is, parallel), in which case an appropriate order shall be superimposed on the sections by the specification writer.
>
> 2.1.3 All subdivisions within a set of specifications shall be consecutively numbered for easy reference.

The decimal system just shown is the one most commonly used to number the microcomponents of a set of specifications.

Quantitative information in specifications is usually given both in symbols and in words:

> The metal in the fan shall not be thinner than sixty-four thousandths (.0064) of an inch.

Repetition of the information two different ways provides an internal referencing system that helps ensure accuracy. The same is true of scale drawings, which provide mathematical information graphically. Many industries require that mathematical descriptions or drawings accompany verbal specifications. Specification writers are usually not required to prepare the drawings, but they are expected to coordinate them with the written documents and to make sure that the two agree. The ultimate responsibility for internal accord lies with the specification writer because, in the case of a discrepancy between drawings and written text, the written text takes legal precedence.

 ## Specification Graphics

Specifications may include any kind of graphic representation appropriate to the subject matter: line drawings, schematic diagrams, cross-sectional drawings, charts (schedules, for instance), exploded diagrams, even photographs. There are two advantages of pictures over words that are of special interest to writers of specifications. First, drawings are exceptionally concise. They convey information very efficiently because they take up relatively little space and say a number of things simultaneously. It would take dozens of pages of text to elaborate all the information provided in a single wiring diagram.

The second advantage of graphic description over verbal description is that graphics are better not only for showing dimensions such as size and shape but also for showing spatial relationships. Words like "above," "below," "beside," and "between" are invariably less precise than pictures. A diagram of a range hood installed above a stove shows *exactly* how far above the stove top the hood is placed and where it is placed in relation to the burners (directly above them, slightly recessed, jutting slightly out, and so on). Words can only approximate the level of exactness of the diagram.

The set of specifications for a timber jetty shown in Figure 7.1 contains both verbal and graphic descriptions. Notice how the two are balanced against one another; words are used to specify the materials and drawings are used to specify dimensions and design features. As you can guess, prescriptive specifications are much more likely to contain graphics than performance specifications are. Words must carry nearly all the meaning in performance specifications, because it is difficult to draw results.

JOHN N. CANDIDE, GENERAL CONTRACTOR
Specializing in Marine Construction
1400 Planet Drive, Sunspot, Maryland

PROPOSAL SUBMITTED TO	DATE
Mr. James L. Yarrison	March 6, 1979
STREET	JOB NAME
Rt.#1, Box 22A, Dutchtown-Zion Rd.	Job No.4-3-79 Timber Jetty
CITY, STATE AND ZIP CODE	JOB LOCATION
Skillman, NJ 08558	Chance, MD

We hereby submit specifications for:

The construction of a timber jetty sixty feet in length on your property at Chance, Maryland. Forty feet of jetty will be constructed channelward of the MHW mark. Jetty will have 2" T & G sheeting on one side and 1" rough sheeting on opposite side. Wales and piles will be sandwiched between the two sheetings.

SPECIFICATIONS: 1st 20' - (that portion landward of MHW and which will also serve as return for bulkhead if it is needed in the future.) 1" sheeting will be placed on one side - nothing on the other side. (This section will not be affected by ice.) 1" x 8" x 5' (rough) sheeting (one side); 2 pcs 6" x 6" x 6' piles on 10' cc.; 4" x 6" top and bottom wales.

Next 15' - 1" x 8" x 5' (rough) sheeting one side; 2" x 8" x 6' T & G opposite side; 1 pc. 6" x 6" x 8' and 1 pc. 6" x 6" x 10' piles on 10' cc.; 4" x 6" top and bottom wales.

Next 15' - 1" x 8" x 6' (rough) sheeting one side; 2" x 8" x7' T & G opposite side; 1 pc. 6" x 6" x 12' and 1 pc. 6" x 6" x 14' piles on 10' cc.; 4" x 6" top and bottom wales.

Last 10' - 1" x 8" x 7' (rough) sheeting one side; 2" x 8" x 8' T & G opposite side; 1 pc. 6" x 6" x 16' pile (last pile); 4" x 6" top and bottom wales.

J. N. Candide

J. N. Candide

FIGURE 7.1   Sample Specifications with Graphics.

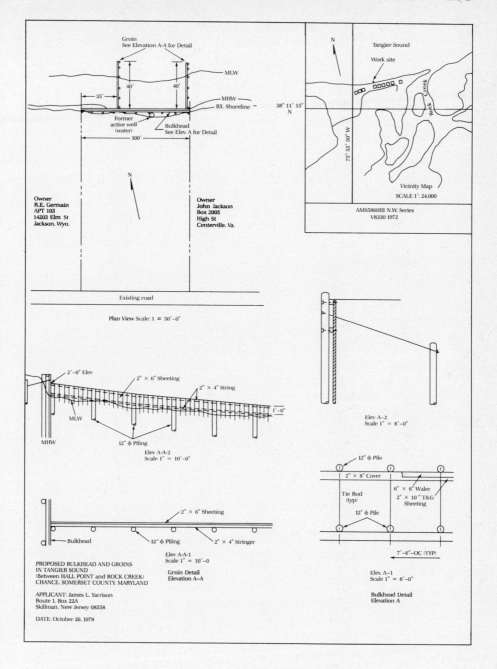

Groin
See Elevation A-A for Detail

MLW

40'   40'

35'

MHW
BX. Shoreline

Former
active well
(water)

Bulkhead
See Elev. A for Detail

100'

N

Owner
R.E. Germain
APT 103
14203 Elm St
Jackson. Wyo.

Owner
John Jackson
Box 2005
High St
Centerville. Va.

Existing road

Plan View Scale: 1 = 50'-0"

N

Tangier Sound

Work site

Rock Creek

38" 11' 15"
N

75° 55' 50" W.

Vicinity Map
SCALE 1': 24.000

AMS5860III N.W. Series
V8330 1972

2'-0" Elev

2" × 6" Sheeting

2" × 4" String

1'-0"

MLW

MHW

12" φ Piling

Elev A-A-2
Scale 1" = 10'-0"

Elev A-2
Scale 1" = 8'-0"

2" × 6" Sheeting

Bulkhead   12" φ Piling   2" × 4" Stringer

Elev A-A-1
Scale 1" = 10'-0

Groin Detail
Elevation A-A

12" φ Pile

2" × 8" Cover

Tie Rod
(typ)

6" × 6" Waler
2" × 10 " T&G
Sheeting

12" φ Pile

7'-6"-OC (TYP)

Elev. A-1
Scale 1" = 8'-0"

Bulkhead Detail
Elevation A

PROPOSED BULKHEAD AND GROINS
IN TANGIER SOUND
(Between HALL POINT and ROCK CREEK)
CHANCE. SOMERSET COUNTY. MARYLAND

APPLICANT: James L. Yarrison
Route 1. Box 22A
Skillman. New Jersey 08558

DATE: October 26. 1978

▶ Conciseness and Clarity in Specifications

Both prescriptive and performance specifications should be written as concisely as possible. They should contain *no* excess words whatsoever; they need not even contain the functional words like definite articles and conjunctions that make reading easier. They need not be complete sentences; fragments are perfectly acceptable as long as they are not unclear:

> Back and all four sides to be insulated with Owens Corning #703 Semi-Rigid low-binder low-organic insulator.

Simple sentences are preferable, so that each individual specification may be given its own sentence:

> Absorbers shall be of 3/4″ I.D. extruded finned tubing with semi-bright nickel and black chrome plating.

Complex sentences (constructions using subordinate clauses) are to be used rarely, most often to state a general specification and then qualify it:

> Flame-resistant cloth shall delay burning for twelve (12) seconds when it is exposed to open flame or heat in excess of 1000° F.

Transitions between individual specifications are unnecessary, for each specification is supposed to stand alone. And the only real transitions used between clusters of related specifications are the numbers used to separate and systematize them. Naturally, the lack of transitions makes specifications sound very choppy, but because a set of specifications is scarcely ever read straight through, its choppiness is not really a problem. In specifications, smoothness is less important than conciseness.

Because conciseness carries a very high priority, it is conventional practice when writing specifications to use a highly technical vocabulary, abbreviating and converting to symbolic language whenever possible. Technical jargon also helps to give specifications an appropriately formal tone. Most importantly, technical jargon is comparatively unambiguous. Thus, it contributes not only to conciseness but also to clarity, the specification writer's fundamental goal.

Unclear specifications are useless. If readers don't know exactly what is being requested, they cannot provide it. Clarifying specifications by issuing addenda or "field change orders" is a complicated and expensive process.

And if specifications are in any way subject to misinterpretation, the fault—and the liability—for their ambiguity rests with their source.

One way in which specification writers avoid ambiguity is by using reference specifications wherever possible. Standard specifications have *become* standard over the years in part because they resist misreading. Another way to preclude misunderstanding is by steering clear of those words and phrases that are known not to survive a court challenge. The most notorious of these equivocal words and phrases is "or equal," but such terms as "equivalent," "satisfactory," "adequate," "normal," "customary," "similar," and "reasonable" are no less problematical. The problem is this: Who decides what is "adequate," what is "normal," or whether or not two products are "equal"?[3] Open specifications cannot always avoid these terms, but they can fix decision-making responsibility in advance. The specification "or equivalent in the opinion of the on-site engineer" may be perilous, but it is not unclear.

Theoretically, of course, equivocal terminology should not be used in the first place, but it is almost impossible for specification writers, given their built-in biases and blind spots, to anticipate which words are going to be misinterpreted by their readers. The specification "The electrician shall replace any defective wiring" *seems* unambiguous, but it does not mean quite the same thing as "The electrician shall replace all defective wiring with new wiring." The word "replace" can mean "put back after removing," or "substitute identical material," or "substitute new material," depending on the context. "With new wiring" is necessary if you don't want the electrician to substitute spare wire from elsewhere. Be especially careful about sentences that require pronouns. With pronouns, there is always a risk of vague pronoun/antecedent reference, as in the following example:

▌ Concrete should not be poured when it is below 30° F.

To what does "it" refer? Is it the air temperature or the temperature of the concrete itself that must exceed 30° F.? Specifying which is worth the extra words.

As this last example illustrates, clarity matters even more than conciseness when the two goals conflict. Consider the following specification:

▌ All water piping, condensate piping and steam piping below the ceiling shall be painted.

What is to be painted? Is it all piping of any kind below the ceiling? All water piping, but only that condensate piping and steam piping that is below the ceiling? All water piping and condensate piping, but only that steam piping that is below the ceiling? This one had to be wrestled out in court. But the litigation could have been avoided if the original sentence had been set up in repetitive, parallel structure, like this:

> All steam piping below the ceiling, all water piping, and all condensate piping shall be painted.

The overall clarity of this specification is enormously enhanced by the revision. That this clarity is achieved at the cost of increased wordiness is secondary.

Unfortunately, learning which words are potentially ambiguous and why is a trial-and-error process, and each industry must find out for itself where the pitfalls in its terminology are. Only when a word or a phrase has been misunderstood once can anyone predict with confidence that it will be misunderstood again (and again). As novice specification writers soon learn, specifications are transcribed rather than written, wherever possible, because previously used words are safe words. Master specifications and guide specifications have been devised in most industries for the express purpose of providing models that are both complete and clearly worded. The writing of specifications is a continual struggle to find words that will accurately represent the properties or performance criteria they stand for.

Review Questions

1. What are closed specifications? Open specifications? Restrictive specifications?
2. What is the difference between prescriptive specifications and performance specifications?
3. What is the goal of a set of specifications? What is its likely audience?
4. How are specifications typically organized?
5. Why are numbers and symbols preferable to words in specifications?
6. What is the role of graphics in specifications?
7. How can a specification writer avoid ambiguity?

Assignments

1. Draft a set of specifications for a personal computer system for your own use.
2. Draft all the specifications necessary to enable a typesetter to set the type that appears on page 149.

Notes

1. The best-known of these official sources are the American Society for Testing and Materials (ASTM), the American Standards Association (ASA), the American National Standards Institute (ANSI), the Construction Specifications Institute (CSI), and the

Occupational Safety and Health Administration (OSHA). Federal specifications (non-military) are set and published by the General Services Administration; military specifications by the Department of Defense.

2. Our thanks to Chesley Ayers of the Detroit Institute of Technology for this example, which appears in his textbook *Specifications* (New York: McGraw-Hill, 1975), p. 35.

3. In 1978, General Motors was successfully sued by Buick owners in whose cars the company had used Chevrolet engines. Clearly, the company and the customers differed over whether or not the engines were "equal."

8 Instructions

As technical readers, you are probably more familiar with instructions than with any other type of technical writing. There are instructions for filling out your income tax forms, for staining bacteria in a biology lab, for setting the alarm on your new digital clock-radio. Some instructions give generalized advice on how to achieve a desired result: how to get a better-paying job, how to decrease heating costs at home, how to improve the gasoline mileage of an old car. However, most of the instructions that technical people read—those found in user's manuals, for instance—are designed to guide them step by step through a specific process or task. These step-by-step instructions are the focus of this chapter.

▶ Goal and Audience

The goal of all instructions is to enable readers to perform a task. Paradoxically, there are two types of instruction writers who regularly fail to achieve this goal: those who don't understand the task well themselves and those who understand it intimately. The first type is obvious. If you don't know exactly what your readers are supposed to do, you can't possibly tell them how to do it. Unfortunately, technical people are sometimes asked to write instructions for processes they have never performed. If you are ever trapped in this way and you can't gracefully decline, get someone to teach *you* how before you try to teach your readers how. If possible, practice the process until you know it thoroughly; at a minimum, talk to someone who does know it thoroughly and take extensive notes.

The problem of understanding the task intimately is more subtle and more pervasive. Steps and details that seem obvious to the experienced writer may be anything but obvious to the inexperienced reader. Examine, for instance, these instructions, printed in *Motor Auto Repair Manual*, for troubleshooting an electronic ignition system:

Measure resistance between terminals "B2" and "B3" of the distributor connector (Fig. 7). If resistance is not 400–800 ohms, replace sensor. If resistance is 400–800 ohms, repair or replace harness between three wire and four wire connector.

In giving these instructions, the writer has made the following assumptions about the readers:

1. That they know what and where the distributor connector is—and where terminals "B2" and "B3" are located.
2. That they know that an ohmmeter should be used to measure the resistance.
3. That they know how to use an ohmmeter.
4. That they know how to replace the sensor.
5. That they know where the harness is and how to repair or replace it.

Because this repair manual is intended for auto mechanics, the writer is probably justified in making these assumptions. For many of us, however, these instructions would be useless without a good deal of supporting information.

With most kinds of technical writing there is a direct relationship between audience expertise and amount of detail: The less your readers know, the less detail you burden them with. This is so because the choice of content depends on both audience and goal, and when the audience is uninformed, the goal is usually modest. But instructions are an exception. Because the goal of instructions is always to enable your readers to do what you are telling them how to do, the less they know to begin with, the more they have to be told, and the more detail you need.

The most common mistake in instruction writing is overestimating what readers know, assuming—often unconsciously—that they possess as much background as the writer. Sometimes, of course, they do. Usually, however, readers know much less. As a writer you must constantly guard yourself against taking the attitude that "any fool knows this"—some very intelligent people just might not. When in doubt, belabor the obvious. Readers won't resent an unnecessary detail nearly as much as a missing one.

You must also guard against omitting a detail by accident. Before attempting to write a set of instructions, gather together all the materials and tools, and do the procedure yourself, taking notes as you go.

 # Making Instructions Clear and Complete

Forcing yourself to see the process you are writing about from your readers' point of view is the key to writing clear and complete instructions. The following eight pointers will help you do just that.

Keep Your Language Simple

Technical terms, abbreviations, special symbols, and mathematical formulas are all kinds of shorthand. They save space for readers who are familiar with them, but they totally defeat readers who are not. Unless you are certain of your readers' expertise, then, avoid these when you can, and define them when you cannot. Remember that an everyday word to you may be a brand-new technical word to your readers. Remember, too, that some common words may have special technical meanings unfamiliar to your readers. Keep a list of all these unavoidable technical terms; you will have to decide whether to define them as they come up in your instructions or to make a glossary, defining them all at once.

Illustrate Your Instructions

All but the most simple instructions are bound to be incomplete without line drawings, diagrams, maps, photographs, or illustrations of some sort. Illustrations are most useful for showing what something looks like, but they are more versatile than that. A labeled drawing is almost essential for helping readers pick out which part of a mechanism you are talking about. A sequence of drawings can demonstrate the steps in a process far more compactly than any written instructions. Designing effective illustrations is a skill in its own right, one we will discuss in detail in Chapter 11. For the moment, bear in mind that your instructions should always be illustrated and should refer to the illustrations often and clearly. However, use technical graphics such as schematic drawings *only* if you are sure your readers will be able to read them. For the beginner, a simple wiring diagram would probably be more useful—and far less intimidating.

Be Precise

How thick is a thick coat of paint? How small is a small screw? This sort of vagueness forces readers to rely on their own judgment, which may be wrong and is bound to be anxious and uncertain. The instructions for using a fabric dye, for instance, state: "Dissolve necessary amount of TINTEX into a quart jar of hot water." But unless you have had some experience with the product, how do you know how much dye is "necessary"—or how hot the water should be? Whenever possible, therefore, use numbers and standard measurements. If you mean a ¾-inch flathead screw, say so; if you mean two tablespoons of dye, say so. If numbers suggest more precision than you need, try a homier measurement—"Heat the water until you can just barely keep your finger in it."

Express Tolerances Clearly

Look back at the instructions for troubleshooting the electronic ignition system, in which the writer tells the reader to see whether the resistance between the two terminals is "400–800 ohms." This is a much more complete instruction than one that said merely "about 600 ohms," leaving the reader to decide how much tolerance was acceptable. The more precisely you spell out what is an acceptable range, the more comfortable your readers will be. On the other hand, you must resist the temptation to "make things easier" by arbitrarily choosing a precise measurement when none is called for—if any reading between 400 and 800 ohms is permissible, don't instruct your readers to seek one of exactly 600! A complete instruction, then, should not only tell readers what to do; it should also tell them—and tell them clearly—how close they have to come.

Anticipate Problems

Every process could go wrong in millions of ways, but each process has only a few ways it *usually* goes wrong. Most of us, for instance, have at one time or another tried to follow instructions that asked us to "insert tabs A, B, and C into precut slots X, Y, and Z," only to find that when two of the tabs were in place, nothing on earth would induce the third tab to fit into its slot. Wise instruction writers advise their readers in advance how to handle these problems: "Insert tab B into slot Y, and secure it behind with a paper clip. Then insert tabs A and C into slots X and Z. Remove the paper clip from tab B." A thoughtful instruction like this can save hours of irritation.

Provide Warnings

Sometimes more than the reader's irritation is at stake. A false move could damage a piece of equipment, ruin the whole procedure, or even cause injury to the reader. If your instructions involve potentially dangerous substances or procedures, be sure to advise your readers of the precautions to take. Though it may seem obvious to you that the step is dangerous, your readers may not realize that, for instance, the transformer is "hot" or the fumes are poisonous. Here, belaboring the obvious is definitely the best policy. Warnings like "Avoid contact with skin" may not be enough; better to say "WARNING!! THIS COMPOUND CAN CAUSE SEVERE BURNS. DO NOT ATTEMPT TO PERFORM THE NEXT STEP UNLESS YOU ARE WEARING PROTECTIVE GLOVES AND LONG SLEEVES." Important warnings deserve every gimmick you can think of to call attention to them—color, capital letters, underlining, asterisks, large type, a box around the word "WARNING," exclamation points, whatever. In addition, you should provide emergency instructions: "IF YOU GET ANY COMPOUND ON YOUR SKIN, IMMEDIATELY FLOOD THE AREA WITH WATER AND CALL YOUR DOCTOR."

Provide Checkpoints Along the Way

Midway through a complex set of instructions, virtually all readers are bound to suspect that something has gone wrong. In fact, everything may be fine, but readers don't know that, and their confidence quickly ebbs as the instructions proceed. Keep in mind that your readers are dependent on you, that they can't do whatever it is they want to do without your instructions, and, consequently, that they are likely to feel insecure about their ability to do what you tell them. Truly complete instructions therefore provide checkpoints along the way, enabling your readers to reassure themselves that they are still on target. "After about five miles you'll pass a big high school on the left" doesn't constitute an instruction at all—but anyone who has ever followed driving directions knows the feeling of relief when the high school comes into view. By the same token, a statement like "At this point, the flywheel may be very hard to turn; however, after you attach the gear section you will be able to adjust both easily" can do wonders for an anxious reader.

Explain the "Why"—But Explain It Briefly

Like the checkpoints just mentioned, explanations of *why* you are telling your readers to do something do not really qualify as instructions, but they are nevertheless important. An endless series of commands—"Do this; do this; then do this"—unrelieved by any explanation of the reasons can leave your readers dispirited and feeling like robots. Brief explanations of the science governing your commands give your readers the sense that they are *understanding* something, not merely doing it. Readers may want to know, for instance, why they are supposed to add three drops of acetone to the solution in step 26. The explanations need not be long or complex (remember that the less your readers know, the less you will be able to explain to them), but even a brief sentence or two can pay large dividends, as you can see by comparing these two versions of an instruction on how to etch a copper plate:

❙ Place the copper plate face-down in the bath of ferric chloride acid.

PLACE THE COPPER PLATE FACE-DOWN IN THE BATH OF FERRIC CHLORIDE ACID. The acid's reaction with copper produces a precipitate that can settle on the plate and thus impair the etching process. Inverting the plate allows the precipitate to sink harmlessly to the bottom of the bath.

The second version provides enough information to satisfy most readers without getting bogged down in chemical equations and theories.

You will have to decide for yourself how much of this kind of background information to include. Keep in mind, though, that there comes a point at which explanations can confuse rather than enlighten an uninformed reader—instructions on how to install a roll of paper towels on a holder don't need to include an explanation of the principle of torque.

The following instructions from a pamphlet demonstrate some of these suggestions. They tell readers how to retrieve the leader of an unexposed roll of film when they have accidentally wound it back into the magazine.

The "why" here makes it more likely that readers will follow this apparently unnecessary step.

1. Cut a strip about 1 inch wide and 5 inches long from a material such as sheet film or acetate. Round both corners on the end which you will insert into the magazine to reduce the chance of scratching your film.

2. Attach double-coated cellophane tape to one side of the strip, covering about 3 inches of one end.

The tolerance for this measurement is obviously wide.

3. Push 2 to 3 inches of the strip through the lips of the magazine. The sticky side of the strip should face the film core.

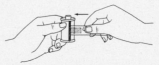

Anticipating a possible problem, the author says what to do if the initial procedure doesn't work. This illustration, together with the preceding one, makes the last two steps absolutely clear.

4. Pull the strip out gently. Usually the leader will attach itself to the sticky tape. If it doesn't, insert the strip again, and rotate the core of the magazine clockwise. This pushes the film against the tape and helps them stick together.

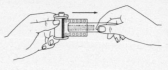

 # Organizing and Writing Instructions

In addition to being clear and complete, instructions must be coherent— that is, the steps must be in the right order—and they must be presented in a format that is easy to follow.

Making Instructions Coherent

Almost without exception, instructions should be written in chronological order. Unfortunately, achieving exact chronological order is trickier than it sounds. Consider this instruction:

> To start this power saw, move the ON–OFF switch to the ON position. Because an unshielded saw blade is very dangerous, be sure that the BLADE SAFETY GUARD is in place. If it is NOT, swing it down until it clicks into position (see Fig. 1).

This is an incoherent instruction. Because the purpose of the BLADE SAFETY GUARD is to protect the operator from the dangerous moving blade, the instruction should make it clear that the BLADE SAFETY GUARD must be in place *before* the machine is turned on. Here is one revision:

> To start this power saw, move the ON-OFF switch to the ON position. *CAUTION*: Because an unshielded saw blade is very dangerous, be sure

that the BLADE SAFETY GUARD is in place *BEFORE* you start the machine. If it is NOT . . .

This version is still incoherent, even though the second sentence now tells the readers to engage the BLADE SAFETY GUARD before turning the machine on. The problem, of course, is that the first sentence tells the readers to do something that they shouldn't do until after they have read—and followed—the second sentence. In other words, merely to say what step to perform first is not enough to ensure coherence; the instructions must be written so that the step that your readers *read* first is the step that they *do* first. Here is a coherent revision:

Because an unshielded saw blade is very dangerous, be sure that the BLADE SAFETY GUARD is in place *BEFORE* you start the machine. If it is not, swing it down until it clicks into position (see Fig. 1). To start the power saw, move the ON–OFF switch to the ON position.

If you could be certain that your readers would always study all the instructions thoroughly before they began to follow them, perhaps the second version of the preceding instruction would be acceptable. Unfortunately, you can't be certain. In fact, most readers perform instructions in the order in which they read them: They read one sentence, do whatever it says; read the second and do whatever it says; and so on. Even if you tell your readers to read all the instructions before they begin, you cannot expect them to impose order on your chaos. Coherence is your responsibility, not theirs.

One implication of this advice is that precautionary comments should be placed where they will do the most good—*before* the step to which they apply. If a particular part might be too hot to handle, for example, the time to warn readers to let it cool is before they touch it. If a particular procedure has to be done differently in very cold climates or at very high altitudes, the place to say so is before you start talking about the procedure. Comments advising readers of exceptions to a particular instruction, likewise, belong near that instruction. It is frustrating to be told to open the valve at the base of the unit when there is no valve at the base, only to learn hours later in a different section of the manual that some models (yours, for instance) have that valve at the rear instead. Of course, if the comment is long, you may have no choice but to put it on another page, but you should still interrupt the instructions long enough to urge readers to check it out before they continue: "Some older models are constructed in a slightly different way; consult the List of Parts Variations on page 49 before you begin." Similarly, steps that should be started in advance must be identified in advance. Suppose step 19 requires an acid that has to sit for a few hours after it is mixed. There are better places than step 18 for telling readers to mix the acid.

Exercise 8.1

The following instructions, which come from the owner's manual for a CB radio, tell how to attach the radio under the dashboard of a car. Reorganize the steps in the proper order. We see nine steps, of which attaching the bracket—the start of the original instructions—is number 5. See also Figure 8.1.

Attach the bracket to the underside of the instrument panel of the vehicle using two machine screws, toothed lockwasher, and hex nuts. Before drilling any holes, however, check the fit and convenience of the unit in the selected location. Also, check behind the dashpanel—wires or vacuum hoses should be moved out of the way. Use the bracket as a template to mark the mounting hole locations. Drill the holes, using a bit size that will allow smooth clearance for the mounting screws. Secure the bracket firmly in position; then attach the CB unit to the bracket, using the appropriate hardware, as shown in the drawing. Note that there are three sets of holes in the bracket for the side screws—use the set that offers the most convenient positioning of the CB unit. Before fully tightening the side screws, tilt the CB unit to the desired angle.

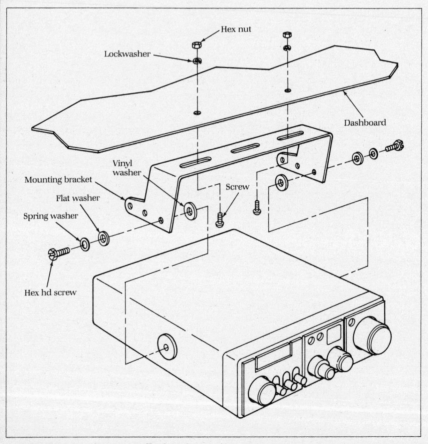

FIGURE 8.1   CB Installation.

Writing the Body

Once your step-by-step instructions are coherent, a numbered list is by far the best format for presenting them. If the procedure is complicated (more than twenty steps, say), try to divide it into stages. Decimal numbers can be used for the steps within a stage, as follows:

3. Third stage
   3.1 First step
   3.2 Second step
   3.3 Third step . . .
4. Fourth stage . . .

This arrangement is easier for the reader to comprehend than a long chain of seemingly unrelated commands. To ease comprehension even more, you should give each major stage an informative heading as well as a number:

3. Collecting the Sample
   3.1
   3.2 . . .
4. Weighing the Sample . . .
5. Testing the Sample for Acidity . . .

Numbered headings like these show readers how the stages are interrelated and also suggest natural stopping points if the procedure is too long to finish in one sitting.

Use layout, typography, and graphics to emphasize the "listy" quality of your instructions. Separate the steps with white space; use italics and capital letters to emphasize warnings and key words. Make sure that your drawings are next to the steps they illustrate and that the terms you use in your drawings are the same terms you use in your text.

The major advantage of the list format is that it is easy to follow. Readers can tell at a glance where they are and where they are going. To get the full benefit of this format, you have to write bite-sized steps. Resist the temptation to combine related steps into a paragraph. Of course, all the information relevant to a *single step* belongs together. But steps in the middle of a paragraph have a way of getting skipped—so give each command its own number.

For similar reasons, keep your sentences as short and simple as you can. Unlike most other writing, instructions are written in the second person: "You should do this." The clearest and shortest instructions, in fact, are written in the imperative mood with the "you" understood: "Do this." Consider these four sentences:

❙ The photographer loads the film in total darkness.

❙ The film should be loaded in total darkness.

❙ You should load the film in total darkness.

❙ Load the film in total darkness.

Of course most instructions include some background explanation written like the first two sentences. And some audiences prefer the formality of the third sentence. As a rule, though, use the imperative in your instructions. Research shows that an average of ten to fifteen words per sentence is ideal for instructions—and as you can see from our four versions, the imperative uses the fewest words to deliver its message.

Don't go too far, however. Stay away from instructions like "mix 8 oz. water, 8 oz. paint in jar; shake 30 secs." Most readers would rather be told, "Pour 8 ounces of water and 8 ounces of paint into a jar. Then shake the jar for about 30 seconds or until the mixture is thoroughly blended."

Inevitably, all those numbered lists of short sentences will give your instructions a chopped-up feeling, the exact opposite of that smooth flow writers usually strive for. That is the price you pay for instructions that your readers will find clear and coherent. Pay it gladly, and save your eloquence for the introduction and conclusion.

Writing the Introduction and Conclusion

A finished set of instructions usually contains an introduction and a brief conclusion to complement its step-by-step procedure section.

Introduction. Logically enough, the introduction should contain information that readers must have before they undertake the procedure. The following items are virtually always furnished in the introduction:

• The procedure to be detailed, and its purpose.
• A list of all the materials called for in the instructions.
• A list of all the tools and equipment needed to perform the procedure.
• Definitions of unfamiliar terms that come up throughout the instructions.

Other topics commonly covered in the introduction include the following:

• Roughly how long the procedure takes.
• The level of expertise required by the procedure and assumed by the in-
  structions.
• Warnings and precautions that apply to the entire procedure.
• Problems that commonly arise and how to cope with them.
• If a mechanism is involved throughout the procedure, a brief description of
  the mechanism and its important parts.
• Why the procedure must be learned.

The last of these is especially important when you are explaining a new procedure to subordinates who are likely to resent the change. As for the second-to-last, we discuss mechanism descriptions in Chapter 10.

Few introductions require all of this information. Confine your introduction to the items your readers really need. The following introduction illustrates some of the points we have been discussing.

Preparing Copper Plates for Photoetching

The procedure is announced, and its purpose is explained. The level of expertise expected by the writer is implied in the terminology: Readers must be comfortable with "regular" intaglio printing.

An overview of the procedure is given. A word that readers might not be familiar with is defined informally.

This warning is set off typographically for emphasis.

Equipment and materials are presented in separate lists. Given their level of expertise, readers are likely to have everything on the equipment list already: It serves as a checklist. However, the chemicals listed under "Materials" will probably have to be purchased.

Described as follows is a technique for transferring a photographic image from a positive film transparency to a copper plate. Once the image has been transferred, the plate can be etched in a bath of nitric acid, Dutch Mordant, or ferric chloride acid, and then processed for regular intaglio printing.

To receive the image from the film transparency, the plate is first coated with a photosensitive solution called a "resist" and allowed to dry. Then the plate is placed under a carbon arc lamp, the transparency is placed on top of it, and the lamp is turned on, exposing the unprotected areas of the resist. Next the plate is placed in a developer that renders the exposed areas impermeable to further chemical action. After the exposed resist has been rinsed off, the plate is ready for etching.

Note
These instructions assume that you have already prepared a positive film transparency of the image you wish to photoetch. If you have not, you should read Section 2.3 of this manual, "Preparing Positive Transparencies," before you proceed.

Darkroom Equipment
Because the resist is photosensitive, the entire procedure must be done in a darkroom under a yellow safe light. Necessary darkroom equipment is listed as follows:

4 stainless steel trays
Exposure table
Carbon arc lamp (or a bank of photofloods or a 250-watt UV lamp)
Vacuum frame (or a sheet of glass larger than the plate)
Funnel
Webril-wipes (or lint-free cloth)

Materials
The chemicals you will need for this procedure are manufactured by Kodak. They include:

1 qt. Kodak KPR3 Photo Resist
1 gal. Kodak KPR Photo Resist Developer
1 qt. Kodak KPR Photo Resist Dye

These chemicals have a shelf life of about one year. Also needed, of course, are:

Copper plates
Positive film transparencies

Approximate times—with allowances for variations—are given.

Each plate will take about 30 minutes to prepare if you use the carbon arc lamp. Using photoflood lamps or a UV lamp will add an additional 20–25 minutes per plate, but plate quality will not be affected.

Conclusion. The conclusion is not the place to bring up new points about the process just completed, but it can provide useful information about problems that may arise now that the procedure is finished. This information might include, for instance, where to go for servicing and spare parts, where to find instructions for performing maintenance procedures, and so forth. Moreover, conclusions have a psychological function. Few things are as comforting to the weary instruction-follower as a conclusion like this: "The assembled radio kit is now ready for use. Simply plug it in, turn it on, and enjoy it."

 # Testing Your Instructions

Ultimately, there is only one way to determine whether your instructions are truly clear, complete, and coherent. Give them to someone who is as familiar with the process as your readers are, and observe silently as he or she tries to follow them. Make a note of anything that causes your test reader trouble, and afterwards discuss specific points where the instructions might be improved. Then revise accordingly.

Review Questions

1. Who is most likely to write inadequate instructions? Why?
2. Characterize the language most appropriate to technical instructions.
3. Why is it important to provide checkpoints and brief explanations in a set of instructions?
4. What order of presentation is virtually always used in instructions? Discuss this order in terms of placing warnings and precautions.

5. What is the imperative mood? Why is it so often used in instructions?
6. Describe the format of instructions—that is, headings, step numbers, and so on.
7. What information belongs in the introduction? The conclusion?

Assignments

1. Choose one of the following, and make a list of everything a beginner would need to know to follow the instructions:

a. Changing the tire on a car.
b. Replacing a plug on an appliance.
c. Balancing a checkbook.
d. Signing on or off a computer.
e. Putting in contact lenses.

2. Write a complete set of instructions for one of the items in the preceding assignment.

3. Write the instructions for some technical process you have learned in another course. Ask someone who is ignorant of the process to follow your instructions; then revise your draft as needed. Hand in both the original and the revised version.

4. Write a set of instructions to entering freshmen on how to drop a course they have registered for and how to add a course they haven't registered for. If the procedure varies according to when the student wishes to drop or add (first week of class, third, tenth), be sure to include all the variants.

9 Process Descriptions

A process is a systematic series of actions. Processes may be linear and directed toward a certain end, like the process of wastewater treatment. They may be linear with no particular end point, like the soil erosion process. They may be cyclical and perpetual, like the process of respiration. The actions that constitute a process may be interdependent and form a cause-and-effect sequence, as in the process of developing and printing a photograph. They may be independent and represent an arbitrary chronology, as in the process of appealing a criminal conviction. They may have a specific duration in time, like the gestation process, or they may have no particular temporal frame. Processes may be continuous, but they need not be. They may recur constantly, or they may have happened only once. *Any* ordered sequence of events is a process; describing a process is describing those events in sequence.

Process descriptions may be complete in themselves, or they may be part of longer technical documents, as is the "Procedures" section of a report or the "System Overview" section of a manual. They may be as complex as a laboratory procedure for the production of synthetic insulin or as simple as: "The lock opens when the key is inserted teeth-down and turned to the right until a sharp click is heard." They may be verbal, graphic, symbolic, or all three, depending on their subject matter and the level of expertise of their intended audience. The more the audience knows about the topic, the more process descriptions tend to use technical symbols. The less the audience knows, the more likely it is that the process description will need to rely on words and pictures to make its point.

▶ Goal and Audience

Process descriptions may have popular or highly technical audiences. All process descriptions, however, have the same goal: to enable readers to *understand* a technical process or procedure. They may have special subgoals: to advise readers of the correct procedure, to recommend a better procedure, to propose a new procedure, to reassure readers about a controversial or unknown procedure, and so forth. But their primary goal is to acquaint readers with the specific steps that comprise a technical process.

Readers of process descriptions do not necessarily need to be able to perform the process themselves. All instructions are, by definition, process descriptions because they are instructions for performing a process, but in-

structions and process descriptions have different goals. Instructions tell readers what to do—"Close the door when the fumes have dissipated." Process
descriptions, on the other hand, tell readers what happens—"When the fumes
have dissipated, the door is closed."

When you profile the audience for a process description, you must concentrate particularly on your readers' knowledge and their needs. If your goal
is to leave them satisfied that they know the process well enough *for their
purposes*, you must know what those purposes are. Readers of procedures
sections in proposals, journal articles, and formal reports may be reading
those procedures so that they can evaluate them or so that they can replicate
them. Obviously, readers who *do* intend to repeat the process themselves need
more detail.

Consider the following description of the process of film development,
adapted from a general-purpose encyclopedia:

As befits a
discussion intended
for nontechnical
readers, technical
terms and details
are avoided here.
How the grains are
altered by the light
is not explained—
most general
readers don't have
the background in
chemistry needed to
understand such an
explanation.

Photographic film is thin, transparent plastic, coated
with an emulsion of grains of silver salt suspended
in gelatin. When light falls on this emulsion, the
grains of silver are slightly altered, and a "latent
image" forms. This latent image is made visible and
permanent by chemical processing, which involves
four main steps. First, the film is placed in a developer, which turns the exposed silver grains dark.
Next, the film is transferred to a "stop bath" that
halts the developer's action. Then the film is placed
in a "fixer," which removes all the silver grains not
altered by the light, thus leaving areas of clear film.
Finally, the film is washed to remove the fixer.

Some process descriptions do no more than outline the sequence of
events in the process ("first A happens; then B; then C . . ."). Others take on
the more ambitious tasks of explaining the "how" or the "why" or both; they
give details of those events and the reasons, theories, and principles behind
them. Goal and audience, of course, determine which you do. Compare the
preceding description with the following excerpts from *Neblette's Handbook
of Photography and Reprography*:[1]

The emphasis here
is on the chemical
process involved
with film
development. The
writer assumes an
audience
comfortable with
advanced chemical
equations; the
vocabulary is

Development is the process of making the photographic latent image visible. It is a catalyzed reduction of silver halide. When hydroquinone is the
reducing agent and the halide is silver bromide, the
reaction may be written

$$C_6H_4(OH)_2 + 2AgBr + 2OH^- \rightarrow$$
$$C_6H_4O_2 + 2Ag^0 + 2Br^- + 2HOH$$

or in more general form,

$$Red + AgX \rightarrow Ox + Ag^0$$

technical. This is a "state of the art" report intended for experts.

Omitted here is a seven-page discussion of the principles and theories associated with various developing agents, complete with diagrams of the chemicals' molecular structures.

The discussion of fixing and washing the film is much more detailed than the discussion in the encyclopedia. Again, the emphasis is on the why and the how of the process.

where AgX is any of the silver halides, *Red* is the reducing agent and Ox is its oxidation product.

A silver halide grain may be thought of as a lattice containing literally millions of silver ions and bromide ions. Upon exposure to light, a number of atoms of reduced silver are formed. . . .

The steps of fixing and washing serve to make the image permanent. Fixation dissolves the remaining silver bromide, and washing removes the remaining silver thiosulfate complexes and the thiosulfate ion. If the thiosulfate ion remains in the film, it may react with the silver image:

$$S_2O_3^{2-} + 2Ag^o \overset{air}{\rightarrow} SO_3^{2-} + Ag_2S$$

In this reaction, the silver has been oxidized, and the image may turn brown. Alternatively, silver sulfide may be formed from silver complexes retained in the processed material. This decomposition stains the highlights of the print or negative with yellow or brown silver sulfide.

A description of the same process in a book written for dedicated photography buffs might be equally detailed, but it would focus on the practical aspects of the process—the mechanical how, rather than the chemical how. Everything depends on the goal and audience.

▶ Organizing and Writing Process Descriptions

Writing a process description involves three steps: analyzing the process, organizing and writing the description, and selecting appropriate graphics.

Analyzing the Process

To analyze a technical process, you must answer two questions: (1) How many steps or hierarchies of steps are there in the process? (2) In what order do the steps take place? You should begin your process analysis by attempting to divide the process directly into steps. You may find that you have too many steps for a convenient list: Eight or ten are about as many as most readers can absorb without losing their bearings. If your process has more than that, group them—if you can—into coherent subprocesses. The nitrogen cycle, for example, is a circular process comprising several subprocesses: (1) ammonification (production of ammonium or ammonia by microorganisms in the soil), (2) nitrification (conversion of ammonium and ammonia to nitrites and then to nitrates), (3) denitrification (reconversion of nitrates to atmospheric nitrogen), and (4) assimilation (conversion of nitrates to amino acids and proteins by plants). Readers can best understand the nitrogen cycle if you describe each of these subprocesses one at a time rather than describing the entire process as a sequence of discrete steps.

The following description of a laboratory procedure for producing high-potency diphtheria toxin in an iron-containing medium differentiates among steps by giving each step its own sentence, and then groups the steps into three numbered paragraphs that represent three subprocesses: (1) preparation of the medium, (2) inoculation, and (3) incubation.

> 1. Twenty ml of Pantothenate Glutamate Tryptophane (PGT) medium was dispensed into each of ten 125-ml Erlenmeyer flasks. The flasks were autoclaved at 115°C for twelve (12) minutes. When the flasks had cooled, .08 ml of maltose solution was added aseptically to each flask. Different amounts of iron solution were added to each flask to provide supplemental iron concentrations ranging between 0 and .50 µg/ml. After incubation overnight to check for sterility, the flasks were ready for inoculation.
>
> 2. The flasks containing 20 ml of PGT medium were each inoculated with two large loopfuls of 48-hour-old *C. diphtheriae* pellicle from tubes of BHI broth.
>
> 3. The flasks were incubated at 34°C for a period of eight days. The flasks were not disturbed during that period except that visibly contaminated cultures were removed from the incubator.

Analyzing a process that is strictly linear is not an arduous task. The pattern is this: A then B then C then D, or, if the process is causal, A causes B, which then causes C, which then causes D, and so on until you've reached the end (or come back around to the beginning). Many processes are like that.

Many processes are not. What do you do when A causes both B and C? B causes D, which causes E, while simultaneously C causes F. Then E and F, working together, cause G, which gets A started over again. The first thing you must do is recognize the problem. Nothing is more likely to confound your readers than a deceptively simple description of a genuinely complicated process. Unless you are certain that the process you are describing is straightforward, make yourself a flow diagram to find out. (See Figure 9.1.) Look for two signs—arrows that go backwards and multiple arrows into or out of the same step. Either is a warning that you are dealing with a complex process, one in which you may not be able to organize the steps in strictly chronological order. Here is how to handle the problem.

Cycles and loops (backwards arrows) aren't too much trouble. If the whole process is a cycle, you can simply enter it at whatever point seems most logical (blood circulation "starts" at the heart; car engines "start" at the intake cycle), follow it around, and end by pointing out that the cycle keeps going. If you have a loop within the process, you will have to pause when you get to it, describe the steps in the loop, and then pick up the linear sequence where you left off. In Figure 9.1, photosynthesis and respiration form a loop within the carbon cycle. When you describe the cycle, you can first narrate

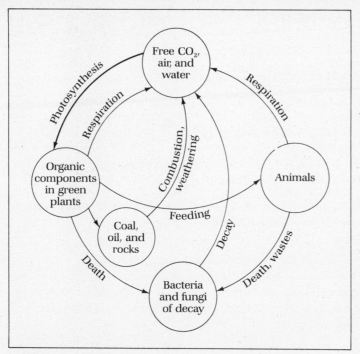

FIGURE 9.1   A Complex Process—The Carbon Cycle.
(From *Life: An Introduction to Biology*, Second Edition, by George
Gaylord Simpson and William S. Beck, © 1965 by Harcourt Brace
Jovanovich, Inc. Reproduced by permission of the publisher.)

that subprocess, and then go on to discuss what happens to the carbon in
plants when they die.

Simultaneous events (multiple arrows) are more challenging. If the arrows converge again after a step or two, stick to chronological order, dealing
with the simultaneous events together: "At this point two things happen, A
and B, which together lead to C." When you describe the carbon cycle, you
can handle combustion and decay in this way. But if A and B each go through
another step before converging at C, you will have to pursue each causal path
separately, bringing A from the point of divergence all the way to the point of
convergence, then going back to pick up B. A process like the formation of
acid rain involves a sequence of industrial events and a series of meteorolog-
ical events occurring simultaneously and affecting each other only at a few
key junctures. To be truly understandable, a description of such a process
would trace the two threads separately, bringing them together only at the
points where they intersect.

Almost always, complex processes with lots of cycles and loops and
simultaneous events have to be dissected into subprocesses before they can
be explained verbally. Before you sit down to write the process description,
decide on an order in which to present the subprocesses (if there are any)
and the individual steps. Even if the process itself is not linear, your description of it will have to be, so choose which of the two or more simultaneous

events you will present first, and number them on your outline. When you have finished arranging all the events in sequence, you are ready to start writing.

Organizing and Writing the Description

When you write a process description, you will probably begin by writing the body—the step-by-step account of the process itself. When you are finished doing that, you will add an introduction and perhaps a conclusion. In this next section, however, we will be discussing the parts of a process description not in the order in which you will doubtless write them but in the order in which your readers will see them.

Writing the Introduction. Here are some of the things you might want to include in the introduction to a process description:

Overview of the Process. Before getting to the first step, readers need to grasp the general idea of the process. Depending on how much your readers already know, the overview may take several pages or just a single sentence. It is often some kind of definition—formal, informal, or expanded. Don't fight to keep it short; a leisurely overview helps ensure that readers won't get lost in the details that follow.

Control of the Process. Saying who or what performs the process seldom takes more than a sentence, but it is sometimes a very important sentence. A description of film developing, for example, needs to specify whether it is talking about home processing, professional processing, or automated processing.

Purpose of the Process. It is essential for readers to know why the process is performed *before* you tell them how it is done. The purpose of coffee percolation may be obvious to most readers, but the purpose of mitosis is not, and deserves explicit attention.

Purpose of the Description. When descriptions of processes are not complete documents, but part of something bigger, readers will appreciate knowing in advance why the process is being described—so they can understand the conclusions or recommendations to come; so they can suggest improvements in the process, or even an alternative; so they can decide whether a certain piece of processing equipment will suit their needs; or whatever.

Description of a Mechanism. If a particular mechanism is involved at just one step in the process, you should give whatever description is needed when you get to that step. But if the mechanism is central to the entire process, the introduction is the best place to describe it. Keep this description as brief as you can, limited to what readers need to know about the mechanism in order to follow the process it controls. (For details on how to write this description, see Chapter 10, "Mechanism Descriptions.")

Underlying Theory. A theoretical principle that applies to just one step in the process should be saved for that step. But a principle that undergirds the entire process deserves explanation at the outset.

Important Steps in the Process. A step buried in the middle may be the most important one in the whole process. For instance, the actual biological process of *in vitro* fertilization comes more or less in the middle of a long and complicated laboratory procedure. When a process contains a buried step that is the nucleus of the entire process or is new or is in any other way especially noteworthy, you can use the introduction to point to the buried step and discuss its meaning.

Equipment and Conditions. Some process descriptions begin with a list of the necessary equipment and the special conditions (location, temperature, cleanliness, whatever) under which the process took place (or should take place). If readers may want to replicate the process or judge its adequacy, they will prefer this information collected at the start rather than spread through the various steps.

List of Steps. Unless your description is very short, a brief list of the steps in sequence should end the introduction. For straightforward processes a numbered list provides the clearest orientation for the reader. The items on the list should be grammatically parallel—all full sentences or all fragments, all starting with gerunds or with infinitives or with nouns, and so on. If the process is sufficiently complex that you have divided it into subprocesses, list those in the introduction, and save the individual steps for the body of your description.

You need not try to work all nine of these suggestions into every introduction to every process description you write. A ten-page introduction to a three-page description defeats its own purpose. Choose from the list those items that best fit your goal, your audience, and your process. The following introduction to the process of atoll formation includes—in two short paragraphs—seven of the nine items we have suggested.

This process description begins with the underlying theory. Most of this first paragraph is an overview of the process.	Over a century ago, Charles Darwin proposed the most probable theory of the evolution of the atoll.
Brief mention of the controlling mechanism; the author assumes that readers know what a volcano is and what it does.	According to Darwin, the process begins with the rising up of a submarine volcano from the ocean floor. When the volcano finally stops erupting, coral colonize the shore of the new island and build a fringing reef around it. Next the volcano
The process is controlled by both the natural behavior of active volcanoes and the natural behavior of coral.	gradually subsides because of crustal movement. The coral then grow upward in response to the subsidence, forming an atoll with a central lagoon. Today, Darwin's theory still remains acceptable; however, to better understand the process of atoll formation, we must add more information.
This sentence, together with the opening sentence, provides the purpose of the description.	

The second paragraph is devoted to a description of the special conditions necessary for the process to take place.

Certain conditions must exist for coral to live, grow, and build atolls: (1) a rocky base for the initial attachment of the coral; (2) water depth between the level of low tide and 600 ft. (122m); (3) water temperature of approximately 68°F (20°C); (4) abundant food supply in the form of plankton; and (5) subsidence of the base, allowing for the upward growth of the coral. The volcanic island serves as the ideal base. It is rocky and extends through a wide range of water depths, thus encouraging the initial attachment of coral. When the water temperature and food supply are suitable for growth, the coral colonize the base.

The author ends the introduction by returning to the opening step, in preparation for a detailed step-by-step analysis.

Writing the Body. Chronological order is really the only sensible one to use in the body of a process description, with due allowances for loops, cycles, and simultaneous events. Once you have established how many steps—or hierarchies of steps—there are in your process and lined up all of the steps in sequence, all that is left for you to do is to decide how much to say about each step. Some steps will contain key terms that will have to be defined; some will depend on mechanisms that will need describing. Sometimes you will want to refer your readers to additional sources of information, to discuss alternative steps, or to add a drawing.

Begin with an overview of the step and any other introductory information that readers need to get themselves oriented to that particular step— purpose of the step, control of the step, underlying theory, or whatever else is appropriate. Then describe the step: "The windward reef front forms as the fringing reef builds upward and outward toward the ocean." If the step has substeps, list those, taking them up in chronological order or in parallel paths if some are simultaneous. ("As the volcanic island gradually subsides, the coral grow upward to compensate for the changing depth.") If the substep is complex, you will have to subdivide again before you move on.

Your description of the step itself may be only one sentence long, or it may take several paragraphs. You may or may not wish to elaborate with supporting details, drawings, analogies, theories, explanations, exceptions, and so forth. Whether or not you do will be governed entirely by how difficult the step is to understand, how much your readers already know about it, and how much they need to know. The amount of detail will be different, of course, for every step in a process description. Some steps are easier to understand than others and require less illustration and elaboration. You are under no obligation to make the degree of detail in the various steps balanced or parallel, so start fresh with each step, asking yourself if a single narrative sentence will make the step clear or if you will need several paragraphs of underlying theory and a schematic diagram.

As you draft the body of your description, be sure to include appropriate chronological and causal transitions. Be sure also to balance active verbs and passive verbs. Verbs in the active voice describe actions that are self-generating or self-propelled:

> As sewage *enters* a plant for primary treatment, it *flows* through a screen, which *removes* large floating objects such as rags and sticks that *may clog* pumps and small pipes.

Verbs in the passive voice describe actions that must be performed by an agent or operator:

> In a well-known type of color film the film *is coated* with three emulsion layers. . . . A yellow filter layer *is interposed* between the blue-sensitive top layer and the two other layers. Positive pictures *are obtained* directly by reversal.

Most processes are a combination of independent steps and operator-dependent steps, so they contain both active and passive verbs. You must know the difference if you are to alternate active and passive verbs successfully in the process descriptions you write.

Exercise 9.1

Following is a process description of the production of beer. List all the verbs in the description, and identify which are active and which are passive.

Production of beer starts with the malting of barley, in which the grain is induced to sprout briefly to produce enzymes that will catalyze the breakdown of starch. The malt is ground and mixed with warm water (and often with other cereals such as corn) before going into the mash tun, where over a period of a few hours enzymes break down the long chains of starch into smaller molecules of carbohydrate. The aqueous extract called wort is separated from the mix and boiled with hops in a brew kettle. The boiling extracts flavor from the hops and stops the enzyme action in the wort. The hops are removed, and the wort is put in a fermenting vessel, where it is pitched or seeded with yeast. After fermentation the beer may go to a lagering tank to mature, following which it is pasteurized and bottled.

Writing the Conclusion. A conclusion is not essential in a process description; the description may simply end with the end of the process. Nevertheless, a conclusion at the end of a long description does help detail-weary readers regain a sense of the overall process, especially if you repeat some of the items from your introduction—the overview, the purpose, the underlying theory, the most important steps. Your readers are now better able to understand them. The conclusion to a process description is also a good place to analyze the advantages and disadvantages of the process, to cite a few of its special applications, to reiterate its importance, to compare it with other processes that aim at the same goal, or to comment on it in any other way that is relevant. Comments on a process do not belong in the description itself; that should be neutral. But putting a description into perspective is an excellent way to conclude it.

Selecting Appropriate Graphics

When you make decisions about how much detail to include in a description of a process, you should consider not only verbal detail but visual detail as well. Photographs, drawings, and diagrams are extraordinarily useful in process descriptions because they can present simultaneous things simultaneously. (We discuss this concept more fully in Chapter 11, "Graphics.") Writing is linear; it cannot express simultaneity. Graphic designs are also better than words for expressing direction and for showing relationships, both of which are important aspects of processes.

The type of graphic design used most often in process descriptions is the flowchart, in which the steps in a process are represented by boxes connected by arrows. However, photographs, drawings, graphs, and diagrams may all be appropriate devices for illustrating an entire process or one of its steps. Drawings and photographs, for instance, are better than words at expressing degrees of change. A series of three photographs showing a bud opening into a flower will describe the process of efflorescence much more concisely than words could describe it.

Good process descriptions contain both verbal and visual illustrations. Pictures can be used very effectively to analyze processes and to compare steps or stages within a process, but pictures are static and two-dimensional, so they are not as good as words at expressing the *quality* or *nature* of the changes that take place between the beginning and the end of a process. Figure 9.2 shows how words can be used to illustrate a drawing. The process by which ethanol is synthesized biologically from a substrate is described in the flowchart; the caption interprets and elaborates on the drawing. A verbal description must necessarily take up various subprocesses one at a time, in sequence. The complex and simultaneous interaction of those subprocesses is best shown, however, in a diagram such as the one in Figure 9.1 on page 203.

 # Process Descriptions and Mechanism Descriptions

As a category of technical writing, process description can be difficult to isolate from mechanism description. When a mechanism manages the process,

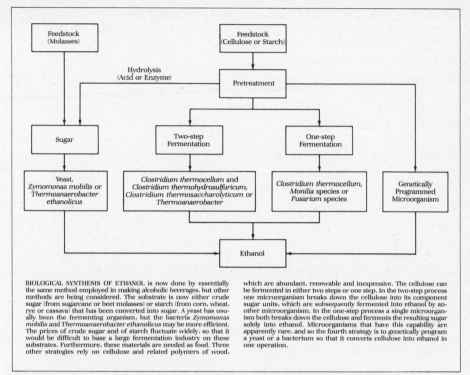

FIGURE 9.2   Process Flowchart with Detailed Caption.
(From "The Microbiological Production of Industrial Chemicals" by Douglas E. Eveleigh.
Copyright © 1981 by *Scientific American, Inc.* All rights reserved.)

as when the heart beats or when a computer performs a complex calculation, describing the process seems indistinguishable from describing the mechanism within which the process takes place. The difference between process descriptions and mechanism descriptions is essentially one of organizational emphasis—process descriptions focus on *how* and *why* things happen rather than *what* makes them happen. Processes are *action*-oriented. The words for processes come from verbs and end, more often than not, with abstract suffixes like "-ion" and "-ing."

writ – ing   erup – tion
swimm – ing   grow – th
defus – ing   revers – al
effloresc – ence   treat – ment
subsid – ence   develop – ment
fertiliza – tion   impeach – ment
abscis – sion

In a process description, an *action* is segmented into its constituent *steps*.

 In a mechanism description, by contrast, a *thing*—a concrete or abstract entity—is segmented into its constituent *parts*. Although the term "process" and the term "mechanism" are rather elusive and hard to define, you may consider anything with discrete steps or operations a "process" and anything with discrete components a "mechanism." When you describe a mechanism, you will describe it not in terms of its stages or steps but in terms of its subunits and parts. And that requires a wholly different organizational strategy, as you will see in the next chapter.

Review Questions

1. What is a complex process? What organizational problems do you encounter when you describe a complex process?
2. What kinds of information should you include in the introduction to a process description?
3. How should you organize the body of a process description?
4. When should you use active verbs in a process description? When should you use passive verbs?
5. What kinds of graphics are commonly used in process descriptions?
6. What is the difference between a process description and a mechanism description?

Assignments

1. Find a description of a technical process in your field. Specify a different goal and audience, and write your own description of the same process. Attach the original and an explanation of why you made the changes you made.
2. Write a 200-word description for a nontechnical audience of an important process in your field. Write an 800-word description of the same process for a technical audience.
3. For the same technical process you used in the preceding question, now write an 800-word description for a nontechnical audience and a 200-word description for a technical audience.

Notes

1. R. W. Henn, "Development and After Process," in *Neblette's Handbook of Photography and Reprography*, ed. John M. Sturge, 7th ed. (New York: Van Nostrand Reinhold, 1977).

10 Mechanism Descriptions

For most people, the word "mechanism" triggers images of something elaborate and mechanical—a clock, for instance, its innards composed of numerous gears, springs, balance wheels, connecting rods, pins, and sprockets. Of course, a mechanism may be a complex device like this, but it may also be as simple as a stick of chalk, as natural as a gallbladder, or as abstract as the law of supply and demand. Anything—or any group of things—acting as a functional unit is a mechanism. Thus, a molecule of ethyl chloride, a dipole antenna, the U.S. Interstate Highway System, and General Motors are all mechanisms.

Having read the previous chapter, you should be clear on the distinction between a mechanism and a process: Processes, you will recall, are action-oriented, whereas mechanisms are thing-oriented. *Circulation*, therefore, is a process, but *the circulatory system* is a mechanism.

Technical people spend a good deal of time writing about mechanisms, and frequently what they write is description. Yet, though an occasional report or journal article will be devoted exclusively to describing some new invention (an improved integrated circuit, say, or a solar wristwatch), mechanism descriptions rarely appear as independent reports. Virtually all reports, however, contain descriptive passages, and often a whole section of a report will be assigned to describing the mechanism mentioned in other sections. User manuals for lab equipment, for instance, usually begin with a detailed description of all the components, especially the controls.

▶ Goal and Audience

The goal of all mechanism descriptions is to give readers a useful mental image of the mechanism. The key word, of course, is "useful." To be useful, a mechanism description must bridge the gap between what the readers already know about the mechanism and what they need to know. Who reads mechanism descriptions? The answer is everyone from experts to laypeople, from technicians to executives. They may read them to get a general overview of a mechanism, to evaluate its technical soundness, to judge its economic potential, to determine how best to manufacture it or repair it, or for any number of other reasons. Thus, a mechanism description may be short and very general, long and highly detailed, or anywhere in between; it may be filled with technical terminology and minute details, or it may be written in words of one syllable; it may focus on a specific mechanism—the IBM Selectric III—or be generic—the typical electric typewriter. But in any event, to bridge the gap between what readers know and what they need to know, the description of a mechanism must address some or all of the following questions:

1. What the mechanism is.
2. What it looks like.
3. What it is made of.
4. What its parts are and how they fit together.
5. What it is for.
6. How it works.

Notice that the second, third, and fourth of these focus on the physical characteristics of the mechanism, whereas the last two address functional characteristics.

We often think of description as concerned exclusively with physical appearance. Here, for instance, is a brief description of the layers of a modern roadbed:

> The new two-lane road from Green Street to Route 222 is composed of the following layers: an 18-inch subbase of crushed stone, a base layer of concrete approximately 10 inches thick, a 2 1/2-inch intermediate layer of rolled asphalt, and a surface of rolled asphalt, which rises 2 inches above the intermediate layer. A concrete haunch separates the road from a shoulder of crushed stone.

But a mechanism description may also concentrate on function. The following description, taken from a desk-top encyclopedia, contains no measurements and few physical details; instead, it focuses on what each part of the camera does:

> The modern camera is a light-tight box with a mechanism that holds a piece of film flat and opposite a lens. The lens focuses onto the film a sharp upside-down image of the scene before the camera. A shutter, located between the film and the lens, prevents light reaching the film until it is opened, usually for only a fraction of a second. The correct exposure is obtained by regulating the relationship between the shutter speed and the diameter of the lens.
>
> The diaphragm controls the amount of light passing through the lens, and the shutter determines how long the film is exposed to the light. . . . All cameras have a viewfinder—from a simple wire frame to a complex optical system—that enables the user to see what the camera "sees."[1]

Most descriptions include both physical and functional characteristics—but not always to the same extent. Descriptions of specific mechanisms are often predominantly physical, especially when they are aimed at knowledgeable readers, who are likely to understand function already. The goal of most generic descriptions, on the other hand, is to give readers an idea of a mechanism's function. Generic descriptions cannot provide details of physical appearance because these vary widely. To see the point clearly, consider the difference between the camera example just given and a description of a Minolta X-700.

▶ Choosing Appropriate Details

In a description of any mechanism there are a nearly limitless number of things you could focus on. Yet your initial attempt at technical description will probably suffer from a lack of detail. And paradoxically, the more complex the mechanism, the more likely you are to have very little to say about it. The sheer number of parts and possible details can be daunting.

Force yourself to *see the mechanism* freshly and completely. Begin by examining the overall mechanism. Study it in operation as well as at rest, noting which parts move—and why. Examine the mechanism from several different perspectives—from above, from below, from the sides—and record your observations. If your readers will be interested in physical description, note the color, shape, size, weight, density, texture, composition, and location of all the exposed parts. And record the model number, the manufacturer, and any other distinguishing features of the overall mechanism.

Once you have studied the mechanism as a whole, *partition* it according to the suggestions we made in Chapter 6. As you take it apart, note which parts work together as subsystems of the mechanism, and examine each subsystem as you did the whole, recording its function and physical characteristics. Also note how these subsystems are attached to each other: by bolts, solder, glue, screws, nails, pressure, and so on. Then take each subsystem apart, and examine the individual subparts, again noting the physical characteristics, function, and method of attachment of each. Accumulate far more detail than you could possibly impose on your readers. Then you can choose.

Which details you choose depends, obviously, on your goal and audience. In describing a certain brand and model of stereo receiver, for instance, the owner's manual would confine itself largely to a description of the things on the exterior: the knobs, buttons, dials, and switches on the control panel in front; the various input and output terminals, hookups, and fuse sockets on the back. Each of these parts would be located (undoubtedly with the aid of a photograph or drawing), and the function of each would be described. A repair manual, on the other hand, would describe the inner workings of the receiver as well, concentrating on the parts that can be replaced or repaired. A manufacturing manual would give precise dimensions, weights, materials, and other specifications and would account for every rivet that goes into the receiver. And a brochure prepared by the manufacturer for salespeople would describe the jazzy new features and latest advances that make this receiver better than its predecessor—and better than the competition. Although some of these descriptions would, of course, be longer than others, each would include some information that the others left out. Most important, each would focus on those details needed by its readers.

If you are writing anything more substantial than a brief overview, always include more details than you think you need—at least in your first draft. You can pare away the excess in revision. It is always easier to see what you have and cut it than to see what you don't have and add it.

▶ Organizing the Mechanism Description

You should organize your description just as you choose your details— with sensitivity to your reader's needs. A random description of the hundreds of individual parts that make up an automobile's ignition system, for instance, would be almost as useless as no description at all. This is why it is so important to partition your mechanism rather than merely cataloging its parts. Because partitioning divides the mechanism into logical and manageable groups of parts and subparts, it will provide the organizational basis of your description. Thus, after an introductory overview, your description will start with one main part or subsystem (using our ignition system example, say, the solenoid), describe it and all its subparts, and then move on to the next main part (the starter motor, for instance). A few main parts, each divided into subparts, will always be easier for readers to follow than a random list.

The problem remains of deciding which of those main parts to describe first, second, and so on, as well as of choosing the best order for the subparts comprising each main part. Any order that makes sense to your readers and follows the logic of the mechanism itself is all right. Following are the most common options.

Spatial Order

By far the most common and generally the most useful order of presentation, spatial order guides the reader through the mechanism from top to bottom, from left to right, from front to back, from inside to outside, from northeast to southwest, clockwise, counterclockwise, and so on. It can be used for any physical mechanism (as opposed to theoretical or abstract mechanisms like the law of supply and demand). It is especially useful when your main concern is to provide physical description of an object that has no moving parts (an armchair, a quartz crystal, a topographical map) or of a mechanism at rest. It is inappropriate for mechanisms in motion; you run the risk of confusing your readers if the moving part that you have located in the rear of the mechanism later turns up in the front.

The spatial orientation of the parts of the mechanism will, of course, determine which spatial order works best. The description of the roadbed we presented earlier in the chapter (p. 213) moves from bottom to top. Oddly shaped mechanisms often require a combination of spatial orders. In fact, even relatively simple mechanisms may require more than one. A description of a common light bulb, for instance, might use an outside-to-inside order for its macrostructure (the metal collar, the glass bell, the layer of inert gas, the filament) and then switch to a top-to-bottom order to discuss the parts of the filament itself.

Operating Order

Describing a mechanism in operating order—the actual sequence in which the parts work—puts the emphasis on function. For those many mechanisms that have only a single sequence of operation, readers find this a logical order of presentation, especially if the description will be predominantly functional. A description of the human digestive system, for example, might start with the mouth, then take up the pharynx, esophagus, stomach, small intestine, and so on, thus following the sequence of digestion. Similarly, a description of a car's braking system might begin with the brake pedal, then move to the master cylinder, the brake lines, the slave cylinder, the brake shoes, and so on. If the mechanism operates in a random or nonlinear sequence, however, operating order won't work. There is, for instance, no single operating order for the parts of the mouth: The teeth grind, the glands secrete saliva, and the tongue manipulates food all at the same time. Thus, even though the overall order of presentation of the digestive system could follow operating sequence, this particular part of the system requires another order—probably spatial.

Order of Importance

Order of importance is a useful organization when readers need to know only some parts of the mechanism in detail. An operator's manual for control room personnel, for example, might start with the major controls (in great detail), move on to the less important controls (in less detail), and conclude with other parts of the mechanism (in very little detail).

Order of Assembly

Occasionally, your readers will want to know how the parts of a mechanism were put together; manufacturers are especially interested in this sort of description. Order of assembly is also useful, obviously, for mechanisms that need to be assembled.

Exercise 10.1

Following are excerpts from two descriptions: the first, of a zinc dry-cell battery; the second, of solar water heaters. Identify the order or orders of presentation used in each.

1.   The zinc dry-cell is essentially a series of concentrically arranged cylinders. The innermost cylinder is the cathode, a core of graphite fitted at the top with a steel collar. Into this collar is screwed a thin threaded

rod, also of steel. This rod is capped with a nut that can be screwed down against the collar to hold a wire. The graphite rod is immersed in a paste of manganese oxide and carbon, which is in turn surrounded by a heavier, moister paste of ammonium chloride, zinc chloride, and chalk, which provides a source of ions to keep the cell electrically neutral.

The cathode assembly and the two chemical pastes fill an enclosing hollow drum of zinc; the top of the cathode collar lies flush with the top of the zinc drum. A small lip encircles the top of the drum. A steel clip, shaped like a horseshoe, is attached to the lip. The clip accepts a threaded rod like the one fitted into the cathode collar, and this rod, in turn, is fitted with a nut that can be screwed down to hold a wire. The zinc drum is the cell's anode.

2. Solar heating systems for residential hot water are available from several manufacturers as prepackaged units. They are pre-engineered and come complete with installation instructions and all necessary hardware. Although system configuration varies by manufacturer, these systems usually consist of:

- approximately 50 ft^2 of flat plate-glass-covered solar collector to absorb the sun's energy
- a heat exchanger to transfer the collected heat to the domestic water supply
- an insulated tank to store the heated water
- an auxiliary source of heat, usually an electric immersion heater element
- controls, pumps, valves, and so on, to operate the system automatically

The Format of Mechanism Descriptions

Once you have decided on an order of presentation, you can begin to write. Descriptions of a mechanism, as we have seen, can be general or detailed, physical or functional, aimed at laypeople or at experts. These differences naturally affect your description. Still, there is a classic model for describing a mechanism. By stressing some items and de-emphasizing others, you can adapt this model to virtually any description. In the outline that follows, the essential items are marked with an asterisk; the others are optional, depending on your goal, your audience, and your mechanism:

I. Introduction
  *A. Define or identify the mechanism.
  *B. Briefly explain its function or purpose.
  C. Identify the purpose of the report and the intended audience.
  *D. Describe the mechanism as a whole (size, shape, materials, and so on), using illustrations if possible.
  E. Briefly discuss the relationship of the mechanism to any larger mechanism of which it is a part.

    F. Explain the principle of operation.
    G. List the principal parts of the mechanism in the order in which they will be discussed.

  II. Detailed Description of Parts
    A. First Part
      *1. Identify (or define) and locate the part.
      *2. Briefly explain its function or purpose.
      *3. Describe the part, using illustrations if appropriate.
      *4. Explain its relationship to other parts and its method of attachment (if applicable).
      5. List subparts, if appropriate, and describe each one following the same outline.
    B. Second Part, Third Part, and so on.
    C. Discuss possible variations, optional features, and so on.

 III. Conclusion
    A. Explain the principle of operation (if you haven't done so in the introduction).
   *B. Describe the entire mechanism in operation, stressing how the parts work together.
    C. Review the major uses, unique characteristics, and overall importance of the mechanism.

This outline works best for highly technical descriptions of complex mechanisms. These sorts of descriptions are usually more physical than functional; often the detailed description of each part occupies nine-tenths of the entire description. But even a very general description for laypeople must cover the same ground, however sketchily; the difference is that the introduction of a nontechnical description may be the longest part.

Most of the items in the outline are self-explanatory. Some, however, deserve special attention.

Many mechanisms have functional names, so that identifying them and explaining their purpose becomes the job of a single sentence: "An electric pencil sharpener, which is activated when a pencil is fed into it, automatically shaves a pencil tip to a smooth, sharp point." If the mechanism's name doesn't clearly suggest its function, you will need to state it explicitly: "An autoclave is a steam chamber that can be regulated to maintain saturated steam at a designated pressure and temperature for a designated period of time. It is used primarily in laboratories and medical facilities to sterilize surgical instruments, laboratory glassware and other equipment, and also certain types of organic substances."

The description of the mechanism as a whole should usually be brief: Give the overall dimensions and weight, identify the basic materials, and point out any identifying characteristics (like a brand name broadly etched across the front). One excellent way to describe the mechanism as a whole is to provide an illustration and refer readers to it.

As you can see from the model outline, the principle of operation—the underlying science that explains the mechanism's function—can go either in the introduction or in the conclusion. If it will help the reader visualize or understand the mechanism and why it is designed the way it is, put it before the detailed description. If it won't, save it until the parts have all been described. For instance, knowing that a car jack works on the principle of leverage might be helpful in advance; on the other hand, the physics involved in the percolation of coffee would probably be easier to grasp once the coffee percolator has been thoroughly described.

The format for each part (and for each subpart) mirrors that for the introduction. Each part must be defined and located, and its function must be explained. Then each part is described. If your description is mostly physical, you will be presenting details on size, shape, materials, weight, color, texture, finish, and so on. You should also explain the relationship of this part to other parts, especially how it is attached to those other parts.

If your mechanism is at all complicated—or even if you are giving a very detailed description of a rather simple mechanism—the detailed description of parts will be long and crammed with information. Use headings and subheadings to break up the text into logical portions and to guide readers to the correct subsection. Use a first-order heading to announce the section, second-order headings for each part, and (if you need them) third-order headings for each subpart (see pp. 93–96 for more information on headings).

Once you have described each piece and its function, conclude with an explanation of how the parts work together to make the mechanism function as a whole. This explanation is usually given the heading "One Complete Cycle of Operation" or "The _____ in Operation." It is actually a process description of whatever action the mechanism performs. If your mechanism requires a human agent, you will probably have to mention what that agent does (and when and where). But keep the focus on the mechanism: This is not, after all, a set of instructions.

The following excerpts from a description of a coffee percolator illustrate most of the points in the outline. Because a percolator is simpler and less "technical" than many of the mechanisms you will be called upon to describe during your career, it makes a useful example, one that enables you to focus on format rather than on content. A good description of a more complex mechanism—a particle accelerator, say, or the homing device of a guided missile—would follow a similar format.

Introduction

An important term is defined.

The mechanism is defined in terms of

Percolation is the extraction of the essence of a dry substance by a downward displacement of a liquid solvent. A coffee percolator is a mechanism that displaces boiling water (the solvent) to extract the essence of coffee from coffee grounds (the dry

its function. The specific mechanism is identified.

substance). An external view of the Apex Percoluxe coffee percolator is shown in Figure 10.1.

FIGURE 10.1   External View of the Apex Percoluxe.

The description given here is quite general.

The Percoluxe is 8½ inches tall and approximately 4 inches in diameter. It is capable of making up to 64 fluid ounces of coffee. Because the Percoluxe is used in conjunction with a stovetop burner, all of its parts are heat-resistant: The exterior is mostly stainless steel; internal parts are all reinforced aluminum.

Only the major parts are presented here. The basis for the organization—order of assembly—is specified.

The major parts of the percolator, in the order in which they are assembled for use, are the following: the pot, the water tube, the coffee basket, the basket lid, and the pot lid with glass plug. . . .

The Water Tube

The part is identified and placed. Its function is explained.

It is described.

Here the order of presentation is spatial. The "raised support ring" was described in the

The water tube (Fig. 10.2) rests inside the coffee pot. It has two functions: It provides a passageway for the boiling water displaced during percolation, and it supports the coffee basket and basket lid.

The water tube resembles an inverted funnel whose wide end has been flattened and whose stem has been stretched. It is made of aluminum and weighs 10 grams. The wide end of the water tube is its base; it is 3 inches in diameter and thus fits just inside the raised support ring at the bottom of

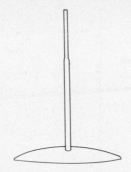

FIGURE 10.2   The Water Tube.

previous (omitted) section.

The relationship to the next part in the description is explained.

the pot. The base is convex, its center rising ½ inch above the bottom of the pot. A small hole near the lower edge of the base allows water to flow into the tube and up its stem. The stem is 8 inches long. From the base to a point 5 inches up the stem, its diameter is ½ inch; from there to its top end, its diameter is ¼ inch. The ridge formed at the point where the diameter drops supports the coffee basket.

The Coffee Basket

The coffee basket (Fig. 10.3) holds the coffee grounds. Made of aluminum, it weighs about 8 grams and is 2¼ inches high. At its bottom edge its diameter is 3½ inches; one inch up from the bottom its diameter increases to 3¾ inches. When the percolator is assembled for use, the water tube slips through a hole in the bottom of the coffee basket; the hole is reinforced with a 1-inch piece of alu-

FIGURE 10.3   The Coffee Basket.

minum tubing to steady the basket on the water tube. The bottom of the coffee basket is perforated with evenly spaced $\frac{1}{32}$-inch holes, 64 holes per square inch. These holes are just large enough to let the percolating water through without allowing the coffee grounds through as well.

The upper edge of the coffee basket has been rolled back to form a small lip. The basket lid fits snugly over this lip. . . .

Preparing the Percolator for Operation

This section shows the relationships of all the parts, and explains how the user of the mechanism would assemble the pieces.

The pot is first filled with the desired amount of water. Then the water tube is placed in the pot so that its base is enclosed by the tube support in the bottom of the pot. The coffee basket is next placed over the stem of the water tube so that it rests securely on the supporting ridge. After the recommended amount of ground coffee has been measured into the basket, the basket lid is fitted over the basket, and the pot lid with the glass plug in place is fitted over the pot. The assembled percolator is then placed over a heat source and the water brought to a boil.

The Percolator in Operation

The science governing the mechanism's operation is briefly explained.

The process of coffee percolation is described.

The process of percolation begins when the water begins to boil. Pressure caused by rising steam and air forces water up the tube. When the water passes out through the end of the tube, it hits the glass plug and is deflected down onto the basket lid. The channels of perforated holes in the lid distribute the water evenly over the coffee grounds held in the basket. The water seeps through the coffee, extracting some of its essence and color, then drips through the holes in the bottom of the basket, mixing with the rest of the water in the bottom of the pot. The small hole at the bottom of the tube continuously supplies water to the system, so that percolation is a cyclical process. The cycle ends when the pot is removed from the heat source—usually when the water splashing against the glass plug turns a dark brown.

Writing Effective Descriptions

Because most mechanism descriptions contain so much highly detailed information, it is easy for readers to get bogged down even if the organization is clear and solid. The following suggestions should help to improve the clarity of the descriptions you write and thus lessen the chance that you will lose your readers in all the necessary detail.

Name Every Part

Because even a simple mechanism is bound to have many similar-looking bolts, plates, bars, rods, pins, and so on, you will help your readers if you give every part a name. If a part *has* a name, you should be able to find it with a little digging. Ask your colleagues; consult a technical dictionary or handbook; call a reference librarian.

If you are satisfied that the part doesn't have an accepted name, feel free to make one up. Give it a name that reflects either its appearance or its function or both. A piece with an appropriate shape might be called a "tongue," another, the "S-shaped connecting plate," a third, the "plastic bridge," and so on. The idea is to think of names that enable the reader to form a concrete image of each piece.

Keep Readers Oriented

Be careful about orientation and point of view—that is, the position of the mechanism with respect to the observer. Some mechanisms have a natural orientation; a car, for instance, has a natural front and back, top and bottom. It does not, however, have a natural left or right. These directions depend totally on where the observer is standing (facing the grille, facing the trunk, and so on). Many mechanisms, moreover, have no natural orientation at all—a can opener, for example, has no immediately recognizable top or bottom, left or right, front or back. If the mechanism you are describing has a natural orientation, referring to it frequently will keep your readers on track. If there is no obvious orientation, you will have to provide one. It might be useful to key the orientation to a prominent part or feature of the mechanism. Instead of saying "top" or "front," then, you might say "on the side with the dull finish" or "on the side without the gears." You should also explain clearly how you are viewing the mechanism: "seen from above," "with the bolt heads resting on the table," and so on. Once you have established where you are in relation to any part of the mechanism, you can use that part as your reference point: "Just above the knurled bolt and 2 mm to its left is the contact flange."

Use transitions to keep readers oriented as you move from one part or subpart to the next. If you have chosen a spatial order, your transitions will be phrases like "below this," "to the right," or "six inches behind the _____."

If you are working with an operating order, use transitional phrases such as "the next part to come into play is the _____" and "when the water passes through the outflow valve, it enters the _____." Often the method of attachment will provide a useful and logical transition between parts. Say, for instance, that you've just finished describing one part of your mechanism—call it the grid plate—and are ready to begin the next—the chassis; you can move smoothly from one to the next like this: "Two rivets fasten the grid plate to the chassis. The chassis is. . . ."

Be wary of transitions that don't provide a clear spatial or functional reference. "The first of these parts" may mean something precise to you, but unless you explain what you mean by "first"—first in importance? first in view?—it might not mean the same thing to your readers.

Illustrate Your Description

For physical descriptions, graphics are virtually indispensable. Photographs or line drawings accompanying the written text can show the orientation of the mechanism and the relative size, shape, and position of its parts much more clearly than words alone. Graphics cannot express function, of course—only words can do that—but they make physical characteristics concrete and thus easier for readers to comprehend. Most descriptions require a series of drawings or photographs—a representation of the entire mechanism in the introduction, representations of the parts and subparts as they come up in the description.

Two kinds of line drawings are especially useful in mechanism descriptions, because they have the ability to show both the whole mechanism and its internal parts simultaneously: cutaway views and exploded diagrams. A cutaway view of the external tank of NASA's space shuttle is shown in Figure 10.4. Note that the tank's outer skin has been made partially "invisible" to reveal the internal parts. Cutaways are typically used in the introduction of a mechanism description. Though a cutaway view need not be geometrically precise, most are.

In an exploded diagram (Fig. 10.5), the parts of the mechanism are arbitrarily disjointed but their relationships within the whole are preserved. Thus, readers can clearly distinguish each of the major parts, and, at the same time, they can "compress" the parts mentally to get a clear image of the assembled mechanism. Like cutaway views, exploded diagrams are often used in the mechanism description's introduction.

With the relative positions of all the major parts established in the introduction, the illustrations within the body of the description can be used to isolate specific parts or groups of parts. Generally, the more illustrations the better. Figure 10.6 shows one group of parts of a wind turbine. Notice how much information is conveyed even without the written text of the description.

We devote the next chapter to graphics. There you will find information on how to label them, how to refer to them in the text, and where to place them.

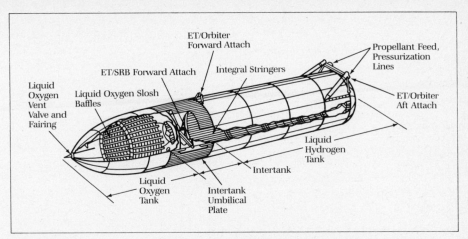

FIGURE 10.4   Cutaway View of the External Tank of NASA's Space Shuttle.
(From "NASA Space Shuttle," U.S. Government Printing Office, 1982.)

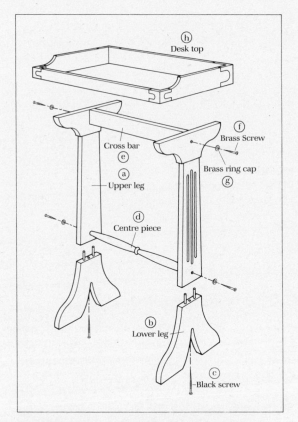

FIGURE 10.5   Exploded Diagram of a Desk.
(Reproduced with permission of The Bombay Company.)

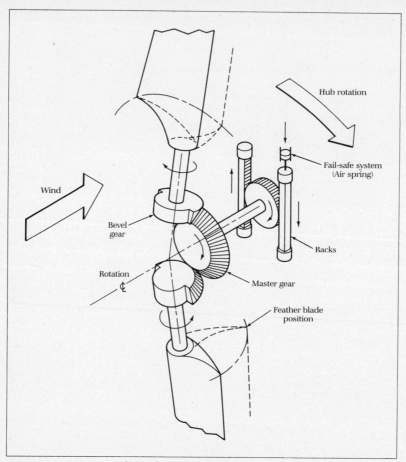

FIGURE 10.6   A Section of a Wind Turbine.
(From "Wind Energy Developments in the 20th Century," NASA, Lewis Research Center, rev. 1979.)

Use Concrete Language

Like your graphics, your writing style should be aimed at making your mechanism description concrete and easy for your readers to visualize. Using action verbs, comparisons and simple words and sentences will help.

Action Verbs. Make an effort to avoid lifeless (but easy-to-write) sentences that endlessly repeat the formula "Next to this, there is the _____. It is a piece...." Instead, use verbs that help the readers see the mechanism and the actions it performs—verbs like "protrude," "encase," "slide," "press," "squeeze," and "radiate."

Three verbs you should be especially wary of are "to use," "to function," and "to serve," as in this sentence: "These screws serve to hold the case to the lid." "Serve" doesn't convey any specific action in this sentence. But even more important, it makes the sentence needlessly wordy; "These screws hold the case to the lid" conveys the same message more concisely. Here is a sentence that misuses "use":

> The Thermos Bottle #2210 consists of a number of layers that use insulation to keep the temperature of the contents relatively constant.

Without "use," the sentence is clearer and shorter:

> The Thermos Bottle #2210 consists of a number of insulating layers that keep the temperature of the contents relatively constant.

Comparisons. Whenever possible, begin your description of an unfamiliar mechanism or part by comparing it with something familiar. This technique is especially valuable in physical descriptions: "The contact disc is approximately the size and shape of a penny"; "This section of the Pearl Paper Punch resembles a mouth: It has an upper and lower jaw, a tongue, and one large tooth—the punch itself." Comparisons of a specific mechanism with the typical mechanism of that type can quickly establish its unique traits. If you are describing an unusually slender ball-point pen, you will help your readers by saying so; to say only that it is "six inches long by ¼ inch in diameter" just doesn't make the point as clearly (though you should probably include the dimensions too).

It is also useful to compare the mechanism you are describing with another that is similar but simpler—a gas turbine with a pinwheel, for example. Beware of extended analogies, however; they usually end up strained and thus defeat their own purpose. (Don't insist on lips and salivary glands for the Pearl Paper Punch's mouth; a steam shovel may be vaguely like an ice cream scoop—but only vaguely.)

The following paragraph makes useful comparisons in its description of a nuclear reactor vessel. In the paragraph that preceded it, the author, Daniel Ford, gave a brief description of steam engines and a short account of their historical importance:

The author is writing to nontechnical readers.

A familiar object is used for comparison.

A nuclear steam-supply system . . . is a machine of great complexity. At the center of the system is a reactor vessel that is connected to one of the most elaborate plumbing systems ever devised. The typical Babcock & Wilcox reactor, to use a specific example, is a steel bottle that stands some forty feet high, measures fifteen feet across, and weighs more than four hundred tons. Inside the reactor's walls—eight and a half to twelve inches of steel—is the uranium core, itself weighing a hundred tons or more, where controlled chain reactions provide the heat that, as in a conventional electric plant, is used to turn water into steam to drive a turbine. The reactor's core is impressively compact: it is only twelve feet wide and twelve feet high.[2]

Simple Sentences; Simple Words. In a mechanism description, keep your writing as simple as you can, both its sentence structure and its vocabulary. Readers trying to absorb all the details of a complicated mechanism should not have to deal with equally complicated sentences. And as for vocabulary, readers who know all the technical terminology you might use to describe your mechanism probably know the mechanism too. They are not usually the people you are writing for. If you do need an unfamiliar term, of course, you must stop everything and define it.

Review Questions

1. What is a mechanism? What aspects of a mechanism are usually addressed in a mechanism description?
2. What is the difference between functional description and physical description? Under what circumstances is each most useful?
3. Which order of presentation emphasizes function rather than appearance? Which emphasizes appearance?
4. The principle of operation can be explained either in the introduction or in the conclusion. How do you determine where it belongs?
5. What belongs in the introduction to a mechanism description?
6. Describe the format of the "Detailed Description of Parts" section.
7. If you must invent a name for a part, how do you name it?
8. How do you keep your readers oriented?
9. What types of graphics are most appropriate in a mechanism description?

Assignments

1. To gain practice in seeing things closely, write a *physical description* of one of the following items. Make your description exhaustively detailed— much more detailed than a formal mechanism description would be:

a soda can	a pencil	a screwdriver
a baseball	a fork	the cover of this book
a hairbrush	a coin	

Make this a specific description of a particular item—your own hairbrush, pencil, and so on.

2. Determine an order of presentation for describing each of the following mechanisms (assume an uninformed audience). Explain why you chose each order.

a cigarette lighter	a bicycle derailleur
a guitar	a pair of toenail clippers
a basketball team	a college fraternity
a coffee grinder	a canoe

a zipper	the U.S. Senate
a tumor	a rock
a desk lamp	a vacuum cleaner
a pine cone	the queen of spades

3. Write a detailed description of a specific mechanism with which you are familiar. Then write a two-paragraph generic description of that type of mechanism. For each description, specify what assumptions you are making about goal and audience and how these assumptions have affected your description.

Notes

1. *The Random House Encyclopedia*, ed. James Mitchell (New York: Random House, 1977) p. 1784.
2. Daniel Ford, "Three Mile Island," *The New Yorker* (19 October 1981) p. 61.

11 Graphics

Pie Charts
Flowcharts
Organization Charts

Technical Graphics as a Part of Technical Style

Most technical writing would be impossible without graphics. Drawings, photographs, tables, and charts not only illustrate technical documents but also convey information that cannot be presented adequately in words. If you have ever tried to follow a set of assembly instructions that came without accompanying diagrams, you know exactly what we are talking about. The words and the images in any technical document complement one another.

There are two types of technical graphics: tables and figures. Tables are linear arrangements of words, numbers, or symbols in parallel columns. Every graphic design that is not a table is a figure. Figures include drawings, photographs, charts, graphs, diagrams, maps, and even actual samples such as a swatch of fabric. We do not have the space in this chapter to give you an example of each of the hundreds of different types of technical figures; your college library has books on scientific and technical illustration that describe the graphical systems used in such fields as medicine and engineering. We will, however, show you a number of widely used graphics that are within even an amateur artist's grasp. Even if you cannot draw very well and lack access to a computer that can generate graphic images, you can learn to create simple graphs and charts that will make your technical reports immensely more readable and professional-looking. And even if you subcontract the art work in a technical document to an illustrator, you will still be the one who has to decide which information is best displayed verbally and which graphically, and it will be up to you to tell the illustrator what you want drawn. You need to be conversant with the various types of tables and figures if you are to choose the most appropriate ones for your particular report.

Tables

Tables are used to give order and coherence to data. They are most often numerical, but they need not be. The table in Figure 11.1, showing vitamin food sources and deficiency diseases, is not quantitative, but it does categorize related items of information systematically.

The Value of Tables

The table in Figure 11.1 is much more concise and readable than the same information would be if it were presented in sentences and paragraphs. Notice how just a small segment of the table makes painfully dreary reading in textual form:

Important water-soluble vitamins include vitamin B_1 (thiamine), which is found in pork, liver, heart, kidney, peas and beans, and wheat germ;

231

vitamin B_2 (riboflavin), which is found in liver, milk, kidney, eggs, and brewer's yeast; niacin, which is found in . . . . A deficiency of vitamin B_1 causes nervous disorders, circulatory disorders, and gastrointestinal disorders; a deficiency of vitamin B_2 causes . . . .

Tables are a concise and efficient way to present large quantities of related or similar information that would be tediously repetitive if it were conveyed as verbal text. Moreover, tables stress the *relationships* among pieces of information visually by arranging related facts in the same column or under the same heading. When information is orderly, tables are bound to show readers the basis for the order more effectively than words.

TABLE 0.   Common Food Sources and Deficiency Diseases of Important Vitamins

Vitamin	Source	Disease
WATER SOLUBLE		
Vitamin B_1 (thiamine)	pork, liver, heart, kidney, peas and beans, wheat germ	nervous disorders, circulatory disorders, gastrointestinal disorders; symptoms include neuritis, rapid pulse, fluid accumulation
Vitamin B_2 (riboflavin)	liver, milk, kidney, eggs, brewer's yeast	lesions on the tongue, lips, and face
Niacin	lean meats	pellagra
Vitamin B_6 (pyridoxine)	yeast, liver, wheat germ, eggs	nervous disorders; symptoms include irritability, seizures, gastrointestinal disturbance
Vitamin B_{12} (cyanocobalamin)	yeast, liver, wheat germ, eggs	pernicious anemia
Vitamin C (ascorbic acid)	citrus fruits, strawberries, green peppers, broccoli, tomatoes, potatoes	scurvy
FAT SOLUBLE		
Vitamin A	fortified whole milk, liver, egg yolk, kidney, deep green and yellow vegetables	night blindness, skin disorders
Vitamin D	fortified whole milk, mackerel, tuna, salmon, sardines, herring (also sunshine)	bone diseases, such as rickets and osteoporosis
Vitamin K	spinach, cabbage, egg yolk	prolonged clotting time, bleeding

FIGURE 11.1   A Nonnumerical Table.

TABLE 0. Grade Distributions for Mechanical Engineering 301, Spring Semester, 1985

Instructor & Section		Grades						
		A	B+	B	C+	C	D	F
C. Chang	MWF	5	6	8	4	2	0	0
	TTh	4	4	8	5	0	1	0
A. Rauch	MTh	4	4	5	5	4	1	0
L. Tichauer	MWF	6	6	9	3	4	0	0
	TTh	7	7	5	3	4	0	1
J. Willard	MW1	4	6	8	2	3	2	0
	MW2	5	7	5	4	4	0	1
B. Yamani	MWF	5	5	5	4	4	1	1
	TTh	7	3	5	5	2	1	0

FIGURE 11.2 A Numerical Table.

Examine the table in Figure 11.2. Here is how the same data would look in narrative form:

> In the spring semester of 1985, C. Chang gave 5 *A*'s in his MWF section and 4 *A*'s in his TTh section. He gave 6 *B+*'s in the MWF section; 4 in the TTh section. He gave 8 *B*'s in the MWF section; 8 in the TTh section. . . .

Not only is this writing excruciatingly redundant and dull, but it also obscures the similarities and differences among the grades given by the five instructors because the instructors must be discussed one at a time, in linear sequence. In a table, all five instructors' grades can be displayed simultaneously, and readers can take in at a glance not only the range of grades given by each instructor but also the number of *B*'s and *A*'s given across all nine sections. Tables are thus an excellent medium for showing comparisons or contrasts. Whenever you want to set up a contrast (advantages/disadvantages, Ford/Chevrolet) by displaying parallel data, choose a table over text. Tables are more concise, more readable, more accessible, and more emphatic.

How much data you display in a table—raw or assimilated—is up to you. The table in Figure 11.2, for instance, could be collapsed into two columns, yielding Figure 11.3. The summary table in Figure 11.3 loses quite a lot of raw data, but it is easier for readers to grasp. It shows much more quickly and clearly that the teachers differed very little in overall grade average. The trade-off is that there is no way to see the gradations in numbers of *A*'s versus *B*'s or *C*'s versus *F*'s. The original permits more sophisticated analysis, but it requires readers to work much harder to see the variations in grading patterns among instructors. Adding the averages in Figure 11.3 to the table in Figure 11.2 would produce a table that met both sets of needs.

Like choices about what words to use, choices about what information to include in tables depend on your goal and audience. The head of the

TABLE 0. Average Grades Awarded in Mechanical Engineering 301	
Instructor	**Average Grade Given (A = 4.0)**
Chang	3.1
Rauch	2.9
Tichauer	3.1
Willard	3.0
Yamani	3.0

FIGURE 11.3   A Summary Table.

Mechanical Engineering Department might prefer Figure 11.3, which would show him or her immediately that no discrepancies in overall grading practice exist across the nine sections. Students interested in enrolling in the course might prefer Figure 11.2, which would tell them which instructors gave the most A's.

Table Formats

The simplest table consists of two columns. (See Fig. 11.4.) The column on the left, headed "Term," contains the *sample* or *independent variable*[1]—in

Term	Abbreviation
Amount	amt
Approximately	approx
Average	avg
Concentration	concn
Diameter	diam
Experiment	expt
Height	ht
Molecular weight	mol wt
Month	mo
Number	no.
Preparation	prepn
Specific activity	sp act
Specific gravity	sp gr
Temperature	temp
Trace	tr
Versus	vs
Volume	vol
Week	wk
Weight	wt
Year	yr

FIGURE 11.4   A Two-Column Informal Table.

this case, the term to be abbreviated. The column on the right contains the *dependent variable*—in this case, the appropriate abbreviation for the given term. The terms are arranged alphabetically; each term is paired with its correct abbreviation.

When you want to display more than one dependent variable in a table, you set up multiple columns like the ones in Figures 11.1 and 11.2, still saving the left-hand column for the independent variable. This creates horizontal *rows*, which permit the table to be read across as well as down. Both columns and rows may be grouped or subdivided. Thus, Figure 11.1 groups the vitamins by solubility, whereas Figure 11.2 divides the teachers by class section. Though it is not essential in a nonnumerical table to assign columns and rows a rational order (alphabetical, chronological, geographical, and so on), it *is* customary, and it makes tables more accessible to readers.

Informal Tables. Informal tables like the one in Figure 11.4 consist simply of two or more columns, with or without headings. Normally, they do not have a title or a table number, but are integrated into the text and introduced by colons. The only rule governing the format of informal tables is that the independent variable always goes in the column farthest to the left.

Formal Tables. When you write a formal proposal, report, or journal article, you use formal tables as part of a formal style. Formal tables differ from informal tables in that they are lined or boxed to set them apart from the text, numbered consecutively, titled, annotated, and set up according to a specific format. Figure 11.5 shows the parts of the standard formal table. The contents

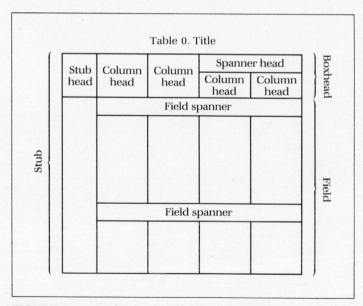

FIGURE 11.5   Major Parts of a Formal Table.

of the stub, the boxhead, the field, the spanners, and so forth are all mandated. Here are the rules:

Stub. The stub (the first column) contains the independent variable—the list of specimens, subjects, or whatever, arranged in numerical or some other rational order. Entries in the stub control information that reads *across* the table—that is, each entry in the stub is the beginning of a horizontal *row*.

Stub Head. The stub head identifies the entries in the stub. Typical stub heads might be these: Type of Vitamin, Species of Rodent, Test Number, Speed (mph), Speed (rpm), or Subject Number. A unit of measure that applies to all the entries in the stub should be placed in the stub head. For instance, if you are tabulating the hatching time for larvae as a function of ambient temperature, you write °F or °C in the stub head and give the numbers (55, 60, 65, 70, or 13, 15, 18, 21) in the stub itself.

Boxhead. The boxhead contains the headings for all the individual columns (including the stub head) and controls information that reads *down* in the table. Like the stub head, the boxhead includes units of measure that apply to all the entries in that column: Time (min), Weight (kg), and so on.

Column Head. Column heads identify the dependent variables—hatching time, number of vertebrae, type of microbial flora, and so forth—and again provide the unit of measure, for example, Length (meters). They should be made as precise as possible, so readers can understand the table more or less by itself, without having to rely extensively on the interpretation provided in the text. The self-sufficiency of a table depends above all on the clarity of the column heads.

Spanner Head and Field Spanner. Spanners are used to group together related variables or to subdivide one variable by another. If a subdivision applies to all the variables, you should use a field spanner, but if it applies in only one column, a spanner head is appropriate. "Grades" in Figure 11.2 is a spanner head; "Water Soluble" in Figure 11.1 is a field spanner. The purpose of spanners is to eliminate redundancy. Consider two columns like the following:

Maximum Speed (kph)	Minimum Speed (kph)
140	60
125	60
135	65

Arranging these columns under a spanner is preferable because it is clearer and more economical:

Speed (kph)	
Maximum	**Minimum**
140	60
125	60
135	65

Field. The field contains the data, neatly gridded. Whole numbers should be lined up symmetrically on the right; decimals, on the decimal point. There should be no blank spaces in the field; a zero, a dash, the word "none," or the letters "NA" (not applicable, not available) should be entered wherever you have no data to record. Units of measure should be placed in the field only if the values in that column or row are being recorded in different units.

The table in Figure 11.6, taken from a corporate research report, conforms to all the requisites for formal tables.

Formal tables are bound not only by format rules but also by stylistic conventions. They are customarily numbered (Table I, Table 1, Table 1.1) and titled (Representative Scale Thicknesses of . . .); the word "Table" is spelled out and the table number and title are usually placed at the top of the table. The title should be precise and comprehensive, so that the table can stand alone without the report in which it will originally appear. If the table has been taken from an outside source, as in Figure 11.6, the source may either be identified in a footnote or be inserted in parentheses after the title (SOURCE: F. Eberle, J. W. Siefert, and J. H. Kitterman, "Scaling of Ferritic Superheater

TABLE 0. Representative Scale Thicknesses Obtained by Eberle et al.[a*]				
	Thickness—One Spot in. (μm)			Range of Total Scale Thicknesses[†] in. (μm)
Material	Inner-Scale Layer	Outer-Scale Layer	Total Scale	
SA-210 C steel	0.0036 (91)	0.0028 (71)	0.0064 (160)	0.0051 to 0.0064 (130 to 160)
0.5%Mo	0.0032 (81)	0.0028 (71)	0.0060 (152)	0.0050 to 0.0060 (127 to 152)
0.5%Cr–0.5%Mo	0.0028 (71)	0.0022 (56)	0.0050 (127)	0.0050 to 0.0065 (127 to 165)
1.25%Cr–0.5%Mo	0.0030 (76)	0.0024 (61)	0.0054 (137)	0.0054 to 0.0065 (137 to 165)
2.25%Cr–1%Mo	0.0030 (76)	0.0024 (61)	0.0054 (137)	0.0049 to 0.0058 (125 to 147)
5%Cr–0.5Mo	0.0028 (71)	0.0020 (51)	0.0048 (122)	0.0044 to 0.0060 (112 to 152)
7%Cr–0.5%Mo	0.0026 (66)	0.0022 (56)	0.0048 (122)	0.0047 to 0.0050 (119 to 127)
9%Cr–1%Mo	0.0026 (66)	0.0022 (56)	0.0048 (122)	0.0047 to 0.0058 (119 to 147)

[a]F. Eberle, J. W. Seifert, and J. H. Kitterman, "Scaling of Ferritic Superheater Steels During 36,999 Hours' Exposure in 980/1030°F Steam of 2350 PSI, with Particular Respect to Scale Exfoliation Tendency," *Proceedings of the American Power Conference*, Vol. 26, 1964, pp. 501–510.
[*]Exposed 36,000 hours (4.1 yr) in steam, 2350 lb/in², 1010°F $\pm$ 25°F 540°C $\pm$ 14°C).
[†]From several measurements.

FIGURE 11.6   A Formal Table.
(From *Controlling Steamside Oxidation in Utility Boiler Superheaters and Reheaters* by Irwin M. Rehn and William R. Apblett, Jr. Courtesy Foster Wheeler Corporation.)

Steels . . ."). You will notice that the table in Figure 11.6 has two annotations, one indicated by an asterisk and the other by a dagger. Notes to tables are always placed at the foot of the table rather than at the bottom of the page, and they are designated by symbols or by superscript letters so that they will not be confused with numbered textual footnotes.

Experts disagree over whether or not horizontal and vertical lines should be used in tables to delineate columns and rows. The sample table in Figure 11.6, from a corporate report, contains both horizontal and vertical lines and is fully boxed. Tables in scientific journals (see Fig. 11.7) have fewer lines, sometimes only at the top and bottom. They may have a horizontal line under column heads or spanners, but they contain no vertical lines at all. When you are preparing a report for publication, model your tables on others previously published in the same source. If your report is to be privately printed—within a company, for instance—check comparable documents and duplicate their table format.

TABLE 0.   Body Weights, Nuclear Numbers, Nuclear Sizes, and DNA Levels in Saccate and Campanulate Females of *A. intermedia*

Variable	Adult Female Saccate	Adult Female Campanulate	P*	Percent Difference
Biomass (μg dry wt/♀)	2.15 ± 0.5†	6.38 ± .07		198
DNA concentration (ng/μg dry wt)	9.3 ± .2	9.2 ± .2		0
Nuclear number				
gastric gland	16.1 ± .2	20.4 ± .2	<0.001	27
vitellarium	43.9 ± .7	47.6 ± 1.2	<0.01	8
Nuclear size (μm)				
gastric gland				
min. dim.	8.7 ± .3	10.5 ± .3	0.001	21
max. dim.	12.4 ± .5	12.7 ± .4	0.63	2
vitellarium				
min. dim.	16.7 ± .7	19.0 ± .5	0.02	14
max. dim.	21.0 ± .9	24.8 ± .8	0.008	18
Nuclear volume (μm³)				
gastric gland	3,931	5,865		49
vitellarium	24,536	37,503		53
Relative DNA/nucleus				
oocyte	2,779 ± 129	3,040 ± 146	0.19	9
gastric gland	9,234 ± 541	14,881 ± 1,347	<0.001	61
vitellarium	56,671 ± 3,443	66,805 ± 4,792	0.09	18

*Probability that the difference between the two means is due to chance.
†Values are means ± S.E.

FIGURE 11.7   A Scientific Table.
(From "Developmental Polymorphism in the Rotifer *Asplanchna sieboldi*" by John J. Gilbert, © 1980 by *American Scientist*. Reproduced by permission of the publisher.)

Using Tables

You may find the idea of crafting formal tables according to a long list of arbitrary rules a little daunting, but once you have gotten some practice, you will discover that tables simplify your job as well as your readers'. It is a waste of effort for you to create elegant sentences to convey information that would be better displayed in a table.

Tables themselves are self-explanatory, but you should always discuss them in the text to show your readers how the information in the table fits into the rest of your discussion. You may wish to point to details in the table that are of particular interest, to give background information on the variables, to tell your readers what the results mean, or to indicate contrasts, comparisons, or trends. Tables show information clearly, exactly, and succinctly, but they place equal emphasis on everything, so if there are interpretations to be made or conclusions to be drawn, you will have to make those interpretations or draw those conclusions in the text. Always make certain the text tells readers why the table is there and what they should look for in it.

To make your tables a coherent part of your report, you should place them as close as possible to the discussion to which they refer. As a general rule, a reader should not encounter a table before it has been mentioned in the text; once you have sent your readers to a table, you should put it in as soon as you reach a convenient stopping place. There are three good locations for a table: (1) at the end of the sentence or paragraph that refers to it for the first time; (2) at the bottom of the page on which it is referred to for the first time or at the top of the following page; or (3) alone on the following page. No table is supposed to spill over onto a second page unless it is longer than a page, so plan ahead when you know a table will be coming up so that you save enough space on the page to fit it in.

You may cite a table in the text either by building an extra phrase or clause into one of your sentences or by inserting a parenthetical reference. Here are some illustrations:

In Table 5.6, we have summarized the results of the bacteria counts.

As you can see in Table 5.6, bacteria count correlated directly with length of exposure.

Bacteria count correlated directly with length of exposure (Table 5.6).

Bacteria count correlated directly with length of exposure. (See Table 5.6.)

Most short tables can be placed vertically on the page, surrounded by ample white space to set them off from the written text and to keep the page from looking too cluttered. Very broad tables, however, may have to go into a report sideways. These tables must be placed on a separate page, with the top of the table along the inside margin. You may also have to carry a very long table over onto a second page; if so, write "continued" at the bottom of the first page and at the top of the second, take subtotals if you will be totaling the columns at the end of the table, and repeat all your headings at the top of the second page so that your readers needn't flip back and forth. Include a table longer than a page only if you absolutely must; long tables constitute a major digression, and readers who are poring over an interesting batch of data may lose your train of thought and forget how those data fit into the point being made in the report. If possible, you should place long tables in an appendix and provide short summary tables in the body of your report.

Although your first textual reference to a table should come before the table itself, save your interpretations of the table until after it has been presented. Readers have to see the data before they can understand a commentary. And feel free to refer back to a table later in your discussion whenever it pertains to the topic at hand. That is a kind of structural repetition—useful for keeping your readers on track.

In a formal technical document, a list of tables should follow the table of contents.

Figures

Every illustration that isn't a table is a figure. We called the seven sample tables we showed you in the last section Figure 11.1, Figure 11.2, and so on, because, for our purposes, they weren't really tables. They were *pictures* of tables, and pictures are one type of figure. Figures include freehand drawings, diagrams such as maps and charts, photographs, printouts, and actual samples; if your illustrations aren't words, symbols, or numbers arranged in columns and rows, they are figures. You would need a course in technical illustration to master the techniques of creating highly specialized figures like schematic diagrams and site plans—and if you plan to be an engineer or a landscape architect, you will probably have such a course in your curriculum. However, with just a drawing pen or two, a ruler, a protractor, an array of colored pencils, and a package of graph paper, you can create figures that will enhance your report's readability, clarity, and conciseness. Moreover, thanks to advances in computer graphics, you may even be spared the vexation of having to create figures from scratch. Even microcomputers now have the capacity to convert mathematical information into symbolic pictures.

Using Figures

In a technical document, never try to say with words what you can say with pictures. If the information you are trying to get across to readers is quantitative or pictorial, graphic designs will express it better than words will.

If your readers can see the information with their eyes as well as in their imaginations, they will comprehend it better and remember it longer. Elementary school mathematics teachers learned long ago that if they want children to understand how a dollar is divided when it is spent, they should present that information in a pie chart (see Fig. 11.8). Information taken in through the senses seems to stick. Furthermore, most people consider pictures easier to understand than words, so they welcome them. Because they make documents easily intelligible, graphics heighten readers' interest and increase the palatability and clarity of any report.

Using figures is very much like using tables; figures should be integrated into the text, placed immediately after the first verbal reference to them, and commented on *after* they have been displayed. Unlike tables, figures are customarily labeled at the bottom rather than at the top, and the word "figure" is often abbreviated, especially in formal documents. The title of a figure— and, if it is borrowed, its source—should be placed below the figure directly after its number.

If you are placing small figures on a page that also contains text, place them to the left and type to the right of them. Use photocopies of figures if you can; also, if you have access to a photocopy machine that has reduction capability, reduce them to save space. (Recommended reduction for figures is generally to 70 percent of original size for photographs and to 50 percent for drawings.) If you cannot avoid including a figure that is larger than a full page, make a foldout for it—preferably one that opens to the right rather than at the bottom.

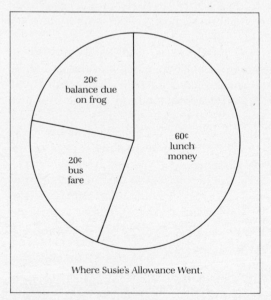

Where Susie's Allowance Went.

FIGURE 11.8   A Pie Chart.

Make all of your figures symmetrical and attractive to the eye, placing labels directly on the figure itself if that does not make it look too cluttered. Otherwise, place them in the surrounding white space, and connect them to the figure itself with short lines. Keep your figures simple; dense and complex graphics are as difficult to understand as dense and complex prose. And make figures, like tables, self-contained. They can be illuminated by in-text explanation, but they should not be incomprehensible without it. Essential explanation belongs in a caption under the figure, not in the text. Distribute figures as evenly throughout a document as your text permits so as to maintain your readers' interest, but make sure that each figure is physically adjacent to the discussion to which it pertains. Splice references to figures smoothly into your sentences, as we have done throughout this chapter. And in a formal technical document, include a list of figures after your table of contents.

Kinds of Figures

There are almost an infinite number of kinds of figures; we will discuss only the most common.

Photographs. If you work as a forensic pathologist, you will take photographs of injuries as a routine part of your daily work. Zoologists take photographs of animals in their natural habitats; archaeologists take photographs of artifacts; geologists take photographs of land outcroppings; paleontologists photograph fossils. Photographs make the perishable permanent; for instance, pictures of a wrecked auto made by a police photographer may be all that exists of the car months later when the case goes to court. Now that advanced technology has made aerial photographs and photomicrographs possible, scientists and engineers have come to regard them as indispensable tools. For accurately recording the surfaces of physical objects, photographs have no equal.

The photographs in Figure 11.11 (pp. 244–245) were used to excellent advantage in a mechanism description of a coffee percolator. (Compare them with the drawings of the percolator on pp. 220–221.) They clearly show both the merits of photography as a communications tool and its built-in drawbacks.

When you include original photographs in a technical document, crop them closely to edit out as much unnecessary background detail as possible. Mount the photos on the pages with rubber cement. You may label the photos underneath, like other figures, but it is also permissible to place a label directly on a photograph in a technical report. You may call attention to details in the picture by drawing lines from the photograph itself to labels or comments that you have placed in the surrounding white space. These notations are known technically as "callouts."

The principal disadvantage of photographs is that they show only surfaces. Another disadvantage is that they faithfully record the unimportant as well as the important. The photograph in Figure 11.9 contains a great deal of superfluous visual clutter. It is not concise. No camera will record only the flower. That requires a drawing.

FIGURE 11.9   A Photograph of a Flower.

Freehand Drawings. Drawings have been a part of technical communication since before human beings could write. The Greek word for "to write" is "graphein," which actually means "to draw." Before there were alphabets, human beings communicated technical information through pictures. Cave drawings, hieroglyphics, and the notebooks of Leonardo da Vinci are all part of the history of technical communication through graphic design. Though photographs have supplanted drawings in some circumstances, they cannot replace drawings entirely. For one thing, drawings can be edited in ways that photographs cannot. In a drawing, the artist puts in and leaves out what he or she wishes. Of course, ethics dictate that drawings be made as faithful as possible to the reality they represent but, because drawings are selective and need contain nothing nonessential, they permit a kind of visual emphasis that is impossible in photographs. Compare the photograph in Figure 11.9 with an artist's rendering of that same flower in Figure 11.10. The drawing concentrates a viewer's attention on the shape of the flower's leaves and petals and

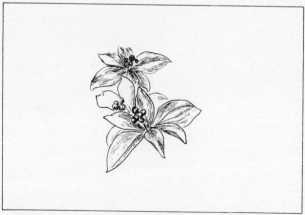

FIGURE 11.10   A Drawing of the Same Flower.

(a) This side view of the fully assembled coffeepot is taken from slightly above, so that an illusion of perspective is created.

(b) The photograph at left of the pot's two subsystems is cluttered by unnecessary background detail. Not only is the background irrelevant but, because the camera lens has limited depth of field, the background appears fuzzy and indistinct. Ideally, the photographer would have used a white backdrop.

(c) Notice how interference from the background is reduced in this photograph. Still, plain white paper, which has no noticeable texture, would have made a less obtrusive background than the white paper towel.

(d) The photographer has chosen to use glossy, high-contrast black-and-white prints. Not only are these prints the easiest to photoreproduce, but also they clearly show every hole and indentation in the basket assembly.

(e) This double exposure functions as a cross-section in the mechanism description. Here, the photographer has successfully gotten around one of the major drawbacks associated with photographs—the fact that they show only surfaces.

(f) This view of the assembled coffeepot is more truthful than the double exposure, but it leaves more to the imagination.

(g) This use of a simulated live-action effect adds verisimilitude and, incidentally, shows the size of the pot relative to the size of the human hand.

FIGURE 11.11 Various Photographic Illustrations for a Mechanism Description of a Coffee Pot.

on the positions of the clumps of berries. The drawing is more concise than the photograph because it contains no underbrush. It is less redundant because it omits the second flower in the lower center of the photograph. But like all abridgments, it is necessarily partial. It tells the truth, but not the whole truth.

Choose freehand drawings over photographs when you wish to emphasize particular attributes of a physical object: its shape, its color, its texture. You should also use drawings when you wish to show the interior of an object as well as its surface.

It is easier to label drawings than to label photographs because you plan a drawing with the labels in mind. The drawing of a mouse in Figure 11.12 shows the standard terminology used by technical illustrators to designate body planes and points of view. As is customary in drawings of animals, the mouse's head is to the left and its tail to the right, but making drawings "read" from left to right is not a hard-and-fast rule. Drawings are as individual as their creators; in fact, if you examine two different drawings of the ruby-throated hummingbird in two different field guides to the birds of North America, you will probably find that they differ in a number of small details. Artists draw what they see, and each of us sees the world a little differently.

Diagrams. Diagrams are abstract, two-dimensional drawings that make no attempt to be representational except in the broadest geometric sense. Diagrams show only the essential exterior or interior properties of objects or systems. As you can see in Figure 11.13, diagrams express lines and contours rather than shading and textures.

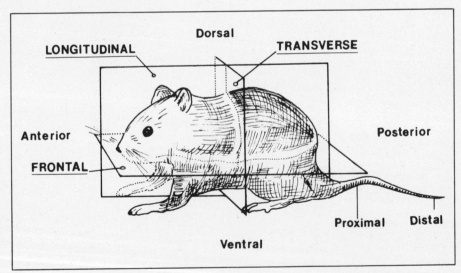

FIGURE 11.12   A Labeled Drawing, Showing the Terminology Used to Designate Body Planes and Points of View.

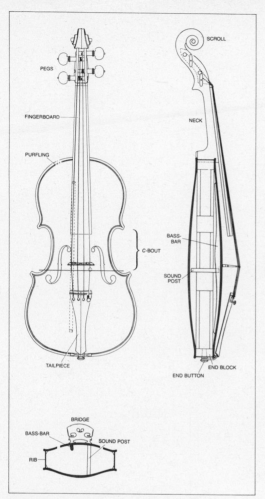

FIGURE 11.13   Three Diagrams of a Violin.
(From "The Acoustics of Violin Plates" by Carleen Maley Hutchins. Copyright © 1981 by *Scientific American, Inc.* All rights reserved.)

Diagrams have many different purposes in technical writing. For instance, the diagram in Figure 11.14 gives concrete visual form to a theoretical structure too small to be seen—a molecule. The flow diagram of a steam-generating system in Figure 11.15 is by no means a complete and accurate rendering of what a steam engine looks like, but it accurately represents the system's principle of operation.

Diagrams, like pieces of writing, vary in accessibility. Schematic diagrams cannot easily be understood by lay readers; you should use them only in reports intended for readers with some engineering background. Dendrograms, or tree diagrams (see p. 162), are self-explanatory, however. Even a

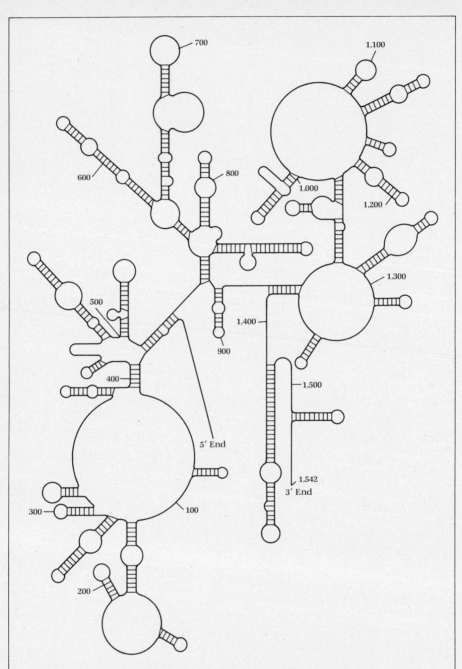

16S Ribosomal RNA. The drawing shows secondary structure of the 16S RNA of the eubacterium *Escherichia coli*, full sequence of which was determined by Harry F. Noller, Jr., of the University of California at Santa Cruz.

FIGURE 11.14   A Diagram of a Theoretical Structure.

(From "Archaebacteria" by Carl R. Woese. Copyright © 1981 by *Scientific American, Inc.* All rights reserved.)

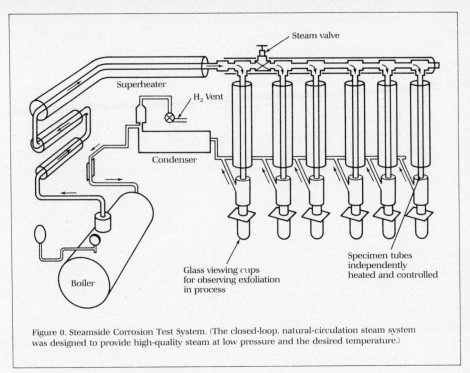

Figure 0. Steamside Corrosion Test System. (The closed-loop, natural-circulation steam system was designed to provide high-quality steam at low pressure and the desired temperature.)

FIGURE 11.15   A Flow Diagram.
(From *Controlling Steamside Oxidation in Utility Boiler Superheaters and Reheaters* by Irwin M. Rehn and William R. Apblett, Jr. Courtesy Foster Wheeler Corporation.)

reader who has never before seen a tree diagram can figure out from the way it is constructed that it is used to divide and subdivide.

One important way in which diagrams differ from drawings and photographs is that they may demand some technical expertise of readers. Because diagrams show relationships and operating principles, they often require an accompanying verbal explanation to make them clear to readers who are struggling to *learn* those relationships and operating principles from the diagrams. Arrows that show direction of flow and clear labels for all the parts may suffice, as they do in the simple flow diagram in Figure 11.15, but if the relationships are particularly complex or if your readers are not technical people, you may need to provide a caption that explains, amplifies, or interprets the diagram.

One specialized type of diagram often used in technical writing is the cross section, a diagram that displays the interior of an object by revealing how it would look if it were sliced in two on a vertical plane. Figure 11.13 contains both a full-front diagram and a cross section of a violin.

All diagrams, from the simple tree to the intricate schematic, are the technical writer's best tool for conveying information about structure and sequence.

Graphs and Charts. Graphs and charts are specialized types of diagrams that are symbolic rather than pictorial. The terms "graph" and "chart" are often used interchangeably, but in this book we will be using the term "graph" to designate lines plotted on horizontal and vertical axes, as shown in Figures 11.16 and 11.17. Other symbolic diagrams we will call "charts."

Graphs. To draw a graph, you plot a set of points on coordinate axes, and then connect those points. The horizontal or x-axis is called the *abscissa*; the vertical or y-axis is termed the *ordinate.* It is traditional to plot the independent variable—in Figure 11.16, the model year—on the x-axis and to plot the dependent variable—in this case, miles per gallon—on the y-axis. It is also traditional, though not obligatory, to plot on the x-axis any values that people tend to perceive horizontally—distance, for instance. When you use a graph in a technical report, you should make sure that you have clearly labeled each axis, giving the unit of measure, and each line. (The graphs in Figs. 11.16 and 11.17 are correctly labeled.) Figure 11.16 shows how solid lines and broken lines can be used in a graph to indicate different values. In a multiple-line graph like the one in Figure 11.17, you may want to use a different color for each line; if color is not feasible, use various sequences of dots and dashes to distinguish the lines. If your graph requires a legend or key, you should box it neatly in a blank area of the graph itself.

Graphs are particularly easy for unartistic technical writers to produce because they can be drawn with a ruler on commercially printed graph paper

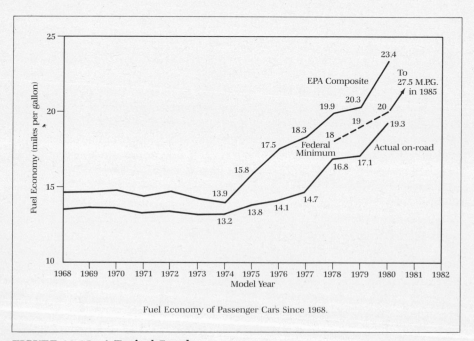

Fuel Economy of Passenger Cars Since 1968.

FIGURE 11.16   A Typical Graph.

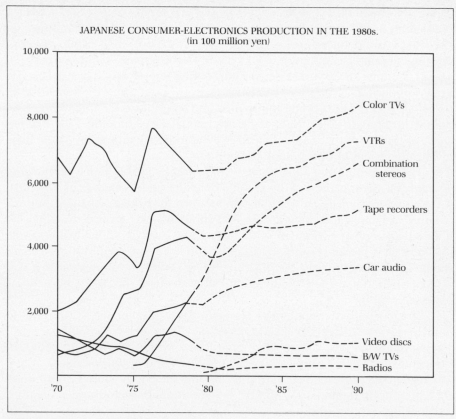

JAPANESE CONSUMER-ELECTRONICS PRODUCTION IN THE 1980s.
(in 100 million yen)

FIGURE 11.17   A Multiple-Line Graph.
(From "Japanese Technology Today," *Scientific American*, October, 1981. Reproduced by permission.)

and then photocopied. (To eliminate the grid on your final photocopy, buy graph paper that does not photoreproduce—called "non-repro blue" graph paper.) The most difficult part of creating a high-quality graph is selecting an appropriate scale. If you make the scale too large, minor changes in the dependent variable will seem huge to readers; and if you make it too small, truly significant changes may appear imperceptible. The American National Standards Institute has determined that readers see a slope of more than 30° in a graph as significant and a slope of less than 5° as insignificant. Try not to condense or exaggerate increments when you are setting your scale.

　　There are three principal techniques for altering the impression created by a graph without actually distorting the data themselves. (1) The illustrator can change the unit of measurement to one that magnifies or hides the result. In designing Figure 11.17, for example, the Japanese Electronics Industry Development Association could have made legitimately dramatic increases in Japan's production of audio and video equipment look astronomical by scaling them in millions of yen instead of hundreds of millions. (2) The illustrator can crowd or extend one of the scales, thus changing the slope—and impact—of the lines in the graph. In Figure 11.16, putting the years closer together on the x-axis would make fuel economy seem to be climbing faster, whereas moving the years farther apart would diminish the apparent improvement. (3) The illustrator can exclude relevant portions of one of the scales or include irrelevant portions. The y-axis in Figure 11.16, for example, goes from 10 to 25 miles per gallon, appropriate minimums and maximums for the averages reported. Running the axis up to 100 miles per gallon would add a lot of white space at the top of the graph; instead of climbing fairly impressively, the lines would then seem to hover along the bottom.

There are no ironclad rules for scaling graphs appropriately; everything depends on the meaning of the results. Consider an increase from 99 to 104. If the measurement is Fahrenheit temperature in a human patient, the increase is alarming, and the graph should show the fever growth dramatically. If the measurement is number of traffic fatalities over two holiday weekends, the increase is probably random variation, and the graph should show a small change. Your goal in every graph should be to create an impression that will be confirmed, not refuted, by thoughtful assessment of the data. Choose your scales accordingly.

The most strictly mathematical of all technical graphics, graphs convey numerical information visually. Moreover, they do it for two variables at once, making them the best technical graphic to use when you want to show relationships between continuous variables. When one of the variables is time, you really have little choice; no other graphic effectively shows continuous change. Graphs demonstrate trends, and multiple-line graphs demonstrate relationships among trends.

Bar Charts. A bar chart displays numerical quantities in the form of horizontal or vertical bars. The lengths of the bars represent the quantities being measured. Most readers can digest comparisons of bar lengths much more easily than comparisons of numbers, making the bar chart a very effective graphic tool.

It is customary to draw a bar chart on two axes, as in Figures 11.18 and 11.19, but actually only one axis gives information. The width of the bars and their distance from one another are only aesthetic matters. (To avoid confusion, make the bars equal in width and equidistant; it is difficult to compare the lengths of bars whose other dimensions are varying as well.) Thus, though graphs can be used to plot two continuous variables, bar charts are useful when only one variable is continuous, and the other is discrete or categorical. To show the progressive effect of an experimental diet on laboratory animals, you want a graph of weight against time. But to compare the final mean weights of different species, you want a bar chart; each bar represents a species, and its length represents the mean weight for that species.

Of course you *can* use a bar chart to show two continuous variables if you round one of them off into discrete units. But a bar chart is the preferred graphic only when you are comparing groups or categories on a single con-

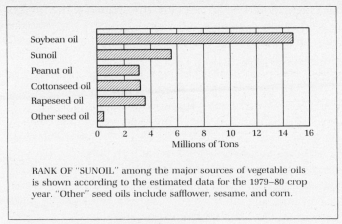

RANK OF "SUNOIL" among the major sources of vegetable oils is shown according to the estimated data for the 1979–80 crop year. "Other" seed oils include safflower, sesame, and corn.

FIGURE 11.18   A Simple Bar Chart.
(From "The Sunflower Crop" by Benjamin H. Beard. Copyright ©
1981 by *Scientific American, Inc.* All rights reserved.)

tinuous variable. Within this constraint, the bar chart is a very flexible device. Each bar can be segmented to show how much of the quantity being measured is accounted for by various subcategories; in a bar chart comparing the budgets of various departments, for example, segments of each bar might show how much of that department's budget went for salaries, equipment, materials, and so on. Or groups of bars can be arranged to show more complex relationships, as in Figure 11.19. In both cases, the different bars or parts of bars should be distinguished by color, shading, or cross-hatching.

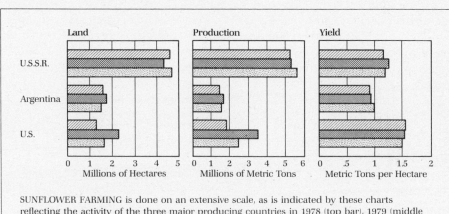

SUNFLOWER FARMING is done on an extensive scale, as is indicated by these charts reflecting the activity of the three major producing countries in 1978 (top bar), 1979 (middle bar), and 1980 (bottom bar). A hectare is 2.47 acres and a metric ton is 2,205 pounds. The figures for production and yield refer to the amount of seed harvested annually. A metric ton of high-quality oilseed will yield about 880 pounds of sunoil.

FIGURE 11.19   A More Complex Bar Chart.
(From "The Sunflower Crop" by Benjamin H. Beard. Copyright © 1981 by *Scientific American, Inc.* All rights reserved.)

The bars in a bar chart should be arranged in the sequence that makes the most sense. In Figure 11.19, for example, each set of three bars is arranged in chronological order. If no other order recommends itself, most bar charts have the bars arranged in ascending or descending order of magnitude.

As with graphs, bar charts can be misleading if the scale is unwisely or unethically selected. Because the lengths of the bars are the reader's only visual cues, scale distortions can make any comparison look puny or gigantic. All the bars in the same chart should use the same scale, of course. Usually the scale should start at zero, as in Figures 11.18 and 11.19—but there are measurements (such as body temperature) where a truncated or "broken" scale is justified. Again, the solution is judgment and honesty, not rigid rules. If one production technique is 94 percent effective and another is 97 percent effective, two long bars of only slightly different lengths would hardly make your point that the second system cuts the failure rate in half.

Histograms. A histogram is easily confused with a bar chart because bars are used in a histogram to express quantitative values. But a histogram like the one in Figure 11.20 is really a graph. Both axes in a histogram represent mathematical scales; that is why the bars have a common line between them. The jagged line formed by the tops of the bars in a histogram is analogous to the curve created in a graph by plotting and connecting points. The difference is that, in a histogram, all values within a certain increment—in this case, five years—are grouped together and plotted as a straight line (the top of the bar). For this reason, histograms are less rounded off and consequently less distorted than bar charts, but they are slightly more distorted than graphs, in which *all* points are plotted.

Pie Charts. A pie chart (see Fig. 11.8) is a segmented circle that represents the partition of a whole—often a monetary whole. Pie charts are simple to draw, but because they are simplistic, you will probably use them mainly in technical documents intended for nontechnical readers. Begin the chart at 12:00 and segment to the right (clockwise), beginning with the largest single segment and proceeding in decreasing order of magnitude. There is one exception: You should save the segment "Other," if there is one, for last.

Flowcharts. Flowcharts show sequence and direction. There are two types of flowcharts—abstract flowcharts, which use boxes to designate the steps in a process (see Fig. 11.21) and diagrammatic flowcharts, which represent those steps with pictures. Because flowcharts most often appear in descriptions of processes, you will find examples of flowcharts in Chapter 9, "Process Descriptions."

Organization Charts. Organization charts show hierarchies—lines of authority and levels of responsibility. As you can see in Figure 11.22, an organization chart is actually a kind of tree diagram; its only special feature is that boxes on the same level represent people or entities that are equal—family members in the same generation, employees at the same salary grade or with equivalent titles, corporate divisions with comparable responsibilities.

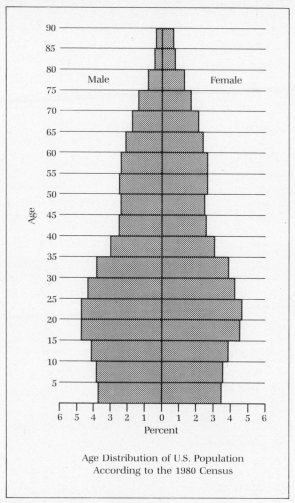

Age Distribution of U.S. Population
According to the 1980 Census

FIGURE 11.20   A Histogram.
(From "The Census of 1980" by Philip M. Hauser. Copyright
© 1981 by *Scientific American, Inc.* All rights reserved.)

▶ Technical Graphics as Part of Technical Style

Graphics in technical writing are seldom merely decorative; they are primary communicators of pictorial and mathematical information. Pictures are universal—that is one reason why they are often more accessible to readers than verbal text. But like all the elements in a technical document, they vary with goal and audience. Logarithmic diagrams don't usually belong in the same document with cartoons. Some graphics are more formal than others, and some are more comprehensible to lay readers than others. Certain kinds of graphics display particular kinds of information better than others—for

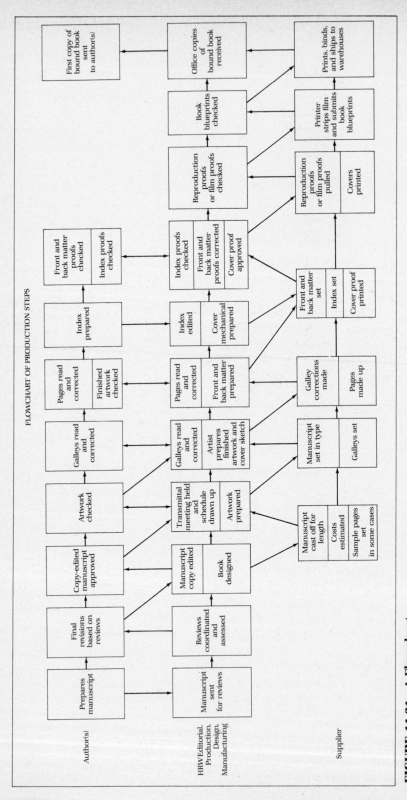

FIGURE 11.21 A Flowchart.

(From *HRW Author's Guide.* Copyright © 1976, 1968 by Holt, Rinehart and Winston. Reprinted by permission of Holt, Rinehart and Winston, CBS College Publishing.)

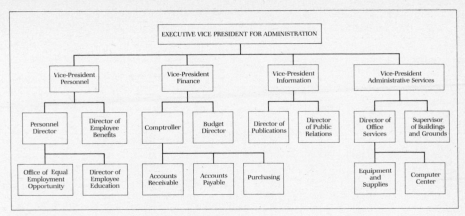

FIGURE 11.22   An Organization Chart.

instance, sharp changes in data are best indicated by a bar chart whereas gradual changes are best indicated by a graph. Selecting graphics means not only choosing the design that is most appropriate for your audience but also choosing the design that is most appropriate to your material. And good technical graphics possess the same attributes as good technical writing: They are clear, concise, coherent, objective, and parallel, with all excess and re-dundancy edited out and all potential ambiguity eliminated.

Review Questions

1. What kinds of information do pictures express better than words?
2. What is the difference between a table and a figure?
3. How do informal tables differ from formal tables?
4. What are the parts of a formal table?
5. Where should tables and figures be placed in the text?
6. What is the primary advantage of photographs as technical illustrations? What is their primary disadvantage?
7. What is the difference between a drawing and a diagram?
8. What is the difference between a graph and a chart? What kind of infor-mation is best expressed in graphs? In charts?
9. What characteristics do technical graphics have in common with technical writing?

Assignments

1. Tables 11.1, 11.2, and 11.3 were published by the U.S. Department of Agriculture (*Agricultural Statistics: 1978*). Using the information they contain, prepare a separate table, graph, or chart to accomplish each of the following:

Table 11·1—Fruits: Production in tons, United States, 1963–77 [1]

Year	Apples, commercial crop [2]	Peaches	Pears	Grapes	Cherries	Prunes and plums (fresh basis)	Apricots	Figs (fresh basis)	Olives
	1,000 tons	1,000 tons	1,000 tons	1,000 tons	1,000 tons	1,000 tons	1,000 tons	1,000 tons	1,000 tons
1963	2,876	1,770	472	3,793	151	490	199	63	57
1964	3,160	1,726	728	3,478	393	647	220	67	54
1965	3,070	1,737	500	4,351	265	606	226	61	50
1966	2,881	1,695	748	3,734	206	492	194	68	63
1967	2,718	1,344	464	3,069	200	584	148	42	14
1968	2,735	1,818	624	3,549	229	534	149	52	86
1969	3,410	1,842	727	3,898	286	496	229	59	70
1970	3,199	1,498	549	3,103	246	787	175	50	52
1971	3,187	1,441	749	3,994	280	582	186	46	55
1972	2,939	1,186	612	2,579	252	359	128	36	24
1973	3,133	1,295	730	4,198	246	783	158	42	70
1974	3,290	1,459	742	4,199	279	659	94	45	59
1975	3,765	1,419	749	4,366	301	656	183	38	67
1976	3,240	1,510	841	4,398	246	643	155	32	80
1977 [3]	3,328	1,496	787	4,298	255	720	147	42	43

Year	Avocados	Nectarines	Cranberries	Bananas	Bushberries	Dates	Papayas	Persimmons	Pomegranates
	1,000 tons	1,000 tons	1,000 tons	1,000 tons	1,000 tons	1,000 tons	1,000 tons	1,000 tons	1,000 tons
1963	61	57	63	3	36	22	8	3	4
1964	37	75	66	5	37	24	12	2	4
1965	61	67	71	4	42	21	10	2	3
1966	80	68	79	4	49	21	9	2	3
1967	52	55	70	4	44	21	11	1	2
1968	74	64	73	3	38	23	12	2	3
1969	46	66	91	3	43	17	10	2	7
1970	86	67	102	3	40	18	12	2	8
1971	45	70	113	3	35	20	10	1	9
1972	89	87	104	3	31	16	13	3	7
1973	74	87	105	4	22	23	16	2	6
1974	127	117	112	3	33	23	19	3	8
1975	87	111	104	3	32	25	20	2	9
1976	141	128	120	3	29	22	25	2	9
1977 [3]	109	150	105	3	29	25	32	2	10

Year	Oranges	Tangerines	Grapefruit	Lemons	Limes	Tangelos	Temples	Strawberries, commercial crop	Total
	1,000 tons	1,000 tons	1,000 tons	1,000 tons	1,000 tons	1,000 tons	1,000 tons	1,000 tons	1,000 tons
1963	4,397	103	1,429	494	16	34	90	255	16,946
1964	3,752	182	1,377	724	18	41	153	274	17,256
1965	4,988	200	1,667	540	22	45	171	216	18,996
1966	5,808	193	1,894	599	17	54	202	232	19,395
1967	7,924	227	2,286	681	17	76	225	237	20,516
1968	5,436	163	1,781	641	29	76	202	263	18,659
1969	7,898	197	2,207	600	28	81	202	243	22,758
1970	8,023	189	2,186	[4] 574	29	113	234	248	21,593
1971	8,205	234	2,473	625	35	122	225	260	23,005
1972	8,237	220	2,627	634	44	162	239	230	20,861
1973	9,737	223	2,676	844	44	140	230	240	25,128
1974	9,386	210	2,692	676	42	167	239	268	24,951
1975	10,241	228	2,504	1,118	44	212	239	272	26,795
1976	10,493	236	2,850	670	43	248	248	286	26,698
1977 [3]	10,595	249	3,029	973	40	216	171	325	27,179

[1] For some crops in certain years, production includes some quantities unharvested on account of economic abandonment.
[2] Estimates of the commercial crop refer to production in orchards of 100 or more bearing-age trees.
[3] Preliminary.
[4] Prior to 1970, growers' marketing season for California lemons was from Nov. 1 to Oct. 31; beginning with the 1970 crop, growers' marketing season is from Aug. 1 to July 31.

Economics, Statistics, and Cooperatives Service—Crop Reporting Board. Data for 1944–62 in *Agricultural Statistics, 1972*, table 299.

Table 11.2—Apples, western: Average auction price per box, all grades, New York, 1963-77

Season beginning July	Golden Delicious	Rome Beauty [1]	Delicious	Yellow Newtown	Winesap	Average [2]
	Dollars	Dollars	Dollars	Dollars	Dollars	Dollars
1963	4.64	3.54	5.15	4.21	5.11	4.97
1964	4.31	4.21	5.33	_____	4.35	5.00
1965	4.47	4.00	5.55	_____	4.66	5.59
1966	4.80	4.11	5.91	_____	5.71	5.62
1967	6.02	3.95	6.75	_____	6.52	6.51
1968	6.10	5.12	7.64	_____	5.20	7.00
1969	4.66	3.14	4.82	_____	4.53	4.33
1970	5.48	2.76	6.45	_____	4.02	6.31
1971	5.76	4.37	6.38	_____	5.08	6.12
1972	6.36	5.04	8.65	_____	6.05	7.85
1973	7.76	5.00	8.56	_____	8.25	8.27
1974	8.81	7.04	9.86	_____	7.44	9.38
1975	8.22	5.16	8.41	_____	6.72	8.11
1976	7.57	3.67	9.85	_____	5.04	7.71
1977 [3]	8.62	7.29	10.70	_____ _____	6.01	9.61

[1] "Rome Beauty" no longer reported. Beginning 1971, prices quoted are for "Red Rome" variety.
[2] Weighted average of varieties shown in this table.
[3] Preliminary.

Economics, Statistics, and Cooperatives Service—Economics. Compiled from New York Daily Fruit Reporter. Prices are weighted by number of boxes sold each day. Data for 1949-62 in *Agricultural Statistics, 1972*, table 307.

Table 11.3—Apples, commercial crop: Production and utilization, by States, crop of 1977 (preliminary)

State	Total production	Utilized production	Fresh [1]	Processed (fresh basis)				
				Canned	Dried	Frozen	Juice and cider	Other [2]
	Million pounds	Million pounds	Million pounds	Million pounds	Million pounds	Million pounds	Million pounds	Million pounds
Calif	480.0	480.0	75.0	102.0	64.0	(3)	197.0	(3)
Mich	570.0	570.0	225.0	112.0	(3)	76.0	143.0	(3)
N.Y.	900.0	900.0	360.0	243.0	35.0	42.0	205.0	15.0
Oreg	147.0	147.0	109.0	21.6	(2)	(3)	14.1	_____
Pa	460.0	460.0	166.1	186.9	(2)	10.4	89.9	(3)
Va	290.0	290.0	90.0	102.9	_____	(3)	71.2	(3)
Wash	2,060.0	2,060.0	1,600.0	87.0	116.0	_____	246.0	11.0
W. Va	195.0	195.0	74.4	65.6	_____	7.1	43.3	4.6
Other States [4]	1,553.6	1,524.0	1,101.1	147.9	5.5	25.4	246.7	88.3
U.S.	6,655.6	6,626.0	3,800.6	1,068.9	220.5	160.9	1,256.2	118.9

[1] Includes "Home use."
[2] Mostly vinegar, wine, and jam.
[3] Data not published to avoid disclosure of individual operations, but included in "Other States" and U.S. totals.
[4] Arkansas, Colorado, Connecticut, Delaware, Georgia, Idaho, Illinois, Indiana, Iowa, Kansas, Kentucky, Maine, Maryland, Massachusetts, Minnesota, Missouri, New Hampshire, New Jersey, New Mexico, North Carolina, Ohio, Rhode Island, South Carolina, Tennessee, Utah, Vermont, and Wisconsin.

Economics, Statistics, and Cooperatives Service—Crop Reporting Board.

    a. To highlight the change in New York auction price (per box) of Golden Delicious, Delicious, Winesap, and Rome Beauty apples between 1963 and 1977.

    b. To show how the apples produced in the United States in 1977 were used.

    c. To compare New York production and utilization to total U.S. production and utilization in 1977.

    d. To display production between 1963 and 1977 of papayas, figs, and dates.

    e. To highlight net or percent growth of production of papayas, figs, and dates.

    f. To compare production of apples and oranges in 1963, 1973, and 1977.

2. You have been asked by the head of the writing program at your school to prepare a report on the students enrolled in technical writing. Using your class as the sample population, gather information on major, career objective, sex, year of graduation, grade point average in major, overall grade point average, and anything else that your audience might find pertinent. After examining your data, construct the graphics that would accompany the report.

Notes

1. An independent variable is a value, function, or entity that determines the value or the characteristics of dependent variables (values that change during a procedure). For instance, in the equation $y = 3x + 1$, the value of y depends on the value of x. Change x, and you change y; x is the independent, y the dependent, variable.

Technical Reports

PART 3

12 Technical Report Style

Textbooks on writing tend to define the elusive term "tone" in engaging but hazy abstractions, calling it "the quality of feeling" that a piece of writing conveys or "the reflection of a writer's attitude" toward his or her subject matter and audience. Luckily, though nailing down a precise definition is difficult, identifying specific tones is not. Virtually all of us can "hear" tone in what we read, and we are quick to label a given piece of writing as "stiff and pompous," "light and witty," "sarcastic and condescending," "intense and emotional," and so on, if that's how it sounds to us.

The distinctive tone of technical reports can be characterized by words like "emotionless," "restrained," "serious," and, especially, "objective" and "authoritative." As readers, you can hear this tone in journal articles and technical books you read for your courses; as writers, you naturally want your own technical work to *sound* technical.

Goal, Audience, and Technical Tone

The first thing to realize is that tone doesn't just happen. The mere fact that a report's subject matter is technical does not guarantee that its tone will be technical too. In fact, tone is only partly controlled by subject matter—a technical piece written for a popular audience, for instance, is designed to convey information *without* sounding very technical. How, then, does a piece of technical writing get its characteristic tone, if not from the subject matter? Tone is the result of the writer's conscious choices about words, phrases, and sentences. And these choices are governed, as we have seen, by the writer's goals and audience.

Because your first goal is always to be understood by your readers, whatever choices you make to get the right tone must be consistent with this goal of precision and clarity. The trick, therefore, is to develop a clear and precise style that also sounds technical, that inspires trust in your abilities and your ideas, that convinces your readers of your objectivity, authority, and seriousness.

The problem with much technical writing is that in trying to sound objective, authoritative, and the rest, technical writers too often end up with passages like this one:

> There is now no effective mechanism for introducing into the initiation and development stages of reporting requirements information on existing reporting and guidance on how to minimize burden associated with new requirements.

This may sound appropriately serious, emotionless, and objective, but what does it mean? The writer's point is lost in a tangle of words and phrases. Pursuing the worthy goal of an objective and authoritative tone, this writer made the fatal mistake of associating appropriate tone with difficult words and sentence structures, as though the harder a sentence is to read, the more important and technical it must be. And to some extent, we have all been taught to feel this way—to measure our intelligence by the length of our sentences. The truth, of course, is that an important idea is no less important if it is written so that readers can understand it—and, conversely, a ponderous stupidity is no less stupid than a breezy stupidity.

In this chapter, therefore, we discuss both aspects of sounding technical. First, we present practical suggestions for giving your written work the objective, serious tone it warrants. Then we discuss how to avoid the many problems that arise from trying *too hard* to sound technical.

Achieving an Appropriate Tone

The serious nature of technical work requires a tone that is equally serious and formal. We discuss below the measures you can take to make your writing suitably formal.

Avoid Colloquial Expressions

Colloquial English, the language of everyday, casual speech, tends to make a technical report sound too chatty. Being too casual is rarely a problem in technical writing; few writers would allow phrases like "foul up" or words like "screwy" to creep into their reports. Yet some colloquial expressions we rely on so heavily in speech that it takes special effort to keep them out of our writing. Two that immediately come to mind are "a lot" and "all right." (When these turn up as "alot" and "alright," they are no longer considered colloquialisms—they are simply errors.) Most good dictionaries will tell you whether a particular usage of a word is considered acceptable ("standard") or not ("nonstandard") for written work. If you have any doubt about a word, look it up.

Avoid Contractions

Contractions—"we'd" for "we would," "that's" for "that is," "won't" for "will not," and so on—are almost never used in formal reports, though they are often used in popular articles and in informal correspondence or wherever a more casual tone is desired. (This textbook, for example, uses contractions—and even an occasional colloquialism, for that matter; this is not, after all, a technical report.)

Use Standard Grammar

Grammatical correctness won't actually improve the tone of your report, but grammatical errors will certainly undermine the tone of authority that you are trying to establish. Though it may be unfair, readers tend to down-grade the ideas conveyed in sentences that contain dangling modifiers, faulty pronoun references, and so on. A report sprinkled with grammatical errors may cause readers to wonder whether the writer is uneducated or just plain sloppy—and either way, the writer loses credibility.

We suggest that you purchase a handbook of English usage—or a hand-book for technical writers—and consult it whenever you have a question about correctness. If consulting the handbook doesn't resolve a question you have about the grammar of a particular sentence, we suggest that you simply revise the sentence to a form you *know* is correct. (See Chapter 4, p. 103.) Just as important is learning from your mistakes. Once your teacher has alerted you to a grammatical error, make an effort to understand what you did wrong—and don't repeat it.

Be Aware of a Word's Connotations

A word may have *connotations*—shades of meaning accrued by usage and context over the years—that are not reflected in its principal dictionary definition (its *denotation*). These connotations may spark positive or negative emotions in your readers.

The words you use in a technical report should be emotionally neutral. A sentence that uses charged language to imply an unstated opinion or atti-tude is a cheat, functioning not to inform readers, but to *prejudice* them. Advocates of a cause, whose main goal is to rally supporters rather than to disseminate information, often rely on connotations to make their points. Thus, an opponent who is *undecided* about a given issue is called *wishy-washy*; one who *acknowledges* a point is said to have *admitted*, *confessed*, or *owned up to* it. Consciously choosing charged words to elicit a calculated response from an audience is the work of propagandists, not of technical writers.

Equally unsuited to technical reports are those words used to sanitize unpleasant references. "Put to sleep" is not an appropriate substitute for "killed"; "passed on" won't do for "died." At best, these cleaned-up words—called *euphemisms*—display a kind of squeamishness that undercuts scientific cred-ibility. At worst, euphemisms come across as imprecise and deceptive. Shortly after the accident at Three Mile Island, the National Council of Teachers of

English cited the corporation that owns the nuclear power plant for purposely using euphemisms to obscure what really happened. The statements released by the corporation referred to an "energetic disassembly" and a "rapid oxidation"—not to an explosion and a fire. At its most extreme, this kind of writing is called *gobbledygook*, and we will have more to say about it later in the chapter.

Even when there is no intent to manipulate reader response, technical writers must be very sensitive to the possible connotations of the words they use. A word chosen carelessly may have unintended overtones. Your readers will think one way about the building contractor who won't lower his or her bid if you describe the contractor as "resolute" or "firm," but quite another way if you call him or her "inflexible." Be especially attentive to words that may have several connotations: the word "cheap" can mean either low in price or low in quality—either "inexpensive" or "inferior" would be a less ambiguous alternative.

To prevent unintended connotations from seeping into your writing, you should try to anticipate how your readers will respond to your words. Some words, though seemingly "neutral" to you, may trigger emotional responses in your readers. The word "exploit" provides a good example. Resource managers use the term to mean "manage wisely," whereas many environmentalists invariably take it to mean "misuse," "ruin," "mistreat—with malice." If you were a resource manager writing to an environmentalist audience, you might try defining the term, since stating explicitly how you are using the word sometimes helps to defuse its emotional charge. But your best bet would be to avoid "exploit" altogether, substituting a neutral paraphrase.

Exercise 12.1

"I am true to my convictions; she is stubborn."
"I am friendly with my instructors; he kisses up to them."

In each of the following spaces, supply a negative, neutral, or positive alternative to the words provided.

Negative	Neutral	Positive
squander	_____	_____
_____	inform	_____
_____	_____	novel
_____	intelligent	_____
knuckle under	_____	_____
_____	young	_____
tightwad	_____	_____
_____	consent	_____
_____	_____	generous
_____	group	_____

Avoiding Tone Traps

The suggestions we have just made will help you achieve the tone you want your writing to have without compromising its clarity and precision. Unfortunately, engineers, scientists, and other technical people often pursue other, less successful ways of making their written work sound technical, mistakenly equating objective, serious, and authoritative writing with writing that is cold, impersonal, abstract, and difficult. What they usually end up with is not a more technical tone, but simply a more forbidding one. And all too often the reports they produce come out boring, imprecise, and obscure.

This brand of writing is characterized by three deliberate choices of style: the choice of the passive voice over the active voice, the choice of impersonal constructions over personal ones, and the choice of vague and abstract words over specific and concrete ones. Probably nothing will more dramatically improve your writing than learning to avoid these bad habits.

The Passive Voice

In everyday speech and in most kinds of writing, the active voice predominates:

I The committee recommended two alternatives.

In the active voice, the subject of the sentence ("the committee") performs the action of the verb ("recommended")—hence the name "active voice"—and the object ("two alternatives") receives the action of the verb. Here are two additional examples of sentences written in the active voice:

I The data indicate a 37 percent increase in reported cases of bone necrosis.

I Polyvinyl alcohol hydrogel, a polymeric material, has safely replaced portions of vitreous humor in albino rabbits.

In the passive voice, on the other hand, the subject doesn't perform the action of the verb; instead it *receives* the action:

I Two alternatives were recommended.

Although this is a complete sentence—the subject ("alternatives") is acted upon by the verb ("were recommended")—the sentence does not indicate who recommended the alternatives. In order to show who or what performs the action of the verb, passive sentences need to append an additional phrase beginning with "by":

I Two alternatives were recommended by the committee.

Compare these passive sentences with the active versions we just gave you:

I A 37 percent increase in reported cases of bone necrosis is indicated by the data.

I Portions of vitreous humor in albino rabbits have been safely replaced by polyvinyl alcohol hydrogel, a polymeric material.

The most important difference between active and passive sentences is that active ones emphasize the person or thing performing the action, whereas passive ones stress the receiver of that action. Passive sentences, as a result, are less dynamic than active ones, tending to de-emphasize not only the actor, but also the action itself. Moreover, passive sentences tend to be a bit longer and a bit clumsier than active ones. Thus, you should use the active voice *unless* you want to draw your readers' attention to the receiver of the sentence's action.

Appropriate Use of the Passive Voice. There are four occasions when the passive voice is preferable to the active voice precisely because it de-emphasizes the actor:

1. *If the performer of the verb's action is unknown, unimportant, or obvious.*

I Gold was discovered in the region sometime in the early 1850s. (Who discovered the gold isn't known.)

I The specimen has been taken to Washington for analysis. (Who or what took the specimen isn't important.)

I Your application for a loan has been denied. (Who has denied the loan is perfectly clear.)

2. *If what is being done is more important than who is doing it.* Here is a description of a process written entirely in the active voice:

> The experimenter exposes a prototrophic strain of *E. coli* to ultraviolet radiation to enhance the mutation rate. The experimenter then transfers the irradiated culture to a nutritionally complete medium. Then he or she transfers the culture to a minimal medium that contains penicillin. Finally, he or she plates out the survivors from the penicillin-containing medium.

In this case, the active voice is less effective than the passive voice would be. For one thing, the repetition of "the experimenter does this" and "the experimenter does that" makes for tedious reading. More importantly, though, the active construction, by drawing the readers' attention to the experimenter, draws it *away* from the meat of the passage—the actual steps of the process. The passive voice, conversely, puts the emphasis on *what*, not *who*:

> A prototrophic strain of *E. coli* is exposed to ultraviolet radiation to enhance the mutation rate. The irradiated culture is transferred to a nutritionally complete medium and then to a minimal medium that contains penicillin. Survivors from the penicillin-containing medium are finally plated out.

Process descriptions and procedures are usually written in the passive voice, whereas introductions, discussions, and conclusions are generally written in the active voice.

3. *If you need to mention both the performer and the receiver, but wish to emphasize the receiver.*

> The Escondido tracking station has been severely damaged by an earthquake. (The writer feels that the damaged station is more important than the cause of the damage.)

> The sample semiconductor superlattice was examined by standard crystallographic techniques.

> Limited production was subsequently begun by the company's Arizona plant.

4. *If you occasionally want to vary your sentence structure.* Because the passive voice reverses the standard "subject-verb-object" order of active-voice

constructions, it tends to slow down a sentence's pace. Thus, in a paragraph of active-voice sentences, the passive voice can be used to stress a point that you especially want your readers to linger over. If you look back through this chapter, in fact, you will see that we have frequently used passives for just this reason—in the previous sentence, for instance.

Abuse of the Passive Voice. Occasions to use the passive voice arise on virtually every page of a technical report. So, unfortunately, do occasions to abuse it. Technical people often seem addicted to passive-voice constructions. Their readers can search in vain for a simple declarative sentence. Overused, the passive voice can deaden a report, make it clumsy to read, or even obscure its meaning.

Consider this passage from a student's progress report on his efforts to find a topic for a research project:

> Early in March a water pollution problem that had been encountered in a previous laboratory exercise was selected as a topic for my proposed project. This topic was rejected after several weeks of library research, when it was revealed that further work by myself was not warranted, given the large amount of progress that has been achieved on the topic. Subsequently, through a brainstorming session with one of my professors in environmental science, a topic was chosen about which comparatively little is known: toxic elements in sewage sludge.
> After reviewing numerous journals and books, it was found that, although there are relatively few major disagreements among the authors, a comprehensive list of toxic elements in municipal sludge has never been agreed upon. . . .

The tortuous and awkward style occasioned by the passive construction is amply demonstrated here. Technical reports are sufficiently taxing without the added encumbrance of such clumsy, leaden sentences.

Even more important, passive constructions are often unclear and imprecise. Look again at the student's second sentence, this time isolated from the others:

> This topic was rejected after several weeks of library research, when it was revealed that further work by myself was not warranted, given the large amount of progress that has been achieved on the topic.

Can you tell with absolute certainty whether the student himself rejected the topic or *had it rejected* by his professor? Chances are he threw it out himself, but the passive construction, because it omits the human agent, causes at least a momentary uncertainty. To what, moreover, does the "it" in "it was revealed" refer? Not to either the subject of the sentence ("topic") or the nearest noun ("research"). Actually, the "it" doesn't refer to anything; it is being used as a dummy subject for the passive verb "was revealed"—and the result is awkward and confusing. The rest of the sentence is even more confusing. Does he mean that further *library* research was not warranted or that further research *of any kind* was not warranted? And whose progress is he talking about—his or other researchers'? Written in the active voice, the sentence would have been much clearer, more precise, and easier to read:

> After several weeks of library research, however, I abandoned the topic, convinced that other researchers had satisfactorily isolated and analyzed the problem.

Yet another problem with passive constructions is that they are vulnerable to the grammatical error of *dangling modifiers*. (See pp. 118–120.) In fact, the student passage we have been examining contains one. Look again at the sentence that begins the second paragraph:

> After reviewing journals and books, it was found that . . .

The "it" didn't review numerous journals and books; the student himself did. Dangling modifiers are among the commonest errors in technical writing. They usually occur in passive-voice sentences where the modifying phrase refers to a word that would be in the sentence if the sentence were in the *active* voice.

Impersonal Constructions

The student whose progress report we just examined probably did not set out to write a string of passive-voice sentences. More likely, he trapped himself into writing in the passive voice when he tried to get around referring to himself as "I." Thus, "I rejected the topic" became "The topic was rejected," and a sea of troubles opened before him.

Reluctance to use the personal pronoun "I" in written work is widespread and long-standing, and contains a kernel of good sense. Certainly, a report in which every other sentence began with "I did such and such" or "I found such and such" or "I concluded such and such" would sound egotistical. Making "I" the subject of the sentence draws attention to the writer and, consequently, draws it *away* from the more important issues of *what* was done, *what* was found, and *what* was concluded. Very often, however, using "I" is perfectly appropriate.

Appropriate Use of "I." In the following sentence, the writer, out of modesty, has found a way not to use "I":

> Although some will disagree, it is believed that further funding is warranted.

This is an ambiguous sentence. Although the writer knows that she means "I believe . . . ," her readers do not. The impersonal "it," coupled with the passive voice, masks the identity of the person or persons who hold this belief. Readers might reasonably conclude that the writer means that someone other than herself—her boss or the board of directors, for instance—believes that funding should continue. They might even decide, on the basis of this sentence, that there is nearly universal agreement that funding is warranted. In cases like this one, where knowing *who* is as important as knowing *what*, the use of the personal pronoun "I" can ensure against possible misinterpretation.

Even when misinterpretation is not an issue, using "I" is often a better alternative than avoiding it. Constructions beginning with "It is . . . ," though not always ambiguous, are almost always less concise than those using personal pronouns. Compare these two sentences for length:

▌ It is hoped that the work will be completed on schedule.

▌ I hope to complete the work on schedule.

Moreover, these impersonal constructions tend to sound sneaky, often making the writer sound as though he or she is trying to evade responsibility for the contents of the sentence. Project directors are likely to use this dodge when they have bad news to report:

▌ It has been decided that the research you are undertaking must be terminated.

What they mean, of course, is "I have decided to terminate your research."

Another common alternative to using "I" is to refer to oneself as "this writer," "the author of this report," and so on. These obtrusive, wordy phrases are, if anything, *less* modest than the much-maligned "I." "We," when the author is an individual, is even more pompous—and misleading.

Although some professors and some professional journals still object to the use of "I" in the papers they receive, demanding that impersonal constructions be used exclusively, the tide is beginning to turn. The *Council of Biology Editors Style Manual*, for instance, now warns "writers who shun the first-person singular" that "'I discovered' is shorter and less likely to be ambiguous than 'it was discovered.'"

Appropriate Alternatives to Using "I." Even if you—or the people you are writing to—are uncomfortable with the first-person singular pronoun, you don't necessarily have to fall back on the impersonal passive. You can, instead, write active sentences that make your *actions*, rather than yourself, the focus of attention. Thus, if you feel that you are drawing unnecessary attention to yourself with a sentence like this one:

❙ I added three drops of alcohol to the solution to prevent it from gelling.

you need not settle for an impersonal passive revision like this:

❙ The addition of three drops of alcohol was made to the solution so that gelling would be prevented.

or even like this:

❙ Three drops of alcohol were added to the solution to prevent it from gelling.

Here are two other possibilities:

❙ Three drops of alcohol added to the solution prevented its gelling.

❙ Adding three drops of alcohol to the solution prevented it from gelling.

Either of these is an appropriate alternative to the original, and both are preferable to the two passive versions. Even though they are written in the active voice, they remain impersonal, keeping the focus on the technical information, not on the writer. Of course, if the point is that *you* rather than anybody else added the alcohol, then the original version is by far the best choice.

Exercise 12.2

Some of the sentences that follow use passives and impersonal constructions improperly. Others are correct. Make any revisions you think necessary (and watch for dangling modifiers).

1. When small traces of smoke are sensed by this detector, an alarm is sounded.
2. The unit is cleaned by wiping with a damp cloth and thoroughly drying.
3. Once the pollutants are removed, the specimen is dried.
4. When a plant is placed on the windowsill, its leaves will turn toward the light.

5. After boiling the specimen for five minutes, the manual should be consulted.
6. It is indicated by our results that a new layer of insulation is needed.
7. Specific qualities are imparted to the clay body by each of the different types of clay.
8. This primitive fish was thought to be extinct until one was caught in 1936.
9. It is suggested by this writer that we begin production immediately.
10. To remove the film from your new Minolta, first the rewind lever is turned up, then the handle is turned counterclockwise.

Vague and Abstract Language

Trying too hard to sound technical often leads writers not only to abandon active-voice sentences for passive-voice ones and personal constructions for impersonal ones, but also to abandon precise, crisp, and expressive language in favor of language that is vague, abstract, and dull. Although this third tone trap can ensnare all technical language, verbs are most susceptible. This section thus begins with a discussion of inert, actionless verbs, called *smothered verbs*, and then turns to the problem of abstract, imprecise nouns and modifiers.

Smothered Verbs. Consider this sentence:

❙ The attending physician vaccinated all the children.

The verb "vaccinated" shows clearly and succinctly the activity performed by the physician. Too many technical writers, however, would write the sentence like this instead:

❙ The attending physician gave vaccinations to all the children.

Here the strong verb "vaccinated" has been turned into a noun (hence "smothered"), and its function has been usurped by the overworked and lackluster verb "gave." Some writers would probably neutralize the sentence's impact further by putting the whole thing in the passive:

❙ Vaccinations were given to all the children by the attending physician.

Writers who smother their verbs by turning them into nouns never *prefer* something; they *have* a preference for it. They never *investigate* a problem;

they *make* an investigation (or *conduct* one). To them, seeds don't *sprout*; instead, sprouting *occurs*. An alloy doesn't *resist* rust; instead resistance to rust *is achieved*. And so on.

   The Dangers of Verb Smothering. Smothered verbs needlessly draw out a sentence's length. They perform two other disservices as well. Verbs can be the most expressive element in a sentence. Strong verbs act; they breathe life into a sentence. But when they are turned into nouns, they lose that power, and the weak verbs that replace them are, by nature, actionless. Long passages composed of such actionless, moribund sentences numb the audience and lull it into inattention—not, obviously, the effect you want.

   Even worse, all-purpose verbs like "have," "make," "perform," "accomplish," "achieve," and so on are altogether too easy to use. If you settle for them, you can come to depend on them—and grow less and less inclined to seek out a more vivid and precise alternative. A general sloppiness and imprecision in your writing may ensue. Expressive verbs, already downgraded to nouns, may disappear entirely. In the following example, the first sentence employs a strong verb, the second transforms it into a noun, and the third omits it completely:

I   In the next phase, workers will bolt the metal plates to the infrastructure.

I   In the next phase, workers will use bolts to attach the metal plates to the infrastructure.

I   In the next phase, workers will attach the metal plates to the infrastructure.

An important piece of information has disappeared by the third version—the method by which the plates will be "attached." Accustomed to writing actionless verbs, you might not remember to include this missing information; accustomed to reading them, your audience might never look for it.

   Correcting Smothered Verbs. Although verb smothering is degenerative, it is also reversible. Once you have completed a draft of your report, reread it with an eye out for dull verbs. Each time you locate one, check to see if the sentence has demoted a sharper, precise verb to the status of a noun. If so, reinstate it. If not, explain to yourself the exact activity you want the subject of the sentence to perform, and then replace the dull verb with that. (Does your new product "have an adverse effect" on mosquitoes, or does it "repel" them? Does it, in fact, "kill" them?)

Exercise 12.3

Unsmother the verbs in each of the following sentences.

1. The turning of the plant is accomplished by the concentration of a growth hormone that collects in the cells on the shaded side of the plant.
2. When the infiltration of muddy water into the soil occurs, it causes the formation of a layer of soil that has a lower rate of infiltration.
3. The eyelashes also give some protection against the entrance of foreign substances.
4. After majority approval is achieved, the measure will get the endorsement of the Board of Trustees.
5. Collection of salamander specimens will be made at two different locations.
6. The crew then performed an examination of the faulty landing gear.

Striving for the right verb will enliven your prose and keep your mind limber. Don't, however, go overboard. In a report on a factory's fire-detection equipment, the alarm need not "shriek" or "shrill"; in an analysis of a new jet engine, the plane need not "purr" or "whoosh." (The alarm *can*, however, "sound" instead of "go off"; the plane can "fly" or "cruise" instead of "be airborne.") To achieve the proper technical tone you should choose verbs on the basis of their preciseness and expressiveness, but not their jazziness.

Abstract Nouns and Modifiers. Verbs are not the only words to fall prey to the distant tone. Nouns, adjectives, and adverbs are also susceptible. Trying to sound properly technical, some inexperienced writers tend as if by instinct to prefer abstract and general words over concrete and specific ones. In their reports, they will pass over the concrete word "truck" in favor of the more abstract "vehicle" or "conveyance." A "box of screws" will appear as a "container of fasteners"—and it will be stored not in a "stockroom," but in a "facility." Writers like these seem to feel that abstract words give their writing a loftier, more dignified tone and that longer, polysyllabic words sound more substantial.

The truth of the matter, however, is that the *sound* of substance and the *fact* of substance are not the same. The more abstract a word is, the harder it becomes for the reader to grasp its exact meaning because the meaning itself becomes less precise. Concrete words, on the other hand, because they refer to specific persons, places, things, and actions that can be verified by the senses, do spark precise images in a reader's mind, and semanticists have long known that visualizing a term is the surest way to comprehend it. Here are some common abstractions you should watch out for:

apparatus	equipment	institution	piece
device	implement	instrument	utensil

Building a Ladder of Abstraction. Abstract and concrete are relative terms, of course. Though "hose" is more concrete than "attachment" or "thing," "the red rubber hose stamped with 'serial #B349-08VXC'" is more concrete still. Linguists call a list of increasingly abstract words a ladder of abstraction. Here is another example:

> entity
> thing
> equipment
> office machine
> duplicating machine
> dry paper copying machine
> Xerox Model IIIxd duplicating machine
> Xerox Model IIIxd machine owned by Acme Chemical Company

The higher up the ladder of abstraction, the more inclusive and less precise the term. You will have to decide how far up the ladder to go. The rule is to be as concrete as your material permits. Thus, don't use "duplicating machine" if you mean Xerox; on the other hand, don't use "Xerox" if *any* duplicating machine will do. If you must use a word that is high up on the ladder of abstraction, the best way to keep your readers on target is to follow the abstraction with concrete examples: "The office needs to purchase a duplicating machine. Canon, Savin, and Xerox, for instance, all make suitable models."

Some terms, moreover, are inherently abstract—"energy," "durability," "momentum," "resiliency," and so on. Abstractions like these are essential to technical writing, and of course experts in your field will have no trouble with them. But they can stop less well-informed readers dead in their tracks. Unless you are absolutely certain that your readers will understand your use of an inherently abstract term, be sure to provide a definition and a concrete example for support.

Avoiding Vague and Empty Modifiers. Although the purpose of adjectives and adverbs is to make the meaning of a sentence more precise, some give only the appearance of precision. Watch out for vague adjectives like "hot," "cold," "sizable," "heavy," "frequent," "small," "short," "deep," and "large," all of which are subject to your readers' interpretation—and therefore misinterpretation. For example, "a large storage facility" is really no more precise than "a storage facility." Do you consider 800 square feet "large"? Or 8,000? Or 80,000? A more precise writer would say, "a warehouse of about 350,000 square feet."

By the same token, be wary of appending adverbs to words that don't need them. The biggest offenders are "somewhat," "quite," "rather," and "relatively," and they often appear in conjunction with an equally vague adjective or adverb:

I   The reaction occurred somewhat slowly.

I   The fuselage will be relatively heavy.

I   The water soon became rather hot.

I   The article is quite informative.

Is a "quite informative" article more or less informative than an "informative" article? The reader cannot tell. The best solution is to expand the sentence with more information:

The article is an excellent summary of freshwater aquaculture research, but it omits the saltwater applications that are of greatest importance to this project.

Once again, the point is to be as precise as your meaning requires.

Exercise 12.4

Construct a ladder of abstraction for two of the following words. Then identify a situation in which each word in the ladder would be appropriate.

elm tree
microscope
star
scalpel
soldering gun
typewriter

 Using and Abusing Jargon

Jargon, the highly specialized vocabulary of a technical discipline, can make your writing sound appropriately technical, but it can also be a disastrous tone trap that makes your writing incomprehensible.

Effective Jargon: Precise and Economical

Jargon is indispensable to technical writing. It is a kind of technical shorthand that uses words, phrases, abbreviations, even acronyms ("SONAR," "B.O.D.") and numbers to name objects, processes, concepts, and qualities that would otherwise require cumbersome circumlocutions—or worse, that would otherwise remain hazily undefined. To get a sense of what it would be like to write a report without jargon, imagine describing the simple process of changing a bicycle tire without using words like "air valve," "rim," "bead," and so on.

In addition to being extremely useful, jargon perhaps more than any other device of style is what gives a report its technical tone. An audience tends to trust those writers who can use jargon appropriately, assuming that because they know the special language of the field, they must also know the field itself. Thus, jargon increases a writer's credibility and authority—it makes the writer sound technical.

Reader Comprehension, Jargon, and Gobbledygook

Despite the usefulness of jargon, many people equate the word with doubletalk, with fancy words for simple things, with terms that seem more Greek or Latin than English—in short, with language that seems designed to exclude them from understanding. It is easy to see why people feel this way. All of us at one time or another have experienced the frustration and anger of running headlong into a solid wall of alien terminology.

It may seem like a contradiction to say that although readers dislike jargon, they trust the writers who use it. In reality, though, what disturbs people is not jargon; rather, it is *other people's jargon*. Consider the following passage:

> Constant-current potentiometry experiments were carried out in an inert atmosphere. The electrochemical cell consisted of a Li-Si alloy positive electrode, a eutectic mixture of LiCl-KCl electrolyte, and two Li-Al alloy electrodes—one negative and one reference electrode.

Electrical engineers could undoubtedly breeze right through this passage, comfortable with the language that they and the writer share, and confident that the writer is an expert in the field. Other readers, however, would give up after the first few words convinced them that the passage was over their heads. And although some of these readers might be impressed by the writer's ability to handle concepts and terms they can't understand, most would be irked that the writer was wasting their time.

Thus, as the specialized and restricted language of people in specific fields, jargon has a legitimate place in technical reports; however, if a writer imposes jargon on readers who can't understand it, they won't perceive it as jargon at all—they'll perceive it as *gobbledygook*, jargon's illegitimate cousin. Gobbledygook is writing that sounds technical, but that *means* less than it says. (The term, incidentally, was coined in the 1940s by Texas Congressman Maury Maverick, who found that much of the language used on Capitol Hill was about as meaningful as the gobbling of a Texas turkey.)

Because one reader's jargon is another reader's gobbledygook, you must choose your words carefully. Feel free to use the jargon of your field if your readers are in the same field. For much of the writing you do—in-house memos and periodic reports to your colleagues and immediate supervisors, lab reports, journal articles—this will indeed be the case. But for reports (or sections of reports) directed to readers outside your specific field, avoid any jargon that isn't essential to your report, and define any jargon that is. Introducing a technical term to an audience that needs to know it may actually have a positive psychological effect: It will make your readers feel part of the select group that "knows the lingo"—that is, it will make them *feel* technical.

Major Sources of Gobbledygook

Even if you decide jargon is appropriate in your report—either because your readers are familiar with it already or because they need to become familiar with it—you must keep a vigilant guard against other sources of gobbledygook. There are several.

Jargon Strings. The great virtue of jargon is its conciseness. Unfortunately, it is sometimes too concise. The zeal to compress vast amounts of technical information into as few words as possible can lead to sentences that are clogged up and choked off by their own terminology. The main culprits are jargon strings, and they can best be explained by example: "microprocessor-based real time data acquisition system." It is common practice in English to use a noun in place of a phrase or clause to modify another noun; for instance, we call the seat of a bicycle a "bicycle seat," and the lamp that sits on a desk a "desk lamp." To describe the lamp in a bit more detail, we might say that it is a "copper desk lamp." And we can increase the string further by adding an adjective: a "new copper desk lamp." So far, conciseness has not interfered with clarity. But if the string of nouns and adjectives gets much longer, problems quickly arise, especially if they are uncommon, technical nouns and adjectives—that is, jargon.

For one thing, jargon strings make for unwieldy sentences: Until readers get to the last word in the series, the word actually being modified, they have no way of knowing where all the nouns and adjectives are leading them. Here is a typical example:

The Technoerg Corporation Production Division Quality Assurance Committee meeting agenda was recently revised by Mr. Mendel.

Before readers get to the word "agenda," the word being modified, they have to slog through eight modifiers—and that, it seems to us, unfairly taxes their memories. Having finally gotten to the end of the string, many readers will have to go back and reread the beginning.

Sometimes, moreover, more than one reading will be necessary just to determine where the end of the string actually is. See if you can follow this sentence:

The continuous readout display terminal monitors any variation in signal frequency.

Chances are good that the first time through, you were so accustomed to seeing one modifier after another that you read the word "monitors" as a noun instead of as a verb. When this reading turned the sentence into non-sense, you read it again and realized that the string ends with "terminal." Obviously, a sentence that is so concise that it requires two readings is *too* concise.

There is still another problem with jargon strings. Some of the words in the string may modify not the last word, but rather one of the other modifiers. In the last example, for instance, the word "continuous" modifies "readout," not "terminal." Readers may find it impossible to sort out exactly which words in the string are modifying which other ones:

The Antelope Valley software manufacturers' convention opened on March 23, 1985.

Does this sentence mean that the convention took place *in* Antelope Valley or that, wherever it took place, the convention was attended by manufacturers *from* Antelope Valley?

The best way to unstring your terms is to turn some of them back into phrases and clauses:

The convention of software manufacturers opened on March 23, 1985, in Antelope Valley.

The terminal, which displays continuous readouts, monitors any varia-tion in signal frequency.

The agenda for the meeting of the Quality Assurance Committee of the Production Division at Technoerg Corporation was recently revised by Mr. Mendel.

Some jargon strings have become so widely used in particular technical fields that knowledgeable readers automatically see them as a single term. Nuclear engineers, for instance, can absorb the following string with one sweep of the eye: "liquid metal-cooled fast breeder reactor." Similarly, astrophysicists and astronomers can handle strings like "hard x-ray imaging spectrometer" and "soft x-ray polychromator" with no trouble. If a compound term already exists and all your readers are familiar with it, you should use it. Don't, however, coin jargon strings of your own—they may be too technical for your readers.

Exercise 12.5

Unstring each of these jargon strings:

1. air quality management district spokesperson
2. single working parent home environment counselor
3. short-term hearing loss restoration techniques
4. ten-ply tennis frame cold steam bending equipment
5. subjectively derived sea surface temperature pattern classifications

Borrowed Jargon. Even worse than language more technical than its audience is language more technical than the ideas it is supposed to communicate. Even a highly technical report inevitably contains sentences—sometimes paragraphs or whole sections—devoted to important nontechnical material. Some writers feel the need to spruce up this nontechnical material so that it sounds as technical as the rest of the report.

The variety of gobbledygook that they end up with often takes the form of vague or loosely applied jargon borrowed from another discipline—physics, psychology, and computer science being among the favorites. Thus, as translated into gobbledygook, two scientists will "interface" instead of merely "talk"; indeed, one of them may even "input" a solution rather than "suggest" it—and then the other will have to "factor in" this solution. With inappropriate words like these at their disposal, writers can swiftly turn a clear, precise, and accurate sentence into a hopelessly garbled one. Hapless readers, not knowing that the writer is merely trying to sound important, may struggle needlessly trying to crack a hard sentence with little beneath its shell. Other readers will realize immediately that the writer is just being pompous—and they will judge the writer accordingly.

Pseudojargon. Another common source of gobbledygook, though it has the general appearance and sound of jargon, can only be called pseudojargon. Consider, for instance, the advertisement for a new type of earplug that labels its product an "intraaural sound reduction device." Despite a puffed-up vocabulary calculated to make us believe that the "device" is a miracle of modern technology, we don't really learn anything more about this product from its fancy name than we would learn from the simple word "earplug." Compare this with a piece of real jargon. "Liquid metal-cooled fast breeder reactor"—though a mouthful—expresses succinctly the characteristics of a particular type of nuclear reactor. Jargon is *elegant*, pared down for maximum efficiency. Pseudojargon is its direct opposite; it dresses up simple concepts with needlessly complex language.

It is easy to create pseudojargon and to pass it off as the genuine article, especially because it shares with real jargon the characteristics of being polysyllabic and based on Latin and Greek words. Virtually any simple object or concept can be disguised. A phone call, for instance, can become a "telephonic communication"; a payment can become "pecuniary remuneration"; a piece of broken lab equipment can become "use-impaired apparatus." Abstract and pompous words lie at the core of this sort of gobbledygook; direct and simple words lose out to big ones, especially those with "-ize," "-tion," "-ism," or "-ate" endings. Here are a few of the most frequent offenders, along with their more straightforward alternatives:

initiate	begin
prioritize	rank
transmit	send
utilization	use (noun)
utilize	use (verb)
terminate	end
analyzation	analysis
be cognizant	know
effectuate	make
ascertain	find out
transpire	happen
encounter	meet
endeavor	try
finalize	finish

This is by no means a complete list. Indeed, you could probably expand it at will. Note, for instance, that a pompous verb can generally be transformed into a pompous noun with a simple change of suffix (see "utilize" and "utilization").

As you might have imagined, this kind of language is practically self-propagating. Phrases like "undertake the endeavor to ascertain" come trippingly to the tongue, and they have the unmistakable *sound* of important technical activity—if only the sound. Often the sole purpose of this kind of writing is to cover up a dearth of ideas with a load of bluster, and writers who have mastered the art can make you wade through pages of it to arrive at a single, simple point.

But even when their intentions are honorable, writers can succumb to the lure of pseudojargon. Many writers intuitively feel that a sentence loaded with necessary high-powered jargon just won't sound right unless the *whole* sentence and the ones around it, too, are consistently high-powered. Thus, if they have found that high concentrations of ammonia suppress *Nitrosomonas* and *Nitrobacter* activity in the soil, they are likely to write that these concentrations should be "utilized" in future experiments; "used" seems too mundane a verb to propel such an important sentence. A similar thought process probably explains why many technical writers prefer words like "optimum" and "maximum" in sentences where "best" and "greatest" would work just as well. If your intuition tells you to go for the big word even though you know an everyday word or phrase that would do just as well (or should we say "that would accomplish the objective in as satisfactory a manner"?), then *override your intuition*.

Exercise 12.6

Provide a simpler alternative for each of the following words and phrases:

eventuate
commence
exhibit
innumerable
a major portion of
transmit
communication

Filler Nouns. Like pseudojargon, filler nouns make a sentence longer and more difficult than its content warrants. The following example is typical:

❚ The migration of these birds is a phenomenon that occurs annually.

The word "phenomenon" has a pleasant rhythm and a technical sound, but it adds nothing to the sentence's meaning. The sentence could be revised like this:

❚ These birds migrate annually.

Note that once "phenomenon" is removed, the smothered verb ("migration occurs" for "migrate") becomes apparent. The revised sentence is less than half as long as the original—and clearer.

The following filler nouns are usually worth eliminating:

area	idea
aspect	nature
case	phenomenon
fact	principle
factor	reason
function	situation

All these words have legitimate uses, but often they are no more than filler, requiring a "that" clause or a prepositional phrase (usually an "of" phrase) to give them some sort of concrete reality. Concise writing dispenses with the empty introductory noun and goes straight to the concrete term that follows.

Exercise 12.7

Eliminate the filler nouns in the following sentences.

1. The annual site visit is a situation where a panel of experts reviews the next year's proposed research.
2. The fact that twenty-four mink escaped during a snowstorm delayed the PCB experiment.
3. The nature of the proposal is such that it requires input from a multi-disciplinary team.
4. Inadequate funding was one factor that contributed to the fact that the research was not completed on schedule.

A Test for Gobbledygook

The characteristic that unites all these sources of gobbledygook is that they sound technical, and it is a characteristic they share with real, useful jargon. Learning to cut inflated, imprecise, or meaningless words that sound right takes patience and practice. You will probably find that until you get the hang of it, the best way to attack the gobbledygook in your writing is through a deliberate and methodical process of revision.

Once you have completed your first draft, go back over it with pen in hand, looking for abstractions and polysyllabic words. Each time you find one—or a string of them—apply this test to determine if it is gobbledygook or legitimate jargon:

1. If the subject matter requires it *and* the readers are familiar with it, use it. It is legitimate jargon, and you will gain credibility by using it.
2. If the subject matter calls for the term but the readers are *not* familiar with it, use it sparingly and define it carefully. The jargon will add to your credibility, but it will also slow your readers down.
3. If the subject matter does not require the term—that is, if you can say the same thing more simply and with as much (or more) precision—revise the sentence. The term is not jargon but gobbledygook.

As your skills develop, you will become adept at recognizing and avoiding both gobbledygook and the tone traps we discussed earlier in the chapter.

Exercise 12.8

We began this chapter by suggesting that gobbledygook is the result of trying too hard to sound technical. A passage we showed you then and two others you haven't seen before are presented now. These three passages, all aggra-

vated cases of gobbledygook, provide a kind of compendium of all the style choices that we've warned you to avoid: They are filled with unnecessary abstractions, euphemisms, and pompous words; they "exhibit a preference for" smothered verbs, indirect constructions, and pseudojargon; and they use the passive voice where the active would be better. First identify, then correct, the problems in each.

1. There is now no effective mechanism for introducing into the initiation and development stages of reporting requirements information on existing reporting and guidance on how to minimize burden associated with new requirements.
2. To give their operations component broader-based relevance from the larger context and at the same time to give the larger context the full benefit of their operational interactions, public administrators must continually keep resonating one against the other.
3. Solar energy for the thermochemical reduction of water into its distinctive components will be utilized in this proposed hydrogen-producing system in a series of controlled chemical reactions. To simplify the construction and understanding of the system, it will be discussed here in three stages. The subsequent actual development of the system will be achieved in an identical order. Obtaining of data from each of the stages will be accomplished in order that the entire system can undergo analyzation.

The style of your reports and the tone that results from that style are said to be appropriate when they are precisely suited to your goals and audience. Each time you replace a vague or abstract word with a concrete one or replace an unnecessary passive construction with an active one, each time you choose to delete some inappropriate jargon or to include some useful jargon, each time you decide how best to express the information in your report, you are one step closer to an appropriate technical tone.

Review Questions

1. Characterize the tone most appropriate to technical reports. How is it achieved?
2. Define "connotation" and "denotation."
3. What are "tone traps," and why do technical writers so often fall into them?
4. Under what circumstances is the passive voice usually a better choice than the active voice?
5. How can you avoid using "I"? When is it appropriate to use "I"?
6. What is the difference between jargon and gobbledygook? Why do they both sound technical?
7. How does pseudojargon differ from real jargon?
8. How can you determine whether something you have written is gobbledygook or appropriate jargon?

Assignments

1. Go to the library, and find two articles in your field. Analyze them carefully for tone, vocabulary, use of the passive voice, and so on. Then rewrite a four- or five-paragraph section of whichever article you decided was worse.

2. To become more aware of the abuse of technical writing, now rewrite a four- or five-paragraph section of the better article from the previous assignment, exaggerating the problems of technical style to produce the worst version you can.

3. Exchange papers from the previous assignment with a classmate, and revise each other's, this time trying for an appropriate technical style.

13 Technical Report Format

Reports can be divided into two groups on the basis of their format: informal reports and formal reports. Informal reports are usually short and highly focused, addressing a single topic to a single reader or group of readers. If you need to tell your employer what you found out on your recent trip to the company's plant in the Southwest or to propose that all new technicians be given one week of instruction in the use of the equipment in your laboratory, you will write an informal report, probably in the form of a memo. If you want to advise your customers of a change in the operating procedure of a device that your company has sold them or to recommend to them that they consider a new line of compatible accessories, you will also write an informal report, this time in the form of a letter. (We discuss letter and memo formats in Chapter 19.) Some informal reports are even written on preprinted company forms: You need only fill in the blanks. Because they are relatively short and simple, informal reports are fairly easy to write. They usually contain an introductory section or paragraph, a body, and a terminal section or conclusion—and often little else.

Generally longer and more complex than informal reports, formal reports also contain an introduction, a body, and a conclusion. But, in addition, they contain title pages, tables of contents, appendixes, and the like. These formal elements are not merely decorative. Most formal reports are complicated. They aim at conveying detailed information about methodology or data or theory or mechanics or policy—or all five—to overlapping audiences of managers, technicians, laypeople, and experts, all of whom have pressing but different needs for this information and most of whom lack the background to understand it quickly or the time to absorb it slowly. Some of these readers will have to master the whole report; they will need frequent reminders of where they have been, where they are, and where they are going in the maze of technical specifics. Many other readers, however, will have to master only those parts of the report that pertain to their particular area of expertise; these readers will need those parts segregated so they are not forced to search through the entire document for relevant information. And all audiences need a context, a big picture, into which they can integrate the individual pieces.

The formal elements of reports are designed to fulfill these varied needs. Some present an overview of the whole report, others provide information needed only by special groups of readers, and still others act as signposts, guiding readers to the places in the report where they can find the information they need. Depending on their function, some formal elements precede the report proper (the introduction/body/conclusion) whereas others follow it. A very formal report might have the following format:

Letter of Transmittal
Title page
Abstract
Table of Contents   These elements are often referred
Lists of Tables and Figures   to as the report's "front matter."
Executive Summary
Lists of Findings, Conclusions, and
    Recommendations

INTRODUCTION
BODY   These constitute the report proper.
CONCLUSION

References (documentation of
    sources used in the report)   These are termed the report's
Glossary   "back matter."
Appendixes

As we discuss each of the formal elements that make up a report's front matter and back matter (except for abstracts and executive summaries, to which we devote the next chapter), bear in mind that although every formal report you write will need to include some of these elements, not all formal reports are equally formal. You might not need a list of figures or an executive summary or even a table of contents for any given report. Unless the format is specified by your employer, therefore, feel free to use as few or as many of these formal report accessories as you need to help your readers. Moreover, though we discuss conventional practices and present typical examples, we are not dictating rigid rules or absolute models. If your company or organization has its own format policy—if it wants the abstract *after* the table of contents, for instance, or if it wants the glossary in the front and the list of recommendations in the back—by all means adhere to that policy.

Note, too, that the report proper is composed of a number of formal elements; the body of a typical laboratory report, for instance, is divided into sections labeled "Materials," "Methods," "Results," and "Discussion." However, because the elements that compose the report proper vary widely from one type of report to another, we don't discuss them here. Later chapters on specific types of reports discuss the elements required in each.

▶ Repetition in Formal Reports

As you will soon discover, many of the formal elements of reports repeat information presented in other formal elements or in other sections of the report. Indeed, one of the distinguishing characteristics of a formal report is that it contains an enormous amount of repetition. Points covered in the abstract turn up again in the executive summary, in the introduction, and, of course, in the body of the report; information presented in the body often turns up again (in greater detail) in an appendix. Likewise, the body itself is filled with summaries, restatements, and other forms of repetition.

Purposeful repetition like this is the key to effective formal report writing. Everything in the front matter of your report is there precisely because it provides the context you want all your readers to keep in mind as they read. If any section of the report goes more than a page or two without referring to that context, readers will start losing track of why it is relevant. Similarly, summaries in the report proper keep your readers on target.

Moreover, because many readers don't read a report from cover to cover, information often has to be repeated in several sections to ensure that all the people who need that information will see it. Although the shop technicians and the legal department presumably won't be reading the same sections of the report, both need to know that a particular chemical is extremely flammable—the former so they will work cautiously, the latter so they will buy the necessary insurance.

Information in a formal report needs to be repeated not only in different places but also in varying degrees of detail and technicality, so that all readers can understand it well enough for their own purposes. Thus, the executive summary, which is designed for administrators, will mention only briefly the technical data discussed in depth in those sections for the technical staff.

Repetition, in sum, is a kind of "understanding insurance," and the formal elements of a report help guarantee readers proper coverage.

 ## Letter of Transmittal

A letter of transmittal usually accompanies a formal report. Either bound at the front of the report or clipped to the cover, the letter of transmittal orients the reader to the attached material. It should do the following:

1. Explain the occasion for and authorization of the report, including why it is being sent to the reader.
2. State the title of the report and the individuals or department that wrote it. This can often be combined with the first item.
3. Briefly summarize the content of the report and its major conclusions and recommendations. If appropriate, indicate a few features of particular interest to the reader and where in the report they can be found.
4. Define the scope of the report and point out any omissions that might confuse the reader—either questions it wasn't intended to answer or questions it tried to answer but couldn't.
5. Acknowledge funding or other assistance received from other organizations, including, of course, the reader's organization.
6. Express the hope that the report is satisfactory. Most letters of transmittal end with an invitation to respond: "Please feel free to write me if you have any questions about the report."

Figures 13.1 and 13.2 present examples of a letter of transmittal. The first, though brief, nevertheless covers most of the points we have mentioned. The second, a letter of transmittal for a competitive proposal, is longer and more persuasive because one of its goals is to arouse interest in the proposal that it introduces.

DONALD NEWTON
CARL W. REH
M. D. R. RIDDELL
RICHARD E. FOERSTER
ELMER F. BALLOTTI
ROBERT M. ZIMMERMAN
PAUL E. LANGDON, JR.
ALLAN B. EDWARDS
PAUL A. KUHN
ARTHUR H. ADAMS
WALLACE A. AMBROSE
RONALD E. BIZZARRI
THOMAS J. SULLIVAN

ASSOCIATES:
WARREN W. SADLER
JACK W. CORMACK
ROBERT DEL RE
JOHN M. SKACH
RICHARD P. MILNE
JOHN C. VOGEL
GLENN R. WENTINK
THOMAS E. WILSON
EDWARD T. KNUDSEN, JR.
EARL L. HECKMAN
WILLIAM C. KRAMER
TERRY L. WALSH
THOMAS W. BURKE
RONALD F. MARTIN
JERRY C. BISH
JOHN R. TROMP
RICHARD S. ERHARDT
JOHN O. ULLINSKEY
KENNETH V. JOHNSON
HAROLD D. GILMAN
R. SRINIVASARAGHAVAN
MELVIN C. LANE
ROGER J. CRONIN
JOHN F. SEIDENSTICKER
CARL M. KOCH
EDWARD M. GERULAT, JR.
FEDERICO E. MAISCH
CLYDE WILBER

CONSULTANTS:
THOMAS M. NILES
SAMUEL M. CLARKE
JOHN F. KAUSAL, AIA
HENRY M. ZUKOWSKI
J. BENJAMIN VERHOEK
T. RUSSELL ALMDALE
JOHN W. HARDIE
PHILLIP B. DENT

GREELEY AND HANSEN

ENGINEERS

1818 MARKET STREET, PHILADELPHIA, PENNSYLVANIA 19103 • (215) 563-3460

October 14, 1982

Mr. Richard E. Roy
Program Manager
Water Pollution Abatement Program
Philadelphia Water Department
1100 Municipal Services Building
Philadelphia, Pennsylvania 19107

Subject: Philadelphia Northeast Water Pollution Control Plant
          Operation and Maintenance Manual - Second Draft Submission

Dear Mr. Roy:

    We are pleased to submit fifteen (15) copies of the second draft of the Operation and Maintenance Manual for the subject project. This submission represents the last draft prior to submission of the final document. The final document will be issued in accordance with the schedule included in the Plan of Operation and as revised through the updates.

    Please return to us any Water Department comments on the submitted draft and notify us if additional copies are required.

                              Yours very truly,
                              GREELEY AND HANSEN

                              Jerry C. Bish

JCB/ao
cc: Elmer F. Ballotti (w/enclosures)

FOUNDED IN 1914

FIGURE 13.1   A Brief Letter of Transmittal.
(Courtesy Greeley and Hansen, Engineers.)

DONALD NEWTON
CARL W. REH
M. D. R. RIDDELL
RICHARD E. FOERSTER
ELMER F. BALLOTTI
ROBERT M. ZIMMERMAN
PAUL E. LANGDON, JR.
ALLAN B. EDWARDS
PAUL A. KUHN
ARTHUR H. ADAMS
WALLACE A. AMBROSE
RONALD E. BIZZARRI
THOMAS J. SULLIVAN

ASSOCIATES:
WARREN W. SADLER
JACK W. CORMACK
ROBERT DEL RE
JOHN M. SKACH
RICHARD P. MILNE
JOHN C. VOGEL
GLENN R. WENTINK
THOMAS E. WILSON
EDWARD T. KNUDSEN, JR.
EARL L. HECKMAN
WILLIAM C. KRAMER
TERRY L. WALSH
THOMAS W. BURKE
RONALD F. MARTIN
JERRY C. BISH
JOHN R. TROMP
RICHARD S. ERHARDT
JOHN D. ULLINSKEY
KENNETH V. JOHNSON
HAROLD D. GILMAN
R. SRINIVASARAGHAVAN
MELVIN C. LANE
ROGER J. CRONIN
JOHN F. SEIDENSTICKER
CARL M. KOCH
EDWARD M. GERULAT, JR.
FEDERICO E. MAISCH
CLYDE WILBER

CONSULTANTS:
THOMAS M. NILES
SAMUEL M. CLARKE
JOHN F. KAUSAL, AIA
HENRY M. ZUKOWSKI
J. BENJAMIN VERHOEK
T. RUSSELL ALMDALE
JOHN W. HARDIE
PHILLIP B. DENT

GREELEY AND HANSEN

E N G I N E E R S

SUITE 233, 233 BROADWAY, NEW YORK, NEW YORK 10279 • (212) 227-1250

February 25, 1983

Mr. Irwin Klein, P.E.
Director of Design
City of New York
Department of Sanitation
Office of Resource Recovery and
  Waste Disposal Planning
51 Chambers Street - Room 801
New York, New York 10007

Subject: Rehabilitation
          West 59th Street Marine Transfer Station, Manhattan
          Proposal for Engineering Services

Dear Mr. Klein:

In response to your February 3, 1983 Request for Proposal, Greeley
and Hansen is pleased to submit six copies of a Technical Proposal and
six copies of a Cost Proposal for engineering services relating to the
complete rehabilitation of the West 59th Street Marine Transfer Station,
Manhattan.

The project approach which we have developed will provide the
City of New York with a facility capable of meeting the Department
of Sanitation's long-term operational needs. The scope of work will
include, but not be limited to, the rehabilitation of damaged and
deteriorated structural members, fendering systems, plumbing, HVAC
systems, electrical systems, exterior architectural treatment, and
personnel facilities. The scope will also include installation of
a communication system throughout the facility, installation of a wider
and more gradually sloped ramp to the tipping floor, and provision for
adequate drainage relief for the facility. Finally, the scope will
include development of dredging requirements, procurement of all
necessary permits, and preparation of an Operations and Maintenance
Manual.

Greeley and Hansen has assembled a project team capable of performing
all work required by the rehabilitation project. This team consists
of Greeley and Hansen, lead consultant; Klein and Hoffman, Inc., structural
engineers; Woodward-Clyde Consultants, Inc., geotechnical engineers;
and Wiss, Janney, Elstner, and Associates, Inc., inspection and testing
engineers. Greeley and Hansen will perform general project engineering
and overall project management, including coordination with the Department
of Sanitation.

F O U N D E D   I N   1 9 1 4

FIGURE 13.2   A More Elaborate Letter of Transmittal.
(Courtesy Greeley and Hansen, Engineers.)

Mr. Irwin Klein, P.E.   -2-   February 25, 1983

    Greeley and Hansen has over 68 years of experience in the engineering
field, and our three supporting firms have, on the average, approximately
30 years of experience in providing specialized engineering services.
Furthermore, all four firms have worked extensively in the New York area,
with Greeley and Hansen's involvement dating back to 1929. All four
have also performed work for the City of New York, and Woodward-Clyde has
worked directly for the Department of Sanitation.

    The fee for engineering services for rehabilitation of the West
59th Street Marine Transfer Station and the estimated construction
cost are shown in the Cost Proposal volume of this submittal. Our
project team is available to begin work immediately upon receipt of
the notice-to-proceed from the Department of Sanitation.

    Greeley and Hansen is proud to have been able to serve the City
of New York over the past 50 years. We look forward to continuing
our association with the City through participation in the rehabilitation
of the West 59th Street Marine Transfer Station.

                                     Yours very truly,

                                     GREELEY AND HANSEN

                                     Richard E. Foerster

                                     Richard E. Foerster

REF/ao
Enclosure

Title Page

The primary function of the title page, of course, is to identify the report, but it often supplies legal or administrative information as well. Thus, in addition to the title, organization, date, and report number, the title page may announce the address of the organization that prepared the report, the place of publication, the names of authors, or the number of pages in the report. It may also include distribution lists, use-control notices, approval signatures, and space for readers' remarks. Sometimes the abstract appears on the title page, too, especially if it is fairly brief.

As you can see, you may have to squeeze quite a bit of information onto a single page. Be guided, therefore, by the principle of readability, and, including only what you absolutely must, strive for symmetry and neatness. Center your title—often set off in all capitals—in the upper third of the page, and space the other information around and below it, leaving as much white area as you can. (If the report is to be bound, be sure to allow a little extra space along the left edge, so that once the report is in its cover, the title page will be centered.)

Though the title of your report will be one of the last things you write, it will be the very first thing your readers read. Word it carefully. It should be informative and specific, indicating the report's topic, purpose, and scope. For instance, if your report analyzes a problem, the title should say so:

Analysis of the Causes of Heat Loss
in the *Sunrise 80* Greenhouse

If it makes recommendations, that point, too, should be included in the title:

Recommendations to Improve Heat Retention
in the *Sunrise 80* Greenhouse

A title that fails to mention the scope of the report (its precise focus) may send readers an erroneous signal:

Recommendations to Improve the *Sunrise 80* Greenhouse

Seeing that title, readers might understandably expect the report to cover topics other than heat retention.

Similarly, watch out for words like "study," "examination," and "exploration." Readers have a right to know the kind of study or examination you are writing about. Consider these three titles:

A Study of Packing Materials in Reverse Phase HPLC

A Study to Determine the Feasibility of Replacing Silica with Reticulated Vitreous Carbon as a Packing Material in Reverse Phase HPLC

Feasibility of Replacing Silica with Reticulated Vitreous Carbon as a Packing Material in Reverse Phase HPLC

The first is vague. The second is specific but wordy. The third, by skipping the word "study" altogether, is as specific as the second and shorter.

A sample title page is shown in Figure 13.3.

Table of Contents

The table of contents not only provides page references to the report's various sections, but also gives readers an overview of the report's scope and a clear indication of its method of development. Experienced report readers know to spend a few minutes studying the table of contents before plunging into a report. Once again, though practice varies slightly from one company to the next, the example we present in Figure 13.4 is fairly representative.

Be sure that the headings in the table of contents are identical to the headings in the report itself. If you have just recently decided to combine the "Sample Collection" section with the "Sample Analysis" section, make certain that both the table of contents and the report itself now read "Sample Collection and Analysis." The easiest way to ensure that you use the same headings throughout, of course, is to type the table of contents directly from the finished report. Put page references for items that precede the body of the report in small Roman numerals. And don't include any items that precede the table of contents itself.

Lists of Tables and Figures

Reports that contain graphs, tables, diagrams, maps, charts, drawings, or photographs customarily list them on a separate page right after the table of contents. It is sometimes useful to make two lists, one for tables and one for figures, especially if you have many of each. A report with only a few tables and figures, on the other hand, might include the list directly below the table of contents, separated by triple or quadruple spacing and, of course, by the heading "List of Tables and Figures."

FIGURE 13.3   A Sample Title Page.

Abstract and Executive Summary

In many reports, the abstract or the executive summary or both immediately follow the list of figures. Just as often, however, an abstract precedes
the table of contents, and sometimes one appears right on the title page.
Because we devote the entire next chapter to abstracts and executive summaries, we won't discuss them here.

TABLE OF CONTENTS

FIGURE 13.4   A Sample Table of Contents.

Lists of Findings, Conclusions, and Recommendations

"Findings" are the results of an investigation. A report on pollutant levels in a particular lake, for example, might *find* that concentrations of heavy metals, PCBs, and organics have decreased by 85 percent over a five-year period. "Conclusions" are generalizations from the findings, usually somewhat more abstract, less tied to the particulars of methodology. The report on the lake might *conclude* that the lake can safely be stocked for sport fishing and the shoreline developed as a recreation area. "Recommendations" are proposed actions based on the conclusions. The report might *recommend* that the lake be stocked with fish.

Often the boundaries among the three are fuzzy. A report may find that a certain product is the cheapest and most effective way to solve a particular problem, conclude that the company would benefit from buying the product, and recommend that it do so—three ways of saying the same thing. Just as often, however, the boundaries are crucial. There can be considerable inference between a finding and a conclusion, and considerable judgment between a conclusion and a recommendation—and a great deal of room for debate over the inferences and judgments. Good report writers keep the three as distinct as possible.

Depending on the topic and the extent of the writer's authority, some reports have findings but no conclusions or recommendations, whereas others have findings and conclusions only, and still others have all three. Report writers often save the findings for the body of the report and make separate lists only for the conclusions and recommendations. This much, at least, is ironclad. If you have conclusions or recommendations at all, list them separately instead of burying them in the body of your report. If you have findings but no conclusions or recommendations, list *them* separately.

Where do your lists belong? Traditionally they have been put at the end of the report, where they follow logically from the evidence in the body. Most organizations now prescribe instead that they come at the front of the report to save busy readers from fumbling to the end to find out what happens. This is a sensible change. Knowing your conclusions as they read will help readers follow your reasoning and assess its adequacy. And many readers, remember, have no intention of tackling the body of your report; your lists of findings, conclusions, and recommendations are the most important thing they will read.

The only problem with putting this information at the beginning is that readers may have trouble following it when they haven't read the report yet. The abstract and executive summary often supply enough of the context of the report to satisfy most readers. If they don't, however, consider putting your lists of findings, conclusions, and recommendations *after* the introduction rather than before it. Because the introduction's main purpose is to identify the problems being addressed in the report and the approach taken to solve them, placing your lists after the introduction should ensure that your readers get the overview they need. In addition, for each finding, conclusion, or recommendation, refer your readers to the section that contains the appropriate evidence and rationale.

Technical writers often try to combine their lists into one master list of findings, conclusions, and recommendations. This has the powerful advantage of showing the reader which findings led to which recommendations. For reports of any complexity, however, you will usually find the reasoning overlaps confusingly—these three findings led to this conclusion and, when added to this fourth finding, led to that second conclusion, which in turn led to these two recommendations, and so on. As a rule, we recommend separate lists, especially because a separate list of recommendations provides a useful checklist for decision makers.

For similar reasons, avoid crowding too much documentation into these lists. Readers who need detail will get it from the body. Try to keep each finding, conclusion, and recommendation to a sentence or two. But don't make them empty sentences. Instead of that old chestnut "further research is needed," for example, consider the following:

> At least two water quality planning agencies with very different problems should be selected as target areas to test the approach reported here, to augment and refine the computer package if necessary, and to suggest additional improvements in the analytic method before it is proposed for nationwide implementation.

The more explicit your findings, conclusions, and recommendations, the better chance they have of getting a fair hearing.

Although your lists of findings, conclusions, and recommendations can't be put into final form until after you have finished the body, it still pays to draft these lists tentatively before you begin the body and to keep them by your side as you write. After all, your findings, conclusions, and recommendations are precisely what the rest of the report aims at proving, justifying, and illuminating. It helps to have a firm grasp of what they are.

 # References

Of the three elements that generally constitute the back matter of a formal report—references, glossaries, and appendixes—the first is by far the hardest to deal with because the type of reference information you give and how you give it depend on the system of documentation that you are using.

When you write any type of report, it is standard procedure to *document* any information that is not original with you. Whoever did the work—or thought of the idea—gets credit for it in your report, whether you are quoting, paraphrasing, or summarizing. There are literally dozens of documentation systems, each one slightly different from the rest, and every discipline has its favorite. We begin, therefore, with a discussion of their chief similarities and differences.

Characteristics of Documentation Systems

All documentation systems have the same purpose: to indicate the information that has been obtained from outside sources and to tell readers where they can find this information in its original source. All documentation systems, therefore, include a method for identifying sources briefly in the text and a method for identifying them extensively somewhere else in the report. They differ, however, in how much information about the source they include in the text itself, how much near the text (in footnotes), and how much at the end of the text (in endnotes, bibliographies, lists of references). In many systems, for instance, the only information supplied in the text is a number indicating more extensive documentation somewhere else:

> Nearly half of all manatees killed in Florida in 1971 were killed by boats or barges.[2]

> Although beach nourishment is an effective method of impeding erosion, it is extremely expensive, with costs often exceeding several million dollars per mile of shoreline (3).

In other systems, however, the author's name, the year of publication, and often the page reference are supplied:

> The small amount of water present in seeds and grains, in addition to the water obtained through metabolic oxidation, is sufficient to sustain the Mongolian gerbil (Schmidt-Nielson, 1951:783).

Again depending on the particular system, information in the text might be supplemented by a note at the bottom of the page (a footnote) or, instead, by a note at the end of the text (an endnote). And the report might or might not also include a separate, alphabetized list of all the sources mentioned in the notes. Indeed, the brief citation in the text might send readers *directly* to this list of sources if the system is one of those that have dispensed with both footnotes and endnotes.

Documentation systems also differ—often radically—in punctuation, capitalization, use of abbreviations, and other matters of format. Examine, for instance, these three footnotes, each from a different system:

> 3. Marvin Rathbun and Christopher Weir, "Some Observations on the Long-Term Systemic Effects of Sustained Nicotine Ingestion in Rats," *New England Journal of Medicine*, 24 (1969), 137ff.

> 4. G. L. Freeman, The relationship between performance level and bodily activity level, *J. Exp. Psychol.*, 26, 1940, 602-608.

5. T. W. Johnston, Can. J. Phys. *41*, 1208 (1962).

As you can see, the author's full name may appear or just initials. The title of the article may or may not be placed in quotation marks, and it may even be omitted, as in the third example, when the volume number and the page number are both given. The name of the journal in which the article appeared may be spelled out; it may be abbreviated. It may or may not be underlined or italicized. The year of publication may appear last or earlier; it may or may not be enclosed in parentheses. And remember that these examples represent only three of the many documentation systems.

Luckily, you don't need to master all the documentation systems that exist. But you do need to master the one used in *your* discipline. Many disciplines therefore publish their own style manuals. Following is a representative selection:

American Chemical Society. *Handbook for Authors.* Washington, D.C.: American Chemical Society Publications, 1978.

American Institute of Physics. *AIP Style Manual.* New York: American Institute of Physics, 1978.

American Mathematical Society. *Manual for Authors of Mathematical Papers.* 6th ed. Providence: American Mathematical Society, 1978.

American Medical Association. *Style Book/Editorial Manual.* 5th ed. Chicago: American Medical Association, 1971.

American Psychological Association. *Publications Manual.* 2d ed. Washington, D.C.: American Psychological Association, 1979.

Council of Biology Editors. *Council of Biology Editors Style Manual.* 4th ed. Arlington, Virginia: American Institute of Biological Sciences, 1978.

Institute of Electrical and Electronics Engineers. *Information for IEEE Authors.* Available from the Editorial Department, Institute of Electrical and Electronics Engineers, Inc., 345 E. 47th St., New York, N.Y.

U.S. Government Printing Office. *Style Manual.* rev. ed. Washington, D.C.: U.S. Government Printing Office, 1973.

University of Chicago. *A Manual of Style.* 13th ed. rev. Chicago: University of Chicago Press, 1979.

These manuals give 20- to 200-page descriptions of how to document—in *that* field—everything from a new reprint of a 1944 translation from the Swedish of an unpublished article to a tape-recorded interview with a now-deceased government official. You should make it a point to learn which style manual is the one accepted in your field—if there is one—and use it.

You can also shop for a documentation system in the scholarly journals. Most journals, on the title page or inside the back cover, supply instructions on how to submit an unsolicited manuscript, usually including a description of the documentation system the editors prefer. (We present an example in Fig. 13.5.) In addition, most large companies and many small ones authorize a documentation format for the reports presented under their auspices. If your company is among these, be sure to follow its format exactly.

REFERENCES

General. The references cited should be to the original source which should be examined in connection with the citation. If this cannot be done, a secondary source (e.g., abstract, book, etc.) may be used, but it should be cited along with the original reference.

Only references actually cited in the paper are to be included in this section.

Use lower case letters a, b, c, etc., following the year to distinguish among different publications of the same author(s) in any one year.

Private communications cited in the text are included in the list of references, the same as unpublished data or publications.

When no author (or editor) is indicated, "Anonymous" is cited in the text as well as in the list of References.

In the text. Cite literature in one of the following ways:

1. With last name of the author as part of the sentence, immediately followed by the year of publication in parentheses:

Example: Smith et al. (1963) reported growth on vinasse. This was demonstrated by Jones (1966).

2. With last name of author and year of publication in parentheses, usually at the end of a sentence:

Example: The starch granules are normally elongated in the milk stage (Brown, 1956).

3. To indicate more than two authors, use et al. after the name of the first author:

Example: This was first observed by Smith et al. (1966) and later confirmed by Smith and Jones (1972).

Complete titles of works should not be cited in the text.

In the list of References. This section is headed References. The entire list MUST be typed double spaced. Each individual citation in the list should begin flush left (with no paragraph indentation). If the citation requires more than one line, indent runover lines about six characters (½ in.).

Arrangement shall be strictly alphabetical, by author(s) LAST name, and where necessary, by order of citation in the text in the case of multiple references to the same author(s) in any one year.

REFERENCE EXAMPLES

EXAMPLES of use in a Reference list are given below. The bold-faced parenthetical type of citation above the example is indicated ONLY for information and is **NOT** to be included in the reference list.

(Anonymous)

Anonymous. 1982. Tomato product invention merits CTRI Award. Food Technol. 36(9): 23.

(Book)

AOAC. 1980. "Official Methods of Analysis," 13th ed. Association of Official Analytical Chemists, Washington, DC.

Weast, R.C. (Ed.). 1981. "Handbook of Chemistry and Physics," 62nd ed. The Chemical Rubber Co., Cleveland, OH.

(Bulletin, circular)

Willets, C.O. and Hills, C.H. 1976. Maple syrup producers manual. Agric. Handbook No. 134, U.S. Dept. of Agriculture, Washington, DC.

(Chapter of book)

Hood, L.F. 1982. Current concepts of starch structure. Ch. 13. In "Food Carbohydrates," D.R. Lineback and G.E. Inglett (Ed.), p. 217. AVI Publishing Co., Westport, CT.

(Journal)

Cardello, A.V. and Maller, O. 1982. Acceptability of water, selected beverages and foods as a function of serving temperature. J. Food Sci. 47: 1549.

IFT Sensory Evaluation Div. 1981a. Sensory evaluation guide for testing food and beverage products. Food Technol. 35(11): 50.

IFT Sensory Evaluation Div. 1981b. Guidelines for the preparation and review of papers reporting sensory evaluation data. Food Technol. 35(4): 16.

(Non-English reference)

Minguez-Mosquera, M.I., Franquelo-Camacho, A., and Fernandez Diez, M.J. 1981. Pastas de pimiento. 1. Normalizacion de la medida del color. Grasas y Aceites 33(1): 1.

(Paper accepted)

Bhowmik, S.R. and Hayakawa, K. 1983. Influence of selected thermal processing conditions on steam consumption and on mass average sterilizing values. J. Food Sci. In press.

(Paper presented)

Takeguchi, C.A. 1982. Regulatory aspects of food irradiation. Paper No. 8, presented at 42nd Annual Meeting of Inst. of Food Technologists, Las Vegas, NV, June 22-25.

(Patent)

Nezbed, R.L. 1974. Amorphous beta lactose for tableting. U.S. patent 3,802,911, April 9.

(Secondary source)

Sakata, R., Ohso, M., and Nagata, Y. 1981. Effect of porcine muscle conditions on the color of cooked cured meat. Agric. & Biol. Chem. 45(9): 2077. [In Food Sci. Technol. Abstr. (1982) 14(5): 5S877.]

Wehrmann, K.H. 1961. Apple flavor. Ph.D. thesis, Michigan State Univ., East Lansing. Quoted in Wehrmann, K.H. (1966), "Newer Knowledge of Apple Constitution," p. 141, Academic Press, New York.

(Thesis)

Gejl-Hansen, F. 1977. Microstructure and stability of freeze-dried solute containing oil-in-water emulsions. Sc.D. thesis, Massachusetts Inst. of Technology, Cambridge.

(Unpublished data/letter)

Peleg, M. 1982. Unpublished data. Dept. of Food Engineering, Univ. of Massachusetts, Amherst.

Bills, D.D. 1982. Private communication. USDA-ARS, Eastern Regional Research Center, Philadelphia, PA.

FIGURE 13.5   Information on Documentation for Contributors to the *Journal of Food Science*.

(From "IFT Style Guide for Research Papers (REV 4) 12/82 5C," *Journal of Food Science*, vol. 47, no. 6, 1982. Reprinted by permission.)

Following the prescribed format *exactly* means just that. You must learn to get all the commas in the right place, all the bibliographical information in the right order, all the right words capitalized or abbreviated, and so on. Every documentation system comes with its own set of technical specifications, some of them quite complex, many of them completely arbitrary. Even within a single system, the format varies according to the type of source you are documenting—a book, a magazine article, a newspaper story, an unsigned pamphlet, and so on. If you don't observe these variations, you may have your report returned to you for correction. Because there are very few logical reasons behind documentation rules, you have to learn them by rote.

Documentation Formats: Reference Citations

The documentation system most familiar to college students is the one prescribed by the Modern Language Association (MLA). It features superscript numbers in the text, either footnotes or endnotes, and a free-standing bibliography. However, virtually all scientific and technical disciplines have developed more streamlined systems for documenting their sources. These systems, called reference citation systems, often include a bit more information in the text than the MLA system, and they combine footnotes and bibliography in a single list titled "References" or "List of Works Cited." Each discipline follows a slightly different format for its reference citations; however, most can be classified as either "author-date" systems or "numerical reference" systems.

Author-Date Systems. Instead of using a superscript number in the text to signal that an outside source is being used, many systems put the author's name and the year of publication right in the text. Sometimes a page reference is included as well:

> Grade-school-aged children who cannot cope with their anxieties and frustrations often display the following symptoms: stuttering, unreasonable fears, temper tantrums, fighting with peers, causing trouble in the neighborhood, truancy, and bed-wetting beyond the age of six (Haberman, 1966:153).

Note that the parenthetical information comes after the sentence but before the period. The page number might refer to the page in the article on which this information appeared, or it might refer to the first page of several; it is customary to give only the initial page.

If the author's name is mentioned in the text, it shouldn't be repeated in the parentheses:

> Kessler (1977) has reported that diabetic patients have no excess mortality from bladder cancer despite their relatively high rate of consumption of saccharin.

If a single sentence refers to more than one source, that, too, can be noted:

> Recent studies of migration patterns (Johnson and Smith, 1981; Lemoine, 1981; Reich, 1982; Vassein et al., 1983) have failed to confirm these findings.

In all author-date systems, the list of references at the end of the report is presented alphabetically. Most systems abbreviate journal titles and cite authors' initials rather than whole first names. Some systems are even more pared down, dispensing with the quotation marks, parentheses, and underlining of the MLA system, and some even omit the title of the article being cited. Some examples from different systems are presented below. (The format used in *your* field, remember, might not be like any of these.)

American Institute of Physics style:

Abercrombie, D. (1967). *Elements of General Phonetics* (Aldine, Chicago).
Dorman, M. F., and Raphael, L. J. (1980). "Distribution of acoustic cues for stop consonant place of articulation in VCV syllables," J. Acoust. Soc. Am. **67**, 1333–1335.
Repp, B. H. (1982). "Perceptual assessment of coarticulation in sequences of two stop consonants," Haskins Lab. Status Rep. Speech Res. **SR-71/72**, 131–166.

American Chemical Society style:

Brand, L., and Witholt, B. (1967) *Methods Enzymol. 11*, 776–856.
Forster, T. (1965) in *Istanbul Lectures* (Sinanouglu, O., Ed.) Part III, Academic Press, New York.
Tanford, C. (1961) *Physical Chemistry of Macromolecules*, Chapter 6, Wiley, New York.

Council of Biology Editors style:

Costello, C. E. Gas chromatographic-mass spectrometric analysis of street drugs, particularly in body fluids of overdose victims. Marshman, J. A., ed. *Street analysis and its social and clinical implications*. Toronto: Addiction Research Found.; 1974: 67–78.
Coughenour, L. L.; McClean, J. R.; Parker, R. B. A new device for the rapid measurement of impaired motor function in mice. *Pharmacol. Biochem. Behav.* 6: 351–353; 1977.

Numerical Reference Systems. Instead of putting the author and date in the text, many systems use numerical references in the text keyed to a numbered list of references at the end of the report:

Serial structure (NAND type cell) was proposed in 1976 as an alternative to the conventional parallel structure (NOR type cell) as a means of increasing bit density for a read-only memory. [5].

In some systems, especially those in the social sciences, a page reference is also included:

Studies have shown that television advertising can have a significant effect on children as young as three years old (5, p. 345; 16, p. 25; 23, p. 79).

Still others use superscript numbers:

Spikes and Pennington[29] discuss the effects of upstream chamfer and orifice thickness. A series of two or more orifices prevented cavitation.[1] This was demonstrated at the inlet of the inner radial-blanket subassemblies of the Fermi reactor,[4] where flow velocity was low but pressure in the inlet plenum was high. Cavitation was a problem in Phenix flow distribution devices.[67] Part of this problem could be attributed to improved performance parameters.[71]

As the references are numbered at the back of the text, the numbers in the text are not in sequence. The entries on the "References" page may be listed alphabetically by author (*and* preceded by a number), or they may be listed in the order in which they first came up in the text. Following are two examples, once again from different systems:

IEEE style:

[1] D. W. Faulkner, R. J. Hawkins, and I. Hawker, "A single chip regenerator for transmission systems operating in the range 2-320 Mbits/s," *IEEE J. Solid-State Circuits*, vol. SC-17, pp. 552–558, June 1982.
[2] J. Davidse, *Integration of Analog Electronics Circuits*. New York: Academic, 1974, p. 252.
[3] IBM ASTAP Program Reference Manual, IBM Corp., 1973.

Council of Biology Editors style:

1. Bailey, D. N.; Shaw, R. F.; Guba, J. J. Phencyclidine abuse: plasma levels and clinical findings in casual user and in phencyclidine-related deaths. *J. Anal. Toxicol.* 2:233–237; 1978.
2. Baselt, R. C. Phencyclidine. *Disposition of toxic drugs and chemicals in man*, vol. 1. Canton, CT: Biomedical Publication; 1978: 163–165.
3. Kase, Y.; Miyata, T. Neurobiology of piperidine: its relevance to CNS function. Costa, E.; Giacobini, E.; Pavletti, R., eds. *First and second messengers—new vistas*. New York: Raven; 1976: 5–16.

Note that the Council of Biology Editors authorizes both the author-date and the numerical reference systems. Most of the style manuals we listed earlier, in fact, point out that whereas some journals and organizations under their aegis prefer one system, others prefer the other. As we mentioned at the beginning of our discussion of documentation, deciding which system *you* should use generally won't be a problem. If your organization requires a certain system, use it. If you have to choose one on your own, adhere to the system used by other writers who have addressed the same audience as you.

 ## Glossary

A glossary lists and defines the technical symbols, abbreviations, and terminology used in your report that your audience might not understand. Of all the elements described in this chapter, it is the one least likely to be needed. It is quite possible that your readers will be familiar with all the technical terms you use or that you use so few that you can define them right in the text. If, however, there are nonspecialists or lay readers among your audiences, a glossary may be not only useful, but indispensable.

Your glossary can be placed either in the front matter or in the back matter. Wherever you place it, two points are worth keeping in mind. First, keep your definitions brief. (If you have legal or stipulative definitions that are long and technical, but necessary, you might consider treating them separately in an appendix.) Second, keep the language simple. If you use technical terminology to define technical terminology, you will only confuse your readers.

 ## Appendixes

Appendixes (or appendices) are storage bins. If you have information vital to one audience but irrelevant to all the others, put it in an appendix ("Appendix II: Construction Specifications"). Appendixes also house either peripheral documents or essential documents that are too cumbersome or complicated to be spliced into the body of the report. Peripheral documents include such things as maps, copies of laws, schedules, résumés, case histories, and raw data (refined data are usually found tabulated in the body of the report). Essential but oversized documents include figures and tables longer than half a page, sample forms (questionnaires, for instance), long mathematical calculations or chemical formulas, and full-length versions of any verbal or mathematical information given in abridged form in the body.

Sometimes it is hard to judge how long a document must be to be exiled to an appendix. Laws are frequently short, but they are so dense that they interrupt the flow. You should ask yourself if your readers can get all the way through to the end of the interruption without losing your train of thought. If they can, you can merge the document into the body. If they cannot, put the document in an appendix. Of course, no matter how long the document is, if the readers need it to understand the rest of the report, then you must at least summarize it in your text; if it cannot be summarized, it must be included in the body, whether it interrupts the flow or not.

Remember, too, that every document in an appendix should be *appended to* some point made in the body of your report. This means that at the appropriate place in your text, you should refer your readers to the relevant appendix:

> The questionnaire (see Appendix I) was randomly distributed to 400 patients of the clinic over a span of two weeks.

> Neither age nor sex could be significantly correlated with performance on the questionnaire. The complete statistical breakdown is presented in Appendix III.

Each document or cluster of documents you append to your report is considered a separate appendix, so you may find yourself with quite a few of them. Number them all (Roman numerals are usual), and give each a short title:

> Appendix I: Marine Mammals Protection Act
> Appendix II: Tons of Yellowfin Tuna Caught in the Eastern Pacific 1971–79
> Appendix III: Incidental Take of Marine Mammals in the Course of Commercial Fishing Operations

You should probably provide each appendix with its own title page at the back, even if the appendix itself is only one page long. This practice is traditional, largely because it facilitates detaching an appendix and circulating it separately.

Your only real concern with appendixes will be order. Nearly always, you will put them in the order in which they are referred to in the body of the report. The first appendix to be cited in the text automatically becomes Appendix I, and so forth. If you refer to several at once, put them in decreasing order of importance or in chronological order if they are dated and follow a recognizable sequence.

Review Questions

1. What are the main purposes of the formal elements of a report?
2. Why is repetition necessary in a long report?
3. What is the purpose of the letter of transmittal?
4. What are the elements of an effective title?
5. What is the difference between a finding and a conclusion?
6. What are the common characteristics of all documentation systems?

7. What is a style manual?

8. What sorts of information belong in an appendix rather than in the body of the report?

Assignments

1. Write a letter of transmittal to accompany a report you have written for this class.

2. Find out which of the style manuals is the accepted one for your field. Compare its policies on documentation, use of appendixes, and format of front matter with those we have discussed in this chapter.

3. Revise the "References" section of a paper that you have already written so that it conforms with the format specified in the style manual of your field.

14 Abstracts and Executive Summaries

An abstract is a very short summary of a technical document, usually 150 words or less and almost never more than 250 words. An executive summary is a longer overview—as long as several pages—aimed at management instead of technical readers. Virtually all formal technical documents begin with an abstract, and most substantial ones include an executive summary as well. On the job, you will write abstracts and executive summaries not only for those reports you prepare entirely on your own, but often for those you write in collaboration with colleagues and superiors. And you may frequently be called upon to summarize reports you have read for the benefit of technical colleagues or managers. The skills needed to write abstracts and executive summaries—the ability to recognize important facts and ideas in a long document, to condense them into a few essential sentences, and to present this distilled information clearly and precisely—will improve the quality of all your written work, technical and nontechnical, in school and out.

Chiefly because they are shorter, abstracts are more difficult to write than executive summaries. Most of this chapter is therefore devoted to abstracts; the final section will discuss the special characteristics of the executive summary.

Goal and Audience

The goal of all abstracts is to condense an entire document into a few pithy sentences that readers can absorb quickly and efficiently. To achieve this goal, the abstract must be completely *self-contained*—if your readers can't understand it until after they have read the report, then obviously it hasn't saved them any time. What makes abstracts difficult to write is that so many different sorts of people read them, each for different reasons and with differing levels of expertise.

Readers of Abstracts

Abstract readers can be divided into three large groups according to what they need from an abstract. Those in the first group read abstracts to see whether they need to read the whole document. A good abstract will give them enough information to decide that the document is just what they have been looking for or that it is absolutely irrelevant to their work or that it just *might* contain that one bit of data they have been trying to find. Members of the second group read abstracts to get a brief overview before they begin a report or article that they know they must read. They are looking for an orientation, an indication both of the direction the document will take and of the sections that will be most relevant to them. The third group reads abstracts with no intention of reading the parent document. This group is becoming larger every day, as articles, reports, books, and journals glut every technical field. More and more, technical people must rely on abstracts to keep at least relatively current on developments in their fields.

Complicating the picture further is the fact that none of these three groups is homogeneous. Each might include management personnel, technicians, students, experts in the field, experts in other fields, and so on. If you pack too much technical terminology and specific detail into your abstract, you are bound to swamp many readers. But at the same time, if your abstract is too general, you will probably exclude some of your most expert potential readers—and they are often the readers you most want.

Balancing Completeness with Comprehensibility

To make your abstract both accessible and useful to its diversified readership, you will have to make compromises about what you include. Although every reader might not be able to understand all of your abstract, even the least informed reader should be able to grasp the main idea of the document being abstracted, as well as its general significance.

Consider this highly technical abstract of an article entitled "Volcanic Contributions of Chlorine to the Stratosphere: More Significant to Ozone Than Previously Estimated?":

> Earlier estimates of the chlorine emission from volcanoes, based upon evaluations of the preeruption magmatic chlorine content, are too low for some explosive volcanoes by a factor of 20 to 40 or more. Degassing of ash erupted during the 1976 Augustine Volcano in Alaska released 525 $\times$ 10^6 kilograms of chlorine (± 40 percent), of which 82 $\times$ 10^6 to 175 $\times$ 10^6 kilograms may have been ejected into the stratosphere as hydrogen chloride. This stratospheric contribution is equivalent to 17 to 36 percent of the 1975 world industrial production of chlorine in fluorocarbons.[1]

Neither a geologist nor a meteorologist would have any difficulty with anything in this abstract. A student doing a paper on the ozone layer could learn enough to decide whether to read the article—or at least to copy it and have someone translate it. Despite its technical terminology like "preeruption magmatic chlorine content" and "degassing of ash," the abstract clearly explains the main point of the article—that the amount of chlorine released into the atmosphere by volcanoes is far greater than previously supposed. And in the last sentence, it suggests the significance of this finding.

Of course, if you can avoid technical terminology without distorting your abstract, then by all means do so. Here is the abstract of an article entitled "Atmospheric Burnup of the Cosmos-954 Reactor":

On 24 January 1978 the Russian satellite Cosmos-954 re-entered the atmosphere over northern Canada. By use of high-altitude balloons, the atmosphere was sampled during 1978 up to an altitude of 39 kilometers to detect particulate debris from the reactor on board the satellite. Enriched uranium-bearing aerosols at concentrations and particle sizes compatible with partial burnup of the Cosmos-954 reactor were detected only in the high polar stratosphere.[2]

With the possible exception of the phrase "enriched uranium-bearing aerosols," this entire abstract could be understood by virtually anyone. Yet the article itself is extremely technical, filled with paragraphs like the following:[3]

Because λ is small, the Poisson probability function can be used to calculate the probability (P_r) of observing a particular number of tracks (r) per particle.

$$P_r = \frac{\lambda^r e^{-\lambda}}{r!}, r = 0, 1, 2, \ldots$$

The authors apparently decided that an abstract covering the reason for conducting the study and the major finding of the study would in this particular case be adequate for most readers, for the article itself deals almost exclusively with sampling procedures and techniques of analysis, drawing no conclusions about the significance of the aerosols in the stratosphere.

 # Types of Abstracts

There are two general types of abstracts, descriptive abstracts and informative abstracts.

Descriptive Abstracts

A descriptive abstract is a sort of table of contents in paragraph form; it announces the topics covered in the article or report. Here is an example:

This report presents the results of a survey of 300 homeowners' attitudes toward converting to solar water-heating. Residents were asked whether they would convert to solar heating if the operating costs averaged a specified amount per month less than heating with fossil fuels, assuming specified initial costs for conversion. Survey responses to questions on availability of solar equipment, perceived reliability of solar systems, and predicted shortages of fossil fuels are tabulated and analyzed, and the implications of the results for home marketing of solar hot water systems are discussed.

This descriptive abstract is admirably clear and jargon-free. It tells readers precisely what topics will be discussed in the report. It states that the report tabulates and analyzes the various data and draws appropriate conclusions. It is silent, however, on what the actual results and conclusions are. In other words, although it describes the *kind* of information to be found in the report, it does not present any of the information itself.

Descriptive abstracts are easy to write. In fact, you could probably write one with only a cursory reading of the actual document. A glance at the introduction and the table of contents will show you the topics covered in the report; then all you need to do to prepare an abstract is append a verb like "is studied," "are discussed," or "are evaluated" to each of the topics. For audiences who read the abstract as a prelude to reading the whole document, the absence of specific information won't be a problem. However, because technical people are relying increasingly on the abstract as their source of information, descriptive abstracts are starting to lose favor.

Informative Abstracts

The alternative to the descriptive abstract is called, appropriately, the "informative abstract." Here is an informative abstract of the report on homeowners' attitudes toward converting to solar heating:

Based on an attitude survey of 300 American homeowners, this study estimates that 7.2 million American homes would convert to solar water-heating if operating costs averaged $20 per month less than heating with fossil fuels and if initial conversion costs were $1,500. Willingness to convert to solar heating increases substantially when conversion costs are reduced to the $500–$1,000 range, but not when monthly savings are increased to $25–$30. Predicted shortages of fossil fuels more than double consumer receptiveness to solar systems at all levels of initial cost and monthly savings. Respondents indicate that unavailability and perceived unreliability of solar systems are major barriers to conversion. The report concludes that a satisfactory home market for solar heating systems requires a reduction in initial costs and an improved image of reliability.

Like the descriptive abstract, the informative abstract covers the purpose of the report and the major topics discussed in it; in addition, however, it summarizes the actual contents—the findings, conclusions, and so on. Informative abstracts are obviously more useful than descriptive ones, especially to readers who do not intend to study the entire document. On the other hand, they are longer and more unwieldy than descriptive abstracts: More space and more jargon are required to report your actual findings than to say that you have findings to report. Thus, the problem of deciding how much technical detail and terminology to include is especially serious in informative abstracts.

Most of your on-the-job writing will be of a length and scope that permit you—with a little effort—to write an informative abstract that is complete and comprehensible to a wide range of readers. Occasionally, however, you will probably have to abstract a real blockbuster of a report, one that is long, complex, and very technical; and especially if it is your own work, you may find it virtually impossible to cram all the essential information into 150 words. An abstract that combines elements of both descriptive and informative abstracts may be the only solution. An "informative-descriptive abstract" includes only the most important findings, then describes the other topics covered in the report. Such a compromise is not nearly so useful as an informative abstract. Whenever you possibly can, therefore, you should write informative abstracts.

Writing Informative Abstracts

Whether you are abstracting your own document or someone else's, the process of abstracting is essentially the same. It involves three major steps.

Read the Document

Read the document carefully several times to get an overall sense of the points covered, the relative importance of each point, and the general direction of the discussion. (If you are abstracting your own work, you will obviously have a good sense of these things already.)

Underline the Important Points

Go through the document again, this time with pencil in hand. Identify and underline these items:

- The *subject* of the report—the problem under discussion.
- The *purpose* of the report—why the author is writing about the subject.
- The *thesis* of the report—the controlling idea; what the author has decided about the subject.
- The *scope* of the report—what it covers and excludes.

These are usually presented in the introduction of the report. Because they are, in essence, what the report is all about, they *must* be in the abstract as well. Remember that the abstract must stand alone, that its function is to reproduce the report in miniature.

Next, underline all the important points, all those pieces of information that support the thesis, that prove or justify what the author is saying in the report. Generally, every paragraph will have one central idea. Termed the "topic sentence," this central idea is usually placed either first or last in the paragraph, and the remainder of the paragraph supports it with examples, explanations, discussion, and so on. In short reports, these topic sentences are usually where you will find the most important points, though not every topic sentence is crucial to the thesis. You will have to decide which ones directly support the main point that the author is making. In long reports, the important points are generally summarized at the beginning or end of each section or chapter. "Topic paragraphs" replace "topic sentences" as the place to look.

What constitutes an "important point" will depend, of course, on the type of report you are abstracting. The important points in a laboratory or field research report, for instance, would include something about the materials and procedures of the experiment (the number and types of subjects involved, the experimental design, the important equipment), the results of the experiment, and the conclusions and significance of the research. The important points in a theoretical or procedural report would be quite different.

Do *not* underline (or put in your abstract) any of the following:

- general background material
- lengthy definitions of terms
- detailed descriptions of standard procedures
- supporting or explanatory examples
- graphs or tables
- footnotes (or references to secondary sources)
- cross-references to other parts of the report
- digressions from the thesis

Put the Information in Your Own Words

Once you have isolated all the important points, write a rough draft of your abstract, putting each point in your own words and following point-by-point the organization of the parent document. After a little practice, you will learn how to condense several points into a single clear sentence.

The following passage contains the opening paragraphs of an article that appeared in the *Journal of Mammalogy*. We have annotated the passage for you:

Alarm Bradycardia in White-tailed Deer Fawns (*Odocoileus virginianus*)[4]

Young of many ungulate species often attempt to conceal themselves ("freeze" or "lie prone") rather than to flee in response to diverse alarm stimuli

This paragraph contains general background information and definitions.

These sentences establish the problem addressed in the study. The last sentence states the purpose of the report (and the scope).

The number and type of subjects are presented.

The methods and equipment are explained.

(Lent, 1974), such as vocalizations and certain movements of the female, noises, and the appearance of humans and other potential predators. An alarmed white-tailed deer fawn (*Odocoileus virginianus*) typically assumes a recumbent posture, producing the lowest body profile. During complete expression of the prone response, the neck is extended so that the ventral surfaces of the head, neck, and body are as close to the ground as possible. The ears are laid back along the neck, and the eyes remain open. The fawn is alert but motionless.

Although prone responses are frequently observed in young fawns, little is known about the physiological aspects of this behavior. A study of the physiology and activity of white-tailed deer fawns from birth to weaning age provided an opportunity to characterize some basic physiological correlates of the prone response. This is a report on decreases in heart rate during responses to alarm stimuli.

Materials and Methods

Two male and three female fawns were acquired from a herd maintained within a 2.4-ha enclosure at the Ithaca Game Farm, New York, during the 1972–1974 fawning seasons. Three of the fawns were separated from their twin siblings and respective dams 4 to 6 h after parturition, and two remained with their tamed dams during the study. The three hand-reared fawns were bottle-fed several times daily a warmed-milk diet formulated to approximate the composition of white-tailed deer milk (Robbins and Moen, 1975). A pelleted diet (horse pellets, Agway, Inc.), water, salt, and herbaceous vegetation were available ad lib. in the 34- by 34-m observation pen. Fawns were weighed on a platform scale ($\pm$.1 kg) at about 4-day intervals.

Heart rates were recorded continuously as an electrocardiogram (ECG) by a Narco Biosystems telemetry system (Jacobsen, 1979). A small elastic harness positioned immediately behind the forelegs of each fawn secured both the surface electrodes and the 18-g transmitter. Fawns wore the transmitter-harness without apparent discomfort or changes in behavior.

Only one fawn was monitored by telemetry during each of the monitoring periods (N = 24) between birth and weaning (485 h of observation). One or two other nontelemetered fawns were present in the observation pen during each monitoring period. The behavioral and social interactions of all fawns were recorded so that physiological data might be correlated with behavioral response patterns. Fawns were observed either from an observation trailer adjacent to the pen or from within the pen at a distance of 5 to 15 m. Observations were aided by binoculars and occasional use of a flashlight at night. Types and intensities of stimuli during each monitoring period were not controlled.

Here is the first sentence of the abstract as it appears in the *Journal of Mammalogy:*

Heart rates of five unrestrained white-tailed deer fawns (*Odocoileus virginianus*) were telemetered during prone behavioral responses ("freezing") to a variety of alarm stimuli.

Notice that the author has managed to include the topic, purpose, scope, number of subjects, and methods all in one sentence. The rest of the abstract is equally good. It briefly and cogently summarizes the results of the experiment and offers an explanation of those results. Here is the abstract in its entirety:

Heart rates of five unrestrained white-tailed deer fawns (*Odocoileus virginianus*) were telemetered during prone behavioral responses ("freezing") to a variety of alarm stimuli. Decreases in heart rates, termed alarm bradycardia episodes, were significant and averaged 38% (range 11–68) below pre-bradycardia heart rates measured during lying-resting behavior in the unalarmed fawn.

Episodes of alarm bradycardia (lasting from 5 to 111 s) were most frequent during the first week of life, when concealment discipline was strongest, but decreased rapidly with growth. Heavier fawns may be further along the continuum of neuromuscular development and associated capacities for "fight and flight" responses, and are less likely to react to disturbances with prone responses and associated bradycardia than lighter fawns of similar age.[5]

Though this abstract contains some technical terminology that might be unknown to certain readers, it keeps such terminology to a minimum and briefly defines terms that it must use.

Preserve the Emphasis of the Original. The "Alarm Bradycardia" abstract says very little about the procedures of the experiment. Apparently, they are well established and thus require little more than a mention. If the purpose of the article had been to announce a new or improved procedure, naturally the abstract would have had to put the emphasis there. The following abstract of an article entitled "Thin Layer Chromatography Method for Analysis and Chemical Confirmation of Sterigmatocystin in Cheese" provides an example:

<table>
<tr>
<td>The purpose of the report is announced.</td>
<td rowspan="4">A semi-quantitative method is described for the analysis of sterigmatocystin in cheese. The method is based on extraction of cheese with MeOH-4% KCL (9 + 1), followed by Florisil and polyamide column cleanup and 2-dimensional thin layer chromatography (TLC). Visualization of sterigmatocystin on the TLC plates is enhanced by an $AlCl_3$ spray reagent. The identity of sterigmatocystin is confirmed by a 2-dimensional TLC test based on reaction with trifluoroacetic acid (TFA) on the plate after first development. The reaction product formed is visualized by spraying with $AlCl_3$. The method allows detection and confirmation of sterigmatocystin in cheese at concentrations as low as 5 μg/kg. The method has been applied to cheese samples ripening in warehouses and naturally molded with *Aspergillus versicolor*.[6]</td>
</tr>
<tr>
<td>The complex procedure is summarized.</td>
</tr>
<tr>
<td>The significance of the new procedure is mentioned.</td>
</tr>
<tr>
<td>An application of the procedure substantiates its significance.</td>
</tr>
</table>

This abstract is, admittedly, quite difficult for the nonspecialist, filled as it is with chemical formulas and references to sophisticated equipment. Given the subject matter, however, the authors could not avoid the dense jargon without turning their abstract into an encyclopedia. What is remarkable is that, despite its terminology, this abstract is quite useful. Its opening and closing sentences—the ones that announce the purpose, significance, and application of the new method—are easily within most readers' grasp; only the technical information is couched in technical terms.

Do Not Judge the Document You Are Abstracting. Because an abstract is a miniaturized version of its parent document, it should not contain *any* information that isn't also in the document. One implication of this is that the abstract should not evaluate anything in the document. Keep your abstract absolutely neutral.

Write Complete, Grammatical Sentences. Although abstracts must be concise, they must also be readable. Don't try to save words by writing sentence fragments or by omitting articles or prepositions ("the," "an," "a," "to," "for," "with," and so on). The information in the abstract is bound to be dense and difficult; compressing the sentence structure may make the abstract impenetrable.

 # Writing Executive Summaries

Right after the abstract in many reports comes the executive summary, another brief overview of the report as a whole (see Fig. 14.1). Like abstracts, executive summaries are often detached from their reports and circulated separately, so they have to be written to stand alone. They have the same three purposes as abstracts, too—to orient the report reader, to help the potential reader decide whether or not to move on to the full report, and to replace the full report for readers who need only the summary.

There are two key differences. First, executive summaries can run a bit longer than abstracts, up to several pages instead of a mere 150 words. And second, executive summaries are written for administrators, especially top management, not for technical people. For almost every report there exists an executive who is high enough in the hierarchy that he or she doesn't need to know the details, but not so high that he or she doesn't need to know the main findings and recommendations.

The characteristics of an executive summary are deducible from the characteristics of its principal readers, busy nontechnical administrators who want to know how your report will affect the organization. Here they are:

1. Focus on your recommendations, perhaps showcasing them with a subtitle and a numbered list (even in a two-page summary). Abstract readers may consider your findings the core of your report, but for administrators the main recommendations are far more important.
2. Explain the purpose and scope of the report cogently but briefly, usually in the first paragraph. Describe your methods in no more than a sentence or two, and describe your key findings in no more than a paragraph or two.
3. Always use an informative rather than a descriptive format.
4. Avoid technical information as much as possible, and avoid technical vocabulary completely. If you have to use a concept or term your lay audience may not understand, define it and explain its relevance.
5. Write the executive summary after you have finished drafting the body of your report so that you are sure it emphasizes the same recommendations as the report itself.

summary

This plan describes what is known about the resources in the Key Largo Coral Reef Marine Sanctuary and summarizes the impacts humans have on the resources. It outlines the research needed to answer specific management questions and describes how the resources are being protected through education and enforcement of regulations.

Coral reefs are the most productive and diverse of all the natural marine communities and the reef areas off Key Largo are particularly luxuriant. The Sanctuary contains four distinct biological zones:

- Coral reef areas consisting of reef banks at the seaward edge of the shelf margin and numerous patch reefs on the landward side;

- Hardground areas veneered with hard and soft corals;

- Marine grass areas, which serve as nursery areas for many important organisms;

- Sand areas interspersed with patches of grass or occurring as haloes around patch reefs.

Populations of spiny lobster and stone crabs, which are commercially important, and recreational species such as grouper, snapper, dolphin, and pompano also are found within the Sanctuary. The area contains many structures and objects such as lighthouses, sunken ships, and associated artifacts, and a bronze statue of Christ of the Deep, significant for their cultural value.

South Florida is one of the fastest growing urban areas in the country. It is unclear what effect dredging and filling, channelization, municipal and industrial discharges, and air pollution associated with this growth have had on the marine environment off Key Largo. Nevertheless, the impact of 400,000 visitors a year is visible: broken and overturned corals, discarded beverage containers and fishing line, and scarred patches on grassy areas all attest to improper operation of boats and carelessness on the part of some visitors.

Overall management of the Sanctuary is the responsibility of the NOAA Office of Coastal Zone Management and day-to-day management at the site has been delegated to the Florida Department of Natural Resources.

Protection of the resources cannot be accomplished without effective surveillance of the area and enforcement of the regulations. These functions are accomplished by the Florida Department of Natural Resources and the U.S. Coast Guard. Cases involving violation of the regulations are handled by the NOAA Office of General Counsel through civil proceedings.

Scientific data on the Sanctuary so far has been restricted to basic hydrographic, environmental, and biological studies. Little research to date has been designed to come to grips with specific management problems or to address any long-term environmental questions. During the next 5 years, therefore, NOAA will give the highest priority to conducting field studies, carrying out laboratory projects, and monitoring biological and sociological aspects of the Sanctuary. Such information will increase the understanding of the resources and assess the impact of human pressures on the Sanctuary environment.

Included will be a biological inventory in 1979 and 1980 that will provide baseline data on the distribution, diversity, and abundance of reef organisms. By comparing these data with the results of inventories to be carried out in 1985 and 1990, managers will be able to determine if conditions in the Sanctuary are getting better or worse, and will be able to restructure their management plans if conditions warrant.

In addition, two other crucial studies are scheduled for the Sanctuary: a reef health assessment (also in 1979 and 1980) to estimate the percentage of live coral versus dead coral cover on the reef and document the extent of coral disease and anchor damage; and a water quality assessment (primarily in 1980) to determine if such factors as organic pollutants, nutrients, temperature, salinity, and turbidity are conducive to coral proliferation. In both instances, the information will provide baseline data to which future studies can be compared.

FIGURE 14.1 An Executive Summary.

In 1979, NOAA, in cooperation with the Harbor Branch Foundation, will carry out a deep-water survey of the Sanctuary. Approximately half of the Sanctuary lies in waters between 100 and 300 feet (30-90 m). Because previous research in the area has been limited to waters no deeper than 100 feet (30 m), NOAA considers it important to fill this significant gap. The agency will examine in detail, using side scan sonar, a magnetometer, and submarines, a deep water reef discovered in 1973. The survey will gather information on the location of this and other deepwater reefs, their associated organisms, and archaeological resources.

Finally, a series of special studies in 1981, 1984, 1987 will be carried out. These will deal with specific rather than general aspects of the Sanctuary and may include a lobster assessment, coral disease studies, anchor damage studies, cultural research and development, and other special items of interest derived from the results of baseline studies.

To supplement this research, NOAA, in cooperation with the Florida Department of Natural Resources, will monitor, on a monthly basis, permanent stations set up at environmentally different locations. The monitoring will include observation of coral diseases, mortality, growth, and recruitment; measurement of sedimentation, temperature, salinity, and turbidity; assessment of anchor damage, diver damage, or other user-related impacts; photographic documentation of representative coral colonies; and life history observations of other invertebrates in the Sanctuary.

Review Questions

1. Why do most reports and articles begin with an abstract?
2. Why is it particularly difficult to balance completeness with understandability in an abstract?
3. What are the differences between an informative abstract and a descriptive abstract?
4. Where in a technical document are you most likely to find points that deserve to be included in the abstract?
5. What information does *not* belong in an abstract?
6. What information *not* in the original document can be included in the abstract?
7. What are the differences between an abstract and an executive summary?

Assignments

1. Go to the library with a classmate. Find an article in your classmate's field, and have him or her find one in yours. Photocopy both articles; then cut out the abstracts and exchange articles. Write a descriptive abstract and then an informative abstract of the article. Turn in the original abstract of your classmate's article when you hand in your own abstracts.

2. Write an abstract and an executive summary for a technical report you have written for another course.

3. Write an executive summary of the New York Bight recommendation report excerpted on pp. 379, 382–383.

4. Turn the executive summary presented on pp. 322–323 into an informative abstract.

Notes

1. David Johnston, "Volcanic Contributions of Chlorine to the Stratosphere: More Significant to Ozone than Previously Estimated?" *Science* 209 (25 July 1980), p. 491.
2. P. W. Krey et al., "Atmospheric Burnup of the Cosmos-954 Reactor," *Science* 205 (10 August 1979), p. 583.
3. Ibid., p. 584.
4. Nadine Jacobsen, "Alarm Bradycardia in White-Tailed Deer Fawns (*Odocoileus virginianus*)," *Journal of Mammalogy* 60:2 (1979), p. 343. Reprinted by permission.
5. Ibid.
6. Hans P. Van Egmond et al., "Thin Layer Chromatography Method for Analysis and Chemical Confirmation of Sterigmatocystin in Cheese," *Journal of the Association of Analytical Chemists* 63:1 (1980), p. 110.

15 Proposals

> **Discussion**
> > **Projected Results**
> > **Limitations**
> > **Evaluation Procedures**
> > **Benefits**

Proposals offer to supply a product or provide a service to an interested client at an equitable price. Research proposals ask for funds to conduct an investigation: to find a cure for cancer, to analyze the causes of declining SAT scores, to learn what is luring American suburbanites back to the cities. Research and development proposals go one step further, asking also for the money to design and construct a prototype. Many engineering proposals fall into this category. Supersonic aircraft, fluoride toothpaste, electronic calculators, and "burpless" cucumbers all began with research and development projects. Sales proposals seek to market a product that has already been developed and tested. When a county contracts for a bridge, it isn't asking the competing firms to reinvent the bridge, only to design one that meets the need at the lowest possible cost. Proposals run the gamut from informal verbal suggestions to elaborate technical documents of 500 pages or more (not counting appendixes). All of them, however, seek approval for the project they describe on the terms they suggest.

 # Goal and Audience

All proposals have the same general goal: to get themselves accepted. They all have, in general, the same audience: people with the power to authorize work and—if necessary—release funds. Proposals are by definition persuasive. They all say: "Choose me" . . . "Award me the contract" . . . "Solve your problems this way." Every proposal is—literally or figuratively—a sales document, organized most carefully so that it creates interest in a product or service and motivates a prospective buyer to select that product or service over competing ones.

To make the sale, a proposal writer must understand *exactly* what the buyer needs. A proposal that pinpoints the needs of its target audience and then addresses those needs conscientiously is very likely to be accepted. This means that knowing your general goal and your general audience is not much use. You must know precisely who will be reading your proposal, how many people will be reading it, which parts will be read by whom, what each reader wants, and whether or not everyone in the audience is looking for the same things. Successful proposal writing requires extensive and detailed audience analysis.

Knowing Your Readers

In Chapter 1, we advised you to talk to a prospective reader whenever possible. When you write proposals, you can do this fairly easily. Your boss will want to review any proposal for new equipment or a new procedure before you submit it. And you will want to discuss such a proposal informally with your colleagues and your supervisors before committing it to writing and sending it up the corporate hierarchy.

For competitive proposals, the practice is much the same, though more elaborate and expensive. Companies spending half a million dollars on a proposal never work in the dark. They assign staff to talk directly with the customer so they can find out in detail exactly what is expected. They want to know the unwritten specifications as well as the written ones. They want to know if the specs are at all negotiable. They want to know whether quality is more important to the contractor than cost or vice versa. Extensive preliminary discussions always precede the writing of a competitive proposal. Sometimes oral presentations or informal written proposals (such as an exchange of letters) are used to test the waters before the company commits large sums of money to the formal paperwork. Always, the people drafting the proposal make it a point to know the people who will be reading it.

Competitive proposals are nearly always *solicited*—that is, they are written in response to a request from a potential sponsor. Such requests are of two sorts: the *request for proposal* (RFP) and the *invitation to bid*. The principal difference between these two is that the invitation to bid is attached to a highly detailed and inflexible set of technical specifications. The issuing agency announces that it wants, for instance, a cargo plane with a specified carrying capacity, a specified number of engines, a specified weight, and so forth. (The federal government publishes such announcements in official organs like the *Commerce Business Daily*, state and local governments often advertise them in the newspaper, and private contractors circulate announcements among subcontractors by mail.) Bidders, or vendors, then propose to meet those precise specifications for a stated price. Because the specifications themselves are not amendable, the cost is the only real variable. Therefore, when you write a proposal to manufacture the plane (it will be a sales proposal), your principal goal will be to justify the price you quote. Your readers will be looking for evidence that you can do the job well, on time, and within budget; they will want to see a convincing relationship between the production details you give and the cost you envision.

Requests for proposals (sometimes called *requests for quotes*) are a great deal looser. They specify a series of performance criteria, then ask the vendors to design equipment or programs or systems to meet those criteria in whatever way they think will work. Requests for proposals are soliciting ideas, not just evidence of production capability. They may or may not contain guidelines. Sometimes they do no more than present a problem and invite proposed solutions. That is why responses to RFPs are the sorts of proposals scientists, engineers, and technicians are obliged to write themselves, usually after talking with their prospective readers.

To the extent that guidelines or specifications are available, you should struggle to make your proposal conform precisely. Among the complaints most frequently voiced by contractors and granting agencies is cavalier disregard of the published specs. If you propose a substitution (the technical term for that is "taking exception"), you must defend it with a convincing rationale; but even so, you will probably be passed over in favor of someone who met the specifications exactly.

Analyze every RFP carefully before you decide to draft a proposal. You want to be sure that it is worth your time and expense to apply for the job. Ask yourself if the project will be profitable. Weigh the cost of preparing the proposal against the probability of success. Evaluate your organization's distinctive competencies. Analyze the competition to see who has the advantage. Consider your knowledge of the client, the scope of the project and the demands it will place on your organization, and the possibility of future work with this client. You may even want to take a loss on this project if you can recover that loss in the next contract. Whether you are acting on your own or as a representative of your organization, you must husband your resources carefully. Decide which contracts you have the best chance of getting, and dedicate your energies to winning those.

Even after they have been submitted, proposals may go through several versions before they reach a final form. Technologies change; prices change. Customers may not decide exactly what they want until after they have read the proposal. They may decide to accept some parts of a proposal and reject others. They may like certain parts of several proposals but not find any one of them totally acceptable. Vendors whose initial proposals seem promising are often asked to submit a revised proposal in response to a more detailed and more clearly focused RFP. Vendors, too, may ask for clarifications of the RFP, and their "queries" may cause the customer to revise the RFP and reissue it. Sometimes the customer may hold a "bidder's conference," to which vendors are invited to go and ask questions about the project. If you are invited to one, go—and have good questions prepared. You want the customer's representatives to notice your articulateness and your interest. And by all means go if you are invited to a meeting held by the customer to clarify its RFP for interested parties. If you do not attend, your organization may not be allowed to submit a proposal.

A proposal writer may need to negotiate directly with the decision-makers at a contracting company or agency. Exchanges of correspondence, phone conversations, face-to-face meetings, and a presentation of the proposal to a group of decision-makers may all take place before the proposal is finally hammered into shape and the customer can issue a purchase order. Few proposals, formal or informal, are approved without an oral presentation of the information contained in the written document. (Chapter 18 gives you the information you'll need to be successful at presenting proposals orally.)

Determining Your Readers' Goals

When you cannot negotiate with a real reader, go back to the *Audience Profile*. (See p. 15.) You may not have filled it out in detail for the mechanism description, for the analysis, or even for the definition, but now is the time to go through the long, involved process of answering every question. Naturally, your readers' degree of technical know-how will be of paramount importance, as will their needs. But in the case of proposals, questions about the audience's professional status suddenly become more relevant than they were in the case of process descriptions or definitions. A proposal addressed to a superior must be modest and slightly deferential; the identical proposal can be put assertively, even aggressively, to an audience of peers. Likewise, questions about probable audience attitudes must be answered as concretely as possible. Most readers approach technical description, specifications, and classifications with comparatively neutral attitudes—at worst, with bored indifference. This is not the case with proposals or, for that matter, with any writing intended to be persuasive. Your readers may be sympathetic or skeptical or bemused, but they are not likely to be uninterested. Estimate their attitudes as precisely as you can, and if you have an opportunity to talk with real readers, note your impressions of their attitudes on the profile. Continue to take readings throughout the proposal-writing process to see if those attitudes change. With proposals, the cost of failing to understand your audience is very high. If your employer loses money on a badly written proposal, you won't be asked to write another.

Once you have completed an exhaustive audience profile, draft a list of the probable goals of the audience you have just characterized. The Department of Health and Human Services, for example, might fund studies attempting to verify damaging effects of nicotine on the circulatory system. The tobacco industry is unlikely to sponsor such research, but it does subsidize the development of low-nicotine cigarettes. Timing is an important consideration, too. In theory, research goals should be valid or invalid regardless of their currency. But research into alternative energy, for example, received little funding until fuel oil became expensive in the 1970s. If the proposal is solicited, you will be able to infer many goals—both overt and implicit—from the wording of the request for proposal.

One problem with defining your audience's goals will arise the moment you discover that your proposal has a multiple audience, as is often the case. Everything we have said about mixed strategies for multiple audiences holds true for proposals as well as for reports. At the very least, you can anticipate that a proposal will be read by managers and by the technical staff to whom it pertains, and these two groups have very different needs. Technical experts are likely to be interested in the intricate details of your proposed procedure, for they will be asked to assess the technical merit of your planned work, to evaluate its soundness, and to recommend acceptance or rejection of the proposal. Most managers do not have the specialized background necessary to understand the procedural technicalities; though they may have begun their careers as biochemists or statistical analysts, they now spend their days struggling with budgets and personnel evaluations rather than with technical information. They will, however, be intensely interested in your rationale for the project and in your budget. If they decide to fund it, they'll have to accept the budget and provide their own superiors with a convincing rationale.

You cannot expect to build enough technical detail into your proposal to satisfy the experts without hamstringing your nontechnical readers. What this means, practically speaking, is that you cannot be content simply to profile the audience for the entire proposal. You must be prepared to analyze the audience for each separate section as well. In the rest of this chapter, we will discuss the major sections of a proposal and how to organize them. As we take up each one in turn, we will talk about who is likely to be reading it and how you can best tailor your organizational strategy and style to suit that particular group of readers. First, though, we would like to make some general comments about organizing the proposal as a whole.

▶ Organizing Proposals

Solicited proposals must be organized according to the specifications provided. Usually, decision-makers supply proposal writers with specifications not only for what's to be in the proposal but also for the proposal itself. Those specifications may be as skimpy as a sentence or two advising proposal writers to stay under five pages and to mention the relationship of the proposed project to the decision-maker's stated goals. Or they may run to pages and pages of forms.

All proposals must conform exactly to published specifications. Within the specifications, however, original and graceful design is a plus. To be persuasive, proposals must be meticulously engineered. Like it or not, your ability to design and draft a proposal will be the yardstick your readers will use to determine whether or not you can carry out your project successfully. You want them to come away from your writing impressed with its careful workmanship. After all, the only evidence they have of your work is the evidence they are holding, and the last thing you can afford is to appear slipshod or slovenly.

Making your readers feel good about your work is essentially a *psychological* goal. It requires an organization that is emotionally satisfying; in fact, it implies four subgoals. You must:

1. Get your readers' attention.
2. Create a need for your product or service.
3. Show how your product or service meets that need.
4. Motivate action on the part of your readers.

The structure of all good proposals is designed to accomplish each of these goals in turn. What that means is that the components of a proposal should be arranged in an interlocking sequence that arouses neutral or unconcerned readers gradually out of their torpor and, in the end, enlists their active support. Theoretically, the sequence looks something like Figure 15.1.

This outline is only a theoretical model, not a pattern or a blueprint. All proposals may have the same *purpose*, but that does not mean they have identical outlines. In the absence of explicit directions, you may include

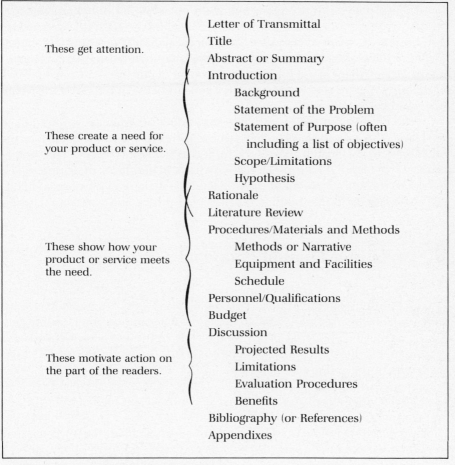

These get attention.
- Letter of Transmittal
- Title
- Abstract or Summary
- Introduction

These create a need for your product or service.
-     Background
-     Statement of the Problem
-     Statement of Purpose (often including a list of objectives)
-     Scope/Limitations
-     Hypothesis
- Rationale
- Literature Review

These show how your product or service meets the need.
- Procedures/Materials and Methods
-     Methods or Narrative
-     Equipment and Facilities
-     Schedule
- Personnel/Qualifications
- Budget

These motivate action on the part of the readers.
- Discussion
-     Projected Results
-     Limitations
-     Evaluation Procedures
-     Benefits
- Bibliography (or References)
- Appendixes

FIGURE 15.1   Model Proposal Outline.

whichever sections you choose from the theoretical model (you are under no obligation to use them all), and you may arrange them in *any* order that creates and holds interest. Mix-and-match versions of these components are fine— often, they are necessary—and if you can create an original and elegant organizational pattern for your proposal, it will work more to your advantage than following the model exactly. It will show that you know not only how to copy but also how to make decisions about design.

Formal proposals require all or nearly all the components listed in the outline. But when you write an informal proposal, you should begin by going down the list of components and asking yourself whether or not you need each one. If you decide you do need a component, then decide where it fits in the persuasive model, and place it there in your draft. Most informal proposals, especially within organizations, do not need all the components. If you are writing a proposal to your division director to suggest that the division purchase its own microcomputer, for example, you will need only a letter or memo of transmittal, a title, an introduction that includes background information, a statement of the problem, a statement of purpose, and a justification or rationale that focuses on the potential benefits of the purchase (cost savings, increased efficiency, and so on).

The proposal in Figure 15.2 (pp. 333–335), a preliminary proposal for a final project in a senior-level biology course, includes an introduction, a methodology section, a qualifications section, and a discussion section. The sample proposal in Figure 15.3 (pp. 336–337), written in the form of an interoffice memo, contains a statement of the problem, a statement of purpose, a rationale, and a discussion that addresses projected results and benefits.

▶ Parts of a Formal Proposal

Formal proposals are not begun at the beginning and written in one long smooth continuous flow of ideas. Like all systems, they are assembled from components *after* the individual components have all been built and tested. If each of the components of a proposal has been skillfully designed and constructed to meet its particular subgoal and if the joints all dovetail, then the whole of the proposal will be greater than the sum of its parts. For that to happen, two criteria must be met. First, each section must be entirely self-contained so that it remains comprehensible even when read out of order, as formal proposals often are. Second, if the proposal *is* read straight through, it must unfold sequentially from beginning to end as if it had been written that way. In this section, we will discuss one by one each of the components commonly found in formal proposals: their goals, their audiences, their characteristic organization, and their customary style. We will discuss them in the order in which they customarily appear in a proposal package.

Letter of Transmittal

We discussed the basic properties of letters of transmittal in Chapter 13; what we want to stress here is that the letter of transmittal for a proposal is anything but perfunctory. It provides you with your one opportunity to say how you feel about your project. Feelings do not belong in a proposal proper; they are irrelevant to the issue of whether or not you can do what you propose. Contractors want to see dispassionate wording and hard evidence in a proposal, not emotional appeals. But the letter of transmittal is an appropriate place to express your enthusiasm for the project. Normally, you will use a three-paragraph structure designed to achieve—in miniature—the four related psychological goals of the proposal.

A Proposal to Compare the Feeding Efficiency of
Laughing Gulls with that of Great Black-Backed Gulls
In New Jersey Salt Marshes

Introduction

The great black-backed gull (Larus marinus) is a large
coastal bird which, until 1970, was rarely seen in New Jersey.
Although native to Canada, great black-backed gulls have steadily
been expanding their range into New Jersey as a result of an
unlimited supply of food from nearby garbage dumps. Most of the
garbage dumps in New Jersey are located in salt marshes, the great
black-backed gulls' natural habitat.

The salt marshes of New Jersey are also the home of the much
smaller laughing gulls (Larus atricilla). Laughing gulls are
native to New Jersey and have been declining in number as a result
of competition with an increasing population of great black-backed
gulls. Both the laughing gulls and the great black-backed gulls
use the salt marshes as a place to feed and breed. Since the two
species of gulls occupy the same territory, or niche, one of the
species populations will decline (competitive exclusion
principle).*

The great black-backed gulls have been invading and taking
over the breeding areas that were formerly occupied by the
laughing gulls. It is generally accepted that this usurping of

*Competitive exclusion principle--No two species can occupy
the same niche in a given environment. Either one must become
extinct or one or both must, through natural selection, diverge
into different niches (Hardin, 1960).

FIGURE 15.2   A Preliminary Research Proposal.

laughing gull territory by the great black-backed gulls is due
solely to the aggressiveness and larger size of the great
black-backed gulls. However, I feel that this invasion is
facilitated, at least in part, by the ability of great
black-backed gulls to feed more efficiently than laughing gulls on
the food available at garbage dumps.

I propose a study to determine if the great black-backed
gulls are more efficient than the laughing gulls in feeding.

Methods

The great black-backed gulls and laughing gulls will be
observed at the Edgeboro Sanitary Landfill in East Brunswick, New
Jersey. Data will be collected in the mornings, when the gulls
are feeding. The following ratios will be recorded:

(1)   Number of food items eaten/number of steps taken -
        for individual gulls
(2)   Number of food items eaten/number of attempts to feed
        - for individual gulls.

The data for each gull will be taken at 60-second intervals
for standardization. The first ratio will indicate how fast (or
efficient) the gulls are at finding food. Since the gulls often
dig for food, the second ratio will illustrate their success rate
in finding food. A comparison of the percentages of gulls that
are feeding versus not feeding will also be used as an indicator
of the total amount of time spent procuring food.

-2-

Qualifications

I have been observing birds on my own for many years. As a student, I have taken courses in ornithology, field identification of birds, wildlife ecology, and animal behavior. I also worked on a project comparing the feeding efficiency of herring gulls with that of ring-billed gulls from September to December of 1979.

Discussion

The practice of filling in salt marshes with mountains of garbage has created problems for both humans and wildlife. Although this practice has favored a few species of animals, it is putting stress on many other species and is increasing their risk of extinction. This is especially true for those animals indigenous to salt marshes. Since humans often create an imbalance in natural systems through environmental alterations such as filling in salt marshes, I feel we have an obligation to study the effects of these alterations. Before making recommendations on how to alleviate problems we have created, we need to acquire as much information as possible on the biology and behavior of animals and their adaptability to habitat changes. This study will help fulfill a part of that obligation.

Literature Cited

Hardin, G., 1960. The Competitive Exclusion Principle. Science 131:1292-1297.

-3-

INTEROFFICE MEMO

October 25, 1984

```
     To:  Dr. J. Doe, Vice-President
   From:  Gerard Gilliam, Plant Supervisor
Subject:  Proposal to improve sterilization by employing
          microwave radiation during the manufacturing
          process
```

Based on our conversation of August 10, 1984, I have evaluated the efficiency of production sterilization procedures. In this preliminary proposal, I will summarize the results of that evaluation.

A more suitable method of sterilizing our products during the manufacturing process is needed. At present, our products have to be sterilized by two different methods. If a single method suitable for all products could be used, we would greatly increase our efficiency and reduce our overhead.

Four sterilization techniques are currently in wide commercial use: (1) steam; (2) gas (ethylene oxide); (3) dry heat; and (4) radiation (Co^{60}). As you know, we now sterilize our products with either steam or dry heat, and both methods have drawbacks. In fact, these methods limit production so severely that if we continue to use them, we will not be able to keep up with future demand.

FIGURE 15.3   A Proposal in Memo Form.

Of the four conventional methods, radiation sterilization is the
most promising from an economic standpoint. It is sufficient for a
wide variety of products because it does not alter the product or
damage the package in any way.

Radiation from a Co^{60} source is quite expensive, however, and to
install and maintain a unit would not suit our purposes. On the
other hand, radiation from a microwave source seems much more
economical and looks as though it could be adapted to our
manufacturing process easily. Installing a microwave unit would
permit all of our products to be sterilized by a single method and
would therefore reduce the time spent in processing. Of course,
safety has to be a major concern whenever radiation is used, but
with recent developments in microwave containment, a hazard-free
environment can be easily maintained.

I have searched the literature to determine if microwave radiation
has been used by other industries as a sterilization technique, but
I have found nothing on microwaves used for this purpose.
Therefore, I am requesting that you authorize me to prepare a
research report to determine:

1. If sterilization by microwave radiation is possible.
2. If so, how it can be applied to our products.
3. The cost of initial investment and estimated return on
    investments.

The proposed report could be completed by December 14, 1984.

I would appreciate your prompt attention to this matter.

In the first paragraph you should announce your title and link your work with the request for proposal to which it responds:

> The accompanying proposal, entitled "The Effects of Crowding in Residential Environments," is being submitted for consideration under your Mental Health Small Grants program for 1985.

(It is customary to use the passive—"is being submitted"—in this first paragraph rather than to say "I am submitting the enclosed proposal.") If the proposal is unsolicited, you will have to create your own link between your work and the mission or mandate of the organization to which you are applying. Use an opening like this one:

> Because the Rockefeller Foundation has in the past funded studies of the effects of energy scarcity on democratic ideals and notions of social justice, I thought that you and your staff might be interested in the enclosed proposal to measure the socioeconomic impact of two pilot programs that provide energy assistance to the elderly and the disabled: the Fuel Stamp Program and the Grant of Restricted Voucher.

("I" is somewhat more appropriate in this sort of direct appeal because you chose this organization deliberately and submitted the proposal on your own initiative.) Keep this first paragraph to one sentence (two at the most), and be sure to get in both a version of your title and a reference to what you know to be the readers' needs. If your title is carefully worded and appropriate to those needs, this first paragraph will get attention and will create a desire among your readers for more details.

The second paragraph, the longest of the three, provides evidence that your project is a sound one that may be worth funding. Here you give select details about your procedures, your target populations, your angle of attack—anything that stresses the value of your project:

> This research is designed to explore the promising field of poetry therapy, a form of art therapy that has already proved successful in the treatment of depressed women. We propose to conduct a poetry therapy workshop with depressed nursing home patients, a group that receives insufficient attention in the area of treatment improvement even though depression is alarmingly common among the elderly. Because the capacity and desire to learn are maintained by most aged people, their need to feel useful and capable—if not met—often leads to profound depression. A humanistic therapy can meet this need and so address the cause of the depression directly.

Because this paragraph seeks to establish a need and to show how your proposed project meets that need, it is the appropriate place to put your "sales" message, and any comments you wish to make about the project. You can predict the likelihood of your success, you can mention your price or your estimate of how soon you expect to get results (if these are favorable to you), or you can hypothesize about what those results might be, provided you keep it general and avoid looking as though you are compromising your objectivity. You may even wish to mention any limitations you anticipate; if you have plans to compensate for those limitations in advance, you appear both competent and farsighted.

If you end up having a great deal you want to say about your project in the letter, you would be wise to expand this central paragraph to two or even three. You don't want a central paragraph that is more than twice as long as the first and third, so subdivide as often as you need to in order to keep the letter looking attractive on the page. Do not exceed one page, however; there *is* a proposal too.

The third paragraph of the letter of transmittal seeks to motivate action by opening a dialogue with your sponsor. You want to end with something upbeat and friendly, but *do* try your best not to say: "If you have any questions, please feel free to contact me." You have no need to put your readers at ease; you are the supplicant here. Instead say: "If you wish to discuss the details of the proposal, please call me at (201) 555–1212." If you want it a bit more formal, say, "I will be happy to discuss the proposal with you in more detail. I can be reached [or "you can reach me"] at (201) 555–1212." One last appeal for funding is not necessarily out of order either, as long as it is dignified and as long as you haven't made a similar one in the second paragraph.

Title

The title of a proposal should be long enough to convey the precise topic and give some clues about approach, methods and procedures, or degree of originality. Somewhere in the range of ten to fifteen words is usually about right. As a rule, it will begin "A Proposal to ..." The more concrete you make the following verb (to construct, to measure, to evaluate), the more your readers will feel confident that they know exactly what they are getting. If you say something vague or evasive (like "A Proposal to Study ..." or "A Proposal to Explore ..."), you may lose both their interest and their trust. To ensure that you have a concrete verb, choose one that suggests either a clear final purpose (for example, "to construct") or a methodology ("to measure").

After the main verb, a good title usually provides the subject of the proposal in the form of a concrete noun phrase with several modifiers, for instance "The Ecological Impact of the Depletion of the Haddock Fishery." Here are some complete titles:

A Proposal to Test the Effectiveness of Poetry Therapy as an Ancillary Treatment for Depression in the Elderly

| A Proposal to Assess the Physiological Effects of Acute Experimental Crowding in Rats

| A Proposal to Determine the Effectiveness of Nonwoven Fiberglass Felt as Belting on a Rotary Vacuum Filter

| A Proposal to Compare the Feeding Efficiency of Laughing Gulls with That of Great Black-Backed Gulls in New Jersey Salt Marshes

Abstract

The only significant difference between report abstracts and proposal abstracts is that in a proposal abstract the verbs are future or conditional, not past tense:

| The proposed study will determine how social expectations affect the way people express hostility. . . . Results from this study will be helpful in assessing the effect of sex-role expectations on individuals' perceptions of the behavioral choices available to them.

| Successful intergeneric crosses between fungi may result in the production of a novel hybrid capable of performing a unique bioconversion. . . . Such a novel hybrid could be used in the production of ethanol for gasohol, in antibiotics production, or in alcoholic fermentations for industrial purposes.

Introduction

Proposals require strong introductions if they are to command attention. Virtually everyone reads the introduction to a proposal (provided it isn't too long) because it contains the statement of purpose. That may be all it contains; for instance, a solicited proposal sent in response to an RFP with detailed specifications need not even include a statement of the problem—the contractor knows what the problem is. However, most formal proposals contain substantial introductions. In unsolicited proposals especially, readers need to be told about the problem and given extensive background information about it before they can understand and appreciate your statement of purpose, your description of scope and limitations, and your tentative hypotheses. This isn't a section to skimp on; the rest of the proposal matters *only* if the introduction makes the case that what you want to do is worth doing.

The most important thing to remember when writing the introduction is that its readers will be using it *to make a decision*. They won't be trying to learn from it, and they won't have to implement it themselves. What they *will* have to do is defend its merits to others—their superiors, their auditors, the runners-up. They want to be impressed with what they read. They want it to contain information they were not already aware of, arguments they cannot fault or refute, and a hint of what's to come—a statement of purpose that implies a procedure, a background section that mentions the relevant literature, a tentative hypothesis that shows you can think intelligently about the material. The following introduction meets those needs.

This first paragraph begins with an implied statement of the problem: Do the benefits of arsenic trioxide outweigh the risks entailed in its use?	Notwithstanding the great commercial benefit of arsenic trioxide as a pesticide, arsenic compounds have recently become a target of public health investigations. Any accumulation of arsenic compounds in nature in a location where they may subsequently be released to the hydrosphere causes great concern. The arsenicals most toxic to mammals and aquatic species are trivalent arsenicals (arsenites) and organic (methylated) arsenicals.
The second paragraph provides a preliminary rationale by elaborating on the problem of risk to human health, then introducing a second problem: Scientists know very little about the biochemical conversion of arsenic compounds within bottom muds and sediments, including their release to overlying waters. These assertions are validated by references to previously published literature; thus, this paragraph also anticipates the literature review.	Bottom muds and sediments have the potential to release hazardous quantities of methylated arsenicals into overlying waters. However, the mechanism of such a release of organoarsenicals is relatively unknown for freshwater systems (2, 7, 13). Evidence of these methylated arsenicals within fresh waters suggests biosynthetic origins. Laboratory experiments have shown that sediment microorganisms can, in fact, biosynthesize toxic dimethyl arsine from inorganic arsenic (26). The biological production of alkyl arsines within bottom muds and sediments requires further study (23).

Next comes the proposal's statement of purpose, coupled with a statement of scope and limitations: freshwater systems, laboratory *and* field studies, New Jersey waters only. The introduction ends with a description of the project's anticipated results—data in these three areas.

We propose to conduct laboratory and field studies into the biochemistry of arsenic compounds in a New Jersey freshwater system. Our major concerns will be the conditions under which arsenic is released into the water phase, the concentrations that are achieved, and the speciation that occurs.

Rationale and Literature Review

In a short proposal, both the rationale and the literature review may be incorporated into the introduction, but in long proposals they are usually combined into a separate section. This section, also called "Survey of Literature" or "Review of Literature" or "Justification" or "Relevance," has a much more limited and hard-to-please audience than the introduction; it is read almost exclusively by other technical professionals. (Nontechnical readers are usually satisfied with the rationale that you provide in your introduction.) The literature review in a formal proposal has two principal goals: to justify your project and to establish your credibility.

Justifying Your Project. Nearly always, a proposal aims to fill a gap, either a gap in knowledge or a gap in services. Before your readers can begin to judge the effectiveness of the plan you wish to employ, they have to be convinced that a gap does, in fact, exist.

You convince them by marshaling relevant facts. In an informal proposal or a proposal that does not require research, a rationale based on nontechnical information may suffice. But where a substantial body of knowledge relevant to your project exists, only a survey of that knowledge can establish the need. Because scientific and technical knowledge is incremental, the existence of previously published research also provides certification that your topic is worthy of study. Before writing a literature review for a research project, you must first conduct an extensive search of all current literature related to the topic you wish to investigate. Then you must present the information you have gathered in a way that effectively demonstrates that the current literature does *not* contain the specific bits of information that your project is aimed at obtaining. You have a delicate mission here: You must describe the existing work thoroughly, yet maintain that there is a need for new work. You want your proposed project to sound new, but not *too* new. The projects most likely to win approval are those that apply a well-established methodology to a new subject or that fill a gap between two established pieces of information. A good literature review shows how what has already been done and what you propose to do will together solve problems and answer questions. To achieve this ambitious goal, you must point out what the literature *doesn't* do in the course of pointing out what it *does* do.

If your project is aimed at improving or instituting services, your approach will be different—less academic, more pragmatic. Use the rationale section to cite *precedents* for your project and to cite hard facts and other proof that, if your project is funded, you will be providing a service that is needed and not currently available. If you want to propose to your town that it institute a recycling program, you will have to demonstrate that other municipalities have done so successfully. If you want to propose that your company set up a flex-time program, you will have to show that the program has worked well in other companies. A proposal for a service without precedents, like a proposal for a research study without relevant forebears, is bound to look very risky to potential sponsors. To the extent that you are walking in others' footsteps, it pays to say so.

Establishing Your Credibility. You establish your credibility by showing your readers that you are thoroughly conversant with what has already been done. Prospective researchers must know what others before them have done if they are to put their own work in context. Project developers must know the details about similar projects—what they cost, where they succeeded, and where they developed problems. Your readers are fellow technical experts; you can expect them to be familiar with the literature themselves. You want to impress them with a knowledge that is both broad and deep, with an ability to summarize and paraphrase your sources accurately, and with a skill at synthesizing the information you take from your sources into a coherent argument in favor of your own project. You should not simply cite a source in a literature review; you should make clear the *reason* for citing it: It proves an assumption fundamental to your own project; it demonstrates a method that may usefully be applied in a new area; and so on. Mastery of both the content of your sources and their relevance to your own plan will establish your credibility.

Organizing a Literature Review. Following is an excerpt from the "Literature Review" section of the arsenical pollutants proposal:

> Epidemiological studies and statistical analyses have strongly related the arsenic content of drinking water to the incidence of skin cancer and other chronic disturbances (5, 10, 20). Inorganic arsenic, which is readily absorbed from the gastrointestinal tract, the lungs, and—to a lesser extent—the skin, becomes distributed throughout the body tissues and fluids with continued ingestion. During chronic exposure, trivalent ar-

senic accumulates mainly in bone, in muscle, in skin, and—to a lesser degree—in the liver and kidneys (14). The U.S. Environmental Protection Agency has set the drinking water standard at not more than 0.05 mg/l of arsenic (4), but prolonged exposure to even this low concentration of arsenic can be very damaging (22). Arsenic compounds have been found to be teratogenic (8, 11, 12); arsenic (especially $AsSO_4^{3-}$) interferes with phosphorus metabolism and uncouples oxidative phosphorylation (9, 17); arsenic has been found to alter the behavior patterns of fish (21). It has been suggested that arsenic interferes with normal DNA repair (14, 18), and in a report to the Occupational Health and Safety Administration, two chemical companies—Dow and Allied—noted an abnormally high rate of deaths due to cancer among workers exposed to arsenic and its inorganic salts (3).

This particular literature review is organized topically, with related studies grouped together. Topical organization is extremely economical because it permits several studies to be discussed jointly and thus saves space. A more leisurely method of organizing a literature review is to make it chronological. A chronological literature survey is, in essence, an annotated bibliography in paragraph form, which gives very abbreviated summaries of previous studies, mentioning the names of the researchers, their results, and their conclusions.

Since its introduction in 1965, the diameter distribution technique has been used extensively to model stand yields. The models differ primarily in the form of the pdf used to quantify the distribution. Early applications used the beta pdf (Clutter and Bennett, 1965; McGee and Della-Bianca, 1967). In 1973, Bailey introduced the Weibull pdf, which has a number of desirable characteristics for modeling diameter distributions. The Weibull is characterized by three parameters, a, b, and c, which have been estimated with maximum likelihood techniques, variable transformations, linear functions, and percentiles (Bailey and Dell, 1973). However, none of these methods accurately describes the relationship between the stand characteristics and the pdf parameters.

Hyink (1980) suggested a method of predicting the pdf parameters using the stand characteristics directly. This method, called parameter recovery, has also been used by Frazier (1981) and Matney and Sullivan (1982) to develop compatible growth and yield models for loblolly pine.

The chronological method of organization is particularly appropriate when your readers are already familiar with the literature. Your closest colleagues will expect you to know not only the salient issues in the field but also the names of the acknowledged experts. Providing a historical overview of the literature demonstrates that you can put it in perspective. You should be aware, though, that "putting it in perspective" means knowing what to omit as well as what to include. Literature reviews don't assess everything that has been written on the topic since Aristotle; they address the most recent work or classic work or work that is closely parallel. If experiments by prior researchers are seminal to your proposed study, or analogous but slightly different in focus, or similar but flawed, your literature review should say so. If you wish to single out the methodology of a particular study for criticism or for special praise, the chronological format is a good one to use.

No matter how you organize your literature review, you will want to end it by carving out a place for your own work in the history you have just outlined. The best strategy for doing this is to remind your readers of the results obtained by your predecessors that are closest in spirit to the results you hope to obtain, or of the gap in knowledge or services that needs to be filled, or of the separate trends that are to be merged in your work—whatever your contribution is to be. This summary of what remains to be done anticipates the procedures section and leads right into it.

Procedures

The procedures section is often the most worrisome to proposal writers, for it is apt to be scrutinized closely by technical experts. Actually, the procedures section (sometimes entitled "Materials and Methods") is just a detailed process description, usually written in the future tense. Whether yours is an experimental procedure, a statistical procedure, a clinical procedure, or a complex schedule of activities, the organizational strategies we outlined for you in Chapter 9, "Process Descriptions," are the right ones to use to describe it. All procedures sections contain a step-by-step narrative of the methods by which you plan to complete the project you are proposing; this is sometimes entitled "Narrative" or "Methods." In addition, procedures sections may contain either of two optional subsections—one devoted to equipment or equipment and facilities (called "Materials" if the other section was called "Methods") and one devoted to schedule.

Methods or Narrative. The overt goal of the methods subsection is to inform readers *how* you plan to arrive at a final product. But there is more to writing a good narrative than making the procedure itself clear. Your description of your proposed procedure has an implicit subgoal: to inspire confidence. Readers want to feel that you know what you are doing and can do it. Detailed working knowledge of the proposed methodology and effortless handling of the appropriate technical jargon both help to induce in readers a feeling of confidence in you. But what separates a mediocre narrative from a truly first-rate one is evidence that you know the procedure so well that you will be able to tinker with it creatively, to correct it if something goes wrong, to experiment with alternative steps (for instance, changing the growth medium to see if that changes your results). Reprinted here is part of the procedures section from the arsenical pollutants proposal.

Specific Objectives:

A. Characterize and identify arsenical compounds in the water, muds, and sediments.
   1. Identify and quantify arsenic species existing in the contaminated surface waters.
   2. Determine, utilizing extraction and leaching techniques, total arsenic content as well as "available" arsenic present in the muds and sediments.
   3. Characterize, whenever necessary, the chemical and physical nature of the muds and waters.
B. Investigate the mechanism(s) by which arsenical compounds are released, retained, or transported by sediments.

Laboratory Studies:

Simulated reservoirs (13) will be prepared using the arsenic-contaminated muds and waters of the Maurice River and the Blackwater Branch. The fate of the arsenical pollutants will be followed in these reservoirs under conditions of varying light, oxygen, and temperature. The effect of these parameters will be examined through arsenic release from muds to water (or the reverse), changes in arsenic species, and losses of arsenic due to volatilization within the system. In addition, both sterile and nonsterile conditions will be employed to differentiate microbial release of arsenic (to overlying waters) from strictly chemical release. Concurrent leaching experiments will be conducted on the various muds to establish "ease" and extent of "availability" of the arsenic present. Variables of pH, D.O., temperature, and ionic strength will be followed within the system over time. . . .

Sampling:

Grab samples of water and bottom muds will be collected on several occasions over a two-year period. Temperature, D.O., and pH will be monitored each time a sample is taken.

A. Water samples will be collected in polyethylene containers (5 gal., 1 gal., 0.5 gal.) which are to be washed, cleaned with dilute nitric acid, and rinsed thoroughly with deionized water. These grab samples will be collected directly from the streams.

B. Mud and sediment samples will be collected in wide-mouthed polyethylene bottles (1 gal.) and Mason jars (1 pt.). These grab samples will be dug with a spade from a depth of four inches.

C. Reservoir samples will be collected in glass carboys (5 gal.). Rubber stoppers will be used to seal the anaerobic jugs.

D. Instrumentation: D.O. Analyzer (Weston and Stack); °C thermometer (Fisher); pH meter (Beckman); Hach (DR-EL) kit to be used for water quality analysis.

Analysis:

Water samples will be analyzed within 24 hours of sampling hour. Mud samples are to be completely dried, pulverized (mortar and pestle), and sized through an 18″ mesh sieve. This dried, pulverized, and sized material will be used directly for leaching and extraction work.

Metal concentrations are to be determined by Flame Atomic Absorption Spectrophotometry using a Perkin Elmer 403 Spectrometer. Standard EPA protocol (4) will be followed for the arsenic determinations. The following two instruments will analyze for arsenic: (a) Beckman DB-GT Spectrophotometer employed for the silver diethyldithiodicarbamate (AgDDC) method (4); (b) Perkin Elmer HGA 2100 furnace on a Perkin Elmer 503 Spectrometer for the flameless atomic absorption procedure.

Summary of Analytical Methods:

All analytical procedures to be utilized in this work conclude with the AgDDC method for arsenic analysis. Eventually, therefore, all arsenic is ascertained spectrophotometrically as a silver-arsine complex. Organic arsenic is obtained by difference. Total arsenic less inorganic arsenic equals organic arsenic (13).

Like its introduction and literature review, the procedures section of this proposal is extremely concrete. Procedures can almost never be too concrete—or too exact; they can only be not concrete or exact enough. If you omit or gloss over any detail because you feel that it is too picayune or too obvious, your reader is more likely to attribute the omission to ignorance on your part than to any other cause. To be safe, be exhaustive. If going into exhaustive detail means that your procedures will be disproportionately long, simply banish the longest and most tortuous sections to an appendix. Copies of a sample survey, for instance, are normally appended at the back of a proposal even though the survey is at the heart of the procedure.

No detailed narrative is needed if your procedure is standard. Borrowing procedures is a common and thoroughly reputable scientific practice so long as you cite their originators (if a procedure is extremely well known, no citation is necessary). If you have added a new twist, cite the standard procedure; then describe only your addendum. If you are deliberately avoiding a standard procedure for some reason, say so, and justify your decision to experiment with an alternative.

Equipment. Equipment may be listed at the beginning of the procedures section (as with instructions), or it may be introduced at the point in the procedure at which it is brought into play. The arsenical pollutants proposal adopted the latter strategy—explaining a step, then indicating what instrumentation would be required for it. If you decide to list all your equipment before you begin outlining your procedures, you will probably want to title the section "Materials and Methods," so that the title reflects the organizing principle. If your equipment list is very long, you may want to put it in an appendix.

Schedule. It is very helpful to readers of procedures if you give them some idea of the time frame you envision for the procedure as a whole and for each step. Estimates of the requisite time for steps are easily interpolated into the procedures narrative itself. Also, if you wish, you may attach a schedule at the end of the procedures section or at the back of the proposal—often in the form of a flowchart showing which steps will take place simultaneously and which consecutively. The author of the arsenical pollutants proposal included such a flowchart. (See Fig. 15.4.) Professional proposals often include an elaborate graphic plan in which the procedure is diagrammed as a network of discrete but sequential tasks. Diagrams show the path from one task to another and permit the time and cost for each subunit of the whole procedure to be clearly displayed. Such graphic project schedules have all sorts of nicknames: critical path method (CPM), program evaluation and review techniques (PERT), critical path scheduling (CPS), critical path analysis (CPA), and so on. Another very common type of project schedule is a task-by-task breakdown

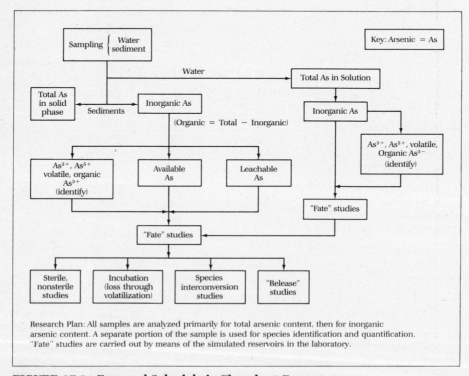

FIGURE 15.4 Proposal Schedule in Flowchart Format.

Steps	Months After Contract Award											
	1	2	3	4	5	6	7	8	9	10	11	12
Purchase materials for gross structure	S			C								
Hire and orient construction crew	S		C									
Contruct gross greenhouse structure				S				C				
Purchase materials for interior design						S			C			
Interior design development								S			C	
Purchase equipment for experiments								S				
Hire student workers								S	C			
Begin collecting data on interior environmental conditions									S			
Set up experiments: plant crops											S	
Submit final report												C

FIGURE 15.5   **Proposal Schedule in Time Line Format.**

of the procedure charted as a series of segmented time lines. (See Fig. 15.5.) This type of schedule stresses completion dates rather than path or flow. Many contractors require that a graphic project schedule accompany a proposal's narrative; if you supply one voluntarily, it will certainly be welcomed by your readers.

Personnel/Qualifications

It is customary in a proposal to present the credentials of your company, of yourself as project leader, and of any collaborators or consultants who will be joining you. Often the simplest way to do this is to append résumés or curricula vitae, but these may contain more information than your readers really need to make their decision. You may wish to write short summaries of the special qualifications you and your colleagues possess for undertaking this project and to group them into a section called "Personnel," inserted directly after the procedures section, placing the résumés in an appendix. Your summaries should include *only* those qualifications that are directly and persuasively relevant to the procedures (for instance, previous experience with the subject matter or the proposed experimental methodology)—that is why the two sections should be adjacent. However, you should name your proposed personnel *only* if they are part of the package you are trying to sell.

Otherwise, write a personnel section that states *specifications* for the staff you will need without naming names; you may wish to label this section "Staff Requirements."

Budget

Because you are seeking funds to carry out your proposed project (as opposed to merely seeking administrative authorization for it), your readers will expect you to account for how those funds will be used. Thus, as part of your proposal package you will need to submit a budget, a detailed breakdown of projected expenditures.

Practice varies as to where in the proposal the budget belongs. For in-house proposals involving only minimal sums, you might be able to work your budget into the justification or combine it with the materials section of the procedures ("Materials and Costs"). For complex in-house proposals and for all external proposals, the budget gets separate treatment, sometimes in a section after the procedures section, sometimes in an appendix, sometimes in an entirely separate document.

No matter where the budget is presented, the key to its success is *detail*. The executive decision-makers whose job it is to scrutinize budgets are interested in "bottom line" projections, of course, but they also need to see how those projections have been obtained so that they can determine whether the budget is reasonable and whether all expenses have been considered. To convince these readers, you will need to be exhaustive: Every nut and bolt, every hour of computer time, everything, in brief, that takes money must be mentioned.

Although budget formats vary, four categories are generally covered:

- Wages
- Equipment costs
- Operating expenses
- Overhead

Wages of project personnel are usually figured on an hourly basis, so you must determine how many hours each person will be working and then multiply by the average hourly wage for that position. If you know who the personnel will be, list them and their positions; if you don't, list the positions.

Equipment costs, sometimes called "capital expenditures," are usually listed by the make and model number of the equipment. If the proposed project doesn't require any major purchases (no linear accelerators, no fleet of hydrogen-powered cars), this section may be incorporated into the section on operating expenses. As its name implies, the section on operating expenses lists such items as project materials; office and laboratory supplies; travel, lodging, and food; telephones; and computer time.

Overhead covers equipment and facility rentals, utilities, and general clerical and secretarial support (but note that clerks or secretaries who will be working exclusively on the proposed project are usually listed under "wages" along with other personnel), as well as a host of other costs: fringe benefits, printing costs, and so on. Overhead is usually expressed as a percentage of the other three categories. (A simplified model budget is presented in Fig. 15.6.)

```
                              Budget

Wages

        Project director       ($00.00/hr)      000.00
        Field crew             ($00.00/hr)     0000.00
        Laboratory assistants  ($00.00/hr)     0000.00
        Programmer             ($00.00/hr)      000.00

        Total                                             0000.00

Equipment

        1st item (name and model)              00000.00
        2nd item (name and model) . . .        00000.00

        Total                                            00000.00

Operating Expenses

        Materials
          1st item                              000.00
          2nd item . . .                        000.00

        Laboratory supplies
          1st item                              000.00
          2nd item . . .                        000.00

        Travel, lodging, food
          1st item . . .                        000.00

        Total                                             0000.00

                              Subtotal                   00000.00

Overhead

        30% of subtotal                                   0000.00

                              Total                      00000.00
```

FIGURE 15.6   A Model Budget.

Budgets for complex and expensive projects are so complicated that they are usually drawn up by a team that includes lawyers and representatives of the firm's accounting department as well as technical staff. Even the budget for a relatively modest project will probably be so problematical that you will wish to seek the help of experts within your organization. And don't forget to check your files—you may be able to model your budget after the budgets of previously submitted proposals.

Discussion

Many proposals include a discussion section after the nuts and bolts of what is going to be done and who is going to be doing it. A discussion section furnishes you space in which to do three things: (1) recapitulate and extend your introduction, (2) address any issues that ought to be brought up in the proposal somewhere but don't fit comfortably into any of the more conventional sections, and (3) bring the proposal in for a smooth landing. The discussion section of a proposal may have any or all of four subsections—projected results, limitations, evaluation procedures, and benefits—which may be identified with headings or which may be fused together into a smoothly flowing series of paragraphs headed simply "Discussion."

Projected Results. The discussion section is the right place to project the results you anticipate (that is really just another way of advancing your hypothesis). The following example predicts results without committing to them:

> Some general trends in the solubility of various coals in aqueous sodium hypochlorite can be predicted from theory. Chemical reaction theory indicates that solubility should increase with time of reaction. Also, as volume of solution per weight of coal is increased, solubility should increase because there are more $-OCl$ molecules available to form carboxylic acids. Solubility predictions can also be made based on previous testing. Studies[4,12] indicate that solubility is greater at higher temperatures, at least in the 25–65°C range. Relative solubilities can be predicted from an analysis of the chemical composition of the various coals. Because aromaticity increases with fixed carbon percentage and because aromatic Sp2 carbons are not reduced by $NaOCl$, it follows that solubility increases as fixed carbon percentage decreases. It would appear, then, that the coals, in increasing order of solubility, are anthracite, bituminous, subbituminous, lignite, and peat.

Limitations. According to the principles of good persuasive organization, any negative information about the project belongs rightly in the discussion section, which comes late in the proposal but not at the very end. Minor limitations can be mentioned in passing in the introduction, but major limitations should be admitted *after* the procedures so that they will seem trivial when weighed against it. Any problems you foresee should be specified and your prospective solutions set forth. Describing what could go wrong is one way of displaying how well you know the procedure; a visible awareness of limitations and potential problems always works to your advantage.

When you point out a weak spot, do so for the express purpose of showing how you plan to get over it:

> We plan to follow the paraffin oil method for pollen treatment as described by M. G. Neuffer for the induction of genetic variability with corn. If, however, the paraffin oil method is unsuccessful for inducing EHS mutation of pea pollen or if we fail to obtain reasonable fertilization, we intend to repeat the procedure using an aqueous medium.

Anticipating problems and objections is a diversionary tactic aimed at displaying your strengths. Therefore, negative information about any feature of your proposal should be presented in a brisk and businesslike tone free of emotionally charged words. It should be subordinated in phrases or dependent clauses within complex sentences, embedded within paragraphs. And it should be adjacent—naturally—to strongly positive information.

Evaluation Procedures. An oddment that belongs in the discussion section for lack of a better home is your self-evaluation plan. Evaluation instruments, like sample surveys, are best placed in an appendix, but a short descriptive paragraph explaining how you intend to evaluate the effectiveness of your work is perfectly in order under "Discussion." The following is a typical example:

> Parents participating in the genetic counseling program will be asked to evaluate the genetic counseling they received and to describe their feelings toward the counseling program. A questionnaire will be used to determine (1) whether or not the counseling program helped them to understand and to face the reality of having an abnormal child; (2) what participants did and did not like about the counseling procedure; and (3) whether or not participants would return for counseling if they intended to have more children and why.

Benefits. The end of the discussion section is effectively the end of your proposal, so you want to end it powerfully. Nearly every proposal ends by looking to the future. Pointing out ways in which your work will supply badly needed information and thus benefit your profession is a common tactic.

> Data from the proposed tests into solubility of coals in aqueous sodium hypochlorite will do more than simply support or disprove theories about relative solubility; they will provide information on the specific amount

> soluble under specific conditions. These raw data will serve as a much-needed foundation on which further experimental work can be based. Further studies should include solubility as a function of particle size and of other variables affecting underground coal seams.

A more pragmatic approach is to discuss the specific benefits that might accrue (and to whom) if the project were funded and implemented.

> Thus, society would gain a new technology. More fuel could be produced from existing coal reserves because more coal could be recovered. There are a large number of coal seams that cannot be mined by present mining techniques. Development of a feasible technology for mining through underground combustion would prove of great significance to our energy-dependent world.

A feeling of satisfaction among readers is your primary goal here. Once you have achieved it, you can end the proposal proper with endnotes or with an alphabetical list of references if you have used reference numbers in the text. After "Notes" (or "References") you will affix your appendixes.

Review Questions

1. What is the goal of a proposal?
2. Why is it more of a disaster if a technical writer produces a bad proposal than if that same technical writer produces a bad report?
3. Why is audience analysis more important for proposals than it is for some other types of technical documents?
4. What differentiates an informal proposal from a formal proposal?
5. What are the four subgoals around which a proposal should be organized?
6. Who reads the introduction to a proposal? The literature review? The procedures? The discussion section?
7. What are the two principal goals of a literature review?
8. What kinds of information should you include in the discussion section of a proposal?

Assignments

1. Write a short proposal (four to six pages, including a cover letter) to the head of your department, in which you propose one of the following: (a) a new course, (b) a specialized program or curriculum not now available, or (c) the purchase of a new piece of laboratory equipment.

2. If you are writing a library research paper as part of this course, write a preliminary proposal for a research topic for the paper. Convince your instructor that the topic is a legitimate and worthy one.

3. Use your library's grant information resources to locate a research or project grant in your field. Send for application materials; then apply for the grant, following the specifications exactly.

16 Progress Reports

Soon most of you will be involved with on-the-job projects in your field that take months or even years to complete. And more often than not, as your project proceeds, you will be required to keep others—employers, administrators, clients, supervisors, department heads, and so on—abreast of your progress.

A progress report presents its readers with an account of the work done on a continuing project during a specific time period. A "continuing project" is one that has been started but not completed; thus, progress reports are those documents you will write between the time your proposal has been approved and the time you submit your final report. The "specific time period" covered by any single progress report will usually be set for you. You may be required to submit reports weekly, monthly, quarterly, or semiannually—or according to some other time scheme. If, for instance, the project involves discrete phases, you may be asked to report at the end of each phase. The type of report just mentioned is sometimes called a periodic report instead of a progress report. Other titles include status, interim, or quarterly (or weekly, annual, and so on) report.

▶ Goal and Audience

Determining who your readers are and what they expect from your report should not present any problems. Progress reports are written for the people who authorized you to do the project in the first place. Therefore, by the time you sit down to write, you already know a good deal about them. In fact, if you had to submit a written proposal for your project, you have already had experience writing to them. Undoubtedly, your readers are just as busy and just as worried about bringing the project in on time as they were when they gave you the go-ahead. One difference, however, is that they now know who you are; your main job is to *maintain* your credibility, not to establish it.

What these readers want from your report is information that will ease their job of executive decision-making. Specifically, they want to know whether the time and money invested in you and your project are paying off—whether you are doing what you proposed to do and doing it at a reasonable pace. They want to know how much work you have completed and how much still remains to be done. And they want to know whether you have encountered any problems that might require them to change the project deadline, alter the project in some way, or even scrap it altogether. In sum, they want evidence that the project is (or is not) coming along and that you are (or are not) fulfilling your responsibilities.

Having seen what your readers expect from the report, you should have no trouble defining your own goals. First, you want your report to inform your readers of the work you have finished, the work in progress, and the work to come. In addition, you want it to inspire in your readers a sense of confidence in your ability to see the project through. If the project is running smoothly, you inspire this confidence by making sure that your report demonstrates your progress persuasively. If you have experienced problems with the project or if you foresee problems, you inspire confidence by showing that you are aware of these problems and by explaining how you are dealing with them or how you intend to deal with them.

The primary function of the progress report is to provide accountability to those who have the power to pull the plug on a project or to let it continue. But a progress report can be useful to its writer as well as to its readers. First, because it must be organized in a coherent and logical way, it forces the writer to rethink the project; to see all the various activities, results, and problems as a whole; to reassess and reaffirm the direction the project is taking. It distances the writer from the project, and sometimes taking that step back will help put a problem in perspective—and thus help resolve it. A progress report has a second and more practical value as well. Once most projects—especially research projects and the like—are completed, they must be reported. The project's findings must be tabulated, analyzed, discussed, justified, and assessed. The progress reports filed when each segment of the project was fresh in the writer's memory will surely make the task of writing the final report a good deal easier.

 # Organizing a Progress Report

Because the essence of progress is that it unfolds in time, organize your report so that it, too, unfolds in time—an introductory section on past work (work already accounted for in previous progress reports), a middle section on work done since the last report, and a final section on work yet to be completed.

Introduction and Summary of Past Work

The purpose of the opening section is to provide the context for the report as a whole. Remember that although this may be the only project that *you* are working on, it is probably not the only project your readers are supervising. In this section, therefore, you should identify your project for them and briefly remind them of the purpose and nature of your work. They will also want to be told right away where this particular report fits in the sequence of reports you have been submitting. Be sure to include the dates covered by this report and to indicate whether this is the first, fifth, or thirty-first in the series. Finally, this section should briefly summarize the information presented in the report that preceded this one. Because the purpose of this summary is only to remind your readers of work you have already reported on, keep it brief; probably the best way to present it is in a list.

Of course, if this is a first progress report, you won't have any previous progress to summarize. To establish a context for the report, then, you might want to summarize the *proposal* you wrote for the project, describing the project's objectives, scope, and methodology a bit more thoroughly than would be necessary in subsequent progress reports.

Present Work

Not surprisingly, the middle section, the discussion of work done since the last report, is usually the longest and most detailed section of the report. It is here that you must give a complete accounting of the facts, figures, and results obtained since the last report. In organizing this section, you must find a design that conveys the information clearly, concisely, and persuasively. Once again, if you think for a moment about the characteristics of *progress*, two organizational patterns recommend themselves. Progress occurs when tasks or problems are addressed and resolved in time. The "present work" section of most progress reports, not surprisingly, is organized either chronologically (by time) or topically (by task/problem). Often, in fact, it is organized both chronologically *and* topically.

Chronological Organization. Suppose that you have agreed to build a dog-house for a neighbor and that you are filing the second report on your progress. Organized chronologically, the present work section might look like this:

Monday	Continued to work on the frame; bought two more two-by-fours, one gallon of white paint, one gallon of sealer, and shingles at Bob's Hardware—cost: $23.00.
Tuesday	Set up table saw for cutting side walls and entrance way; worked some more on frame (completed all but front section); went to library to review techniques for shingle application.
Wednesday	Finished the frame; cut plywood for side walls, roof, and floor; mounted floor to frame.
Thursday	Mounted walls and roof to frame; applied undercoat.
Friday	Began shingles; painted walls; applied insulation.

The exact chronological order emphasizes movement in time: did this first, did this second, did this third. Because you merely record everything you have done in the order in which you did it, the report is easy to write and natural to read—these are its advantages. One disadvantage, however, is that this pattern of organization tends to put equal emphasis on each time period, regardless of the quantity or quality of work done in that period. Thus, if you have had a bad day or two, this pattern will draw attention to that fact. Furthermore, if you have done lots of relatively unimportant things in one time period, your report will be forced to pause over them unnecessarily, perhaps drawing attention away from more important matters. Look at our example: Thursday and Friday look less productive than the rest of the week. In fact, Friday's work probably took longer to complete and was more complicated than Monday's, but because Monday's entry is longer, it commands more attention.

Another disadvantage is that chronological order becomes hard to read if you are reporting work on several different phases of the project all done simultaneously, because each may be at a different stage of development. In a microbiology experiment, you might be growing some cultures, starting to grow some others, centrifuging some that you grew a while ago, and staining some from the same batch so that you can examine them tomorrow. If you merely listed these activities in the order in which you did them, you would very likely confuse your readers. A topical pattern of organization will solve this problem.

Topical Organization. Organized by tasks instead of time, our doghouse example might look like this:

Task 1:   *Buying the materials*. Bought two more two-by-fours, one gallon of paint, one gallon of sealer, and shingles at Bob's Hardware for $23.00. All materials have now been purchased.

Task 2:   *Building the frame*. Having worked on sides and back on Monday and Tuesday, completed frame on Wednesday.

Task 3:   *Building walls and floor*. Completed walls; worked . . .

Task 4:   *Applying shingles*. Reviewed techniques, but barely began application.

Task 5:   *Painting*. Completed undercoating and applied first coat of white . . .

With this kind of organization, the reader can see exactly how much progress has been made on each of the separate phases of the project. This, of course, can be a disadvantage as well as an advantage. If one of the tasks isn't going very well (applying shingles), this pattern will make it seem even worse by devoting an entire section to its lack of progress.

Topical organization does, nevertheless, offer another advantage over chronological. It provides you some leeway to choose the topics to be discussed. If you use chronological order, you can't very well leave out Thursday and Friday, for instance, without drawing attention to their absence. With a topical order, however, you can combine Tasks 3 and 4 and call them Task 3: Building walls, floor, and roof. Such a grouping would minimize the attention drawn to the problem of shingling. Of course, you can't always cover up weak spots by combining tasks; they must fit together logically. Moreover, you must decide to what extent you really *want* to cover up. If you are actually having that much trouble with shingles, you might be better off acknowledging the fact and seeking advice now, before you start falling behind schedule.

Depending on the project itself, there might be still another advantage to a topical organization. It allows you some freedom to arrange the order in which you discuss the topics. The various tasks in the microbiology experiment might be arranged in a number of ways. If the most important task at the moment is the centrifuging, you could start with that and work down to the less important ones (Task 34: grew more *E. coli*).

Combined Order. For complex progress reports, you will usually need to combine chronological and topical patterns to organize the present work section. There are two choices:

 I. First time period
    A. First topic/task
    B. Second topic/task
    C. Third topic/task
    D. . . .
 II. Second time period
    A. First topic/task . . .
 III. Third time period . . .

Or this:

 I. First topic/task
    A. First time period
    B. Second time period
    C. Third time period
 II. Second topic/task
    A. . . .
 III. Third topic/task . . .

The plan you adopt will depend on your goal. If you want to highlight the progress you have made on each discrete task, organize the section by task first, and then divide each task chronologically. If you don't wish to highlight individual tasks, reverse the order: Divide the section into time periods, and then discuss the progress you have made, task by task, during each time period.

Future Work

Because their subject is work in progress, most progress reports end with a reference to the future—specifically, a brief statement of the progress that will be made in the next time period and recorded in the next progress report.

Although this section (sometimes labeled "Work to Be Completed" instead of "Future Work") is usually fairly short, in a first progress report it may actually be longer than the section on present work. With little completed work to discuss, you will use this section to detail your project strategy; discuss the order in which you will undertake your tasks; hypothesize contingencies; prepare your readers for probable changes in scope, method, schedule, or cost; and outline your predictions and expectations about the project.

Whether you are writing a first report or a subsequent one, however, this final section's primary purpose is to be persuasive. As you discuss your future work, you want to convey the sense that because your past work has been satisfactory, you feel confident to predict the amount of work you will be able to complete before the next report is due. Obviously, if you promise more than you can deliver, you will lose the credibility that this section is designed to establish. So be moderate in your predictions; promise only what you realistically believe you will be able to accomplish.

Also use this section to anticipate future problems and to solicit any help you might need in solving them. Forewarning your readers of possible problems builds your credibility by demonstrating that you have thought the project through. It also covers you in case the problems cause a serious delay in progress, and it gives your readers the opportunity to take whatever action they find appropriate to help you solve the problems. A timely, rational warning thus serves both your interests and the interests of your readers.

In addition to a look into the future, this section occasionally includes a summary of the recommendations or conclusions (or both) reached in the present report. If the evidence suggests that the project plan needs modification or rethinking, this is the place to say so.

 Including Adequate Detail

Once you have decided on a suitable organizational pattern, you must decide how detailed to make your progress report and how best to present the details you choose to include.

Why Detail Is Important

When it comes to progress reports, most writers feel a strong and immediate impulse toward fast, readable generalizations that state, "Everything is coming along fine." Unfortunately, you must learn to resist this impulse.

Unsupported generalizations just aren't very convincing, and a report that strings them together unrelieved by hard data is bound to convey the impression that you haven't made much progress—or worse, that you are trying to conceal something. Consider the effect of a report composed exclusively of sentences like these:

> All the design problems have now been rectified, and we're making good progress in the experimental phase. Preliminary results are very encouraging.

Such writing is too vague to be convincing; it doesn't inspire confidence in your work because it doesn't show *your work* at all—these sentences might be about anybody's work, anywhere, at any time; they sound as though they've been applied through a template, not written by a thinking, responsible professional. Thus, even if the generalizations are absolutely accurate, they are, if presented alone, ineffective and potentially disastrous.

There is a second reason to avoid loading your report with generalizations. Because it is the job of your readers to make ultimate decisions about the project as a whole, a report without facts and details gives them precious little to do their jobs with. Unless you are on unusually friendly terms with your readers, your word alone—and that is what unsupported generalizations are—won't satisfy them.

How to Present Detail

Is it possible to go to the other extreme and include too much detail? If you try to account for every nut and bolt, your readers may exhaust their patience before they have exhausted your report. Paragraph after paragraph of hard data, numbers, and results can make for hard reading. But the problem isn't really one of including too many details; rather it is one of including them ineffectively. The truth is that most readers want to *have* the data more than they want to *read* the data—that is, they want the data to be there so that they can refer to them as the need arises, but they don't necessarily want to read all the data with equal care.

The solution to this apparent problem is simple: use graphic aids—tables, charts, graphs. These will compress your data and make them accessible so that your readers can use them if they want to. At the same time, using graphics allows you to devote your paragraphs to highlighting those data that you think are most worthy of attention. Moreover, because they provide hard evidence, graphics allow you to make convincing generalizations:

> Preliminary results are encouraging. We have consistently managed to recover about 18 percent of the H_2SO_4 from the test samples. Detailed results are recorded in Table 1.

TABLE 1   Recovery of H₂SO₄ from Test Samples

Sample	Location	Description	Sample Size (in mg)	% Recovered
1	P3	Type I	500	14
2	P4	Type I	435	08
3	Q1	Type III	750	19
4	Q3	Type II	600	18
5	Q4	Type III	730	18
6	R1	Type I	500	18

If you have many tables and graphs, you can safely put them in an appendix; they will have the same credibility-building effect, and they won't clutter up your report.

 # Writing Effective Progress Reports

Technical writers often have a great deal of trouble with the style and tone of their progress reports. The following suggestions should help.

Use Concrete, Concise Language

In earlier chapters we talked at great length about the need to keep the language of technical reports concrete and concise. Progress reports, unfortunately, are often repositories for the worst sort of flabby, murky language, especially when the writer feels that there isn't much progress to report. Figure 16.1 presents an example that we hope you will find amusing. (The author of this mimeographed parody is unknown; before landing in these pages, it made the rounds of various government agencies and scientific organizations.)

The first thing to notice, of course, is the report's total lack of information. Quite simply, it presents no data. But its lack of information goes beyond this. Even the *generalizations* are uninformative—more than that, they are hieroglyphic, rendered obscure by flabby constructions and abstract words. Look again at a piece of one sentence:

> . . . it was deemed expedient to establish a survey and to conduct a rather extensive analysis of comparable efforts in this direction . . .

Does the sentence mean anything more than "we decided to study projects similar to ours"? And what do you make of "the functional structure of components of the cognizant organization"? None of these words conjures up a concrete picture in the reader's mind; they are all abstract. Don't expect

STANDARD PROGRESS REPORT

(For Those with No Progress to Report)

During the report period that encompasses the organized phase, considerable progress has been made in certain necessary preliminary work directed toward the establishment of initial activities. Important background information has been carefully explored and the functional structure of components of the cognizant organization has been clarified.

The usual difficulty was encountered in the selection and procurement of optimum materials, available data, experimental data, and statistical analysis, but these problems are being attacked vigorously, and we expect that the development phase will continue to proceed at a satisfactory rate.

In order to prevent unnecessary duplication of previous efforts in the same field, it was deemed expedient to establish a survey and to conduct a rather extensive analysis of comparable efforts in this direction, to explore various facilities in the immediate area of activity under consideration, and then to summarize these findings.

This committee held its regular meeting and considered rather important policy matters pertaining to the overall organization levels of the line and staff responsibilities that develop in regard to the personnel associated with the specific assignments resulting from these broad functional specifications. It is assumed that this rate of progress will continue to accelerate as these necessary broad functions and functional phases continue further development.

FIGURE 16.1   Parody Progress Report.

pompous, inflated, abstract language to give you a protective screen to hide behind if you feel that you have nothing to report. Similarly, don't delude yourself into thinking that you can improve a good report by dressing it up with "impressive" words and sentences.

Find the Appropriate Tone

Even if your project isn't going very well, you will have something to report. Be specific and use concrete language, but be as positive as you can. You do not have to say:

> We have not had any success with the new formula, but we will keep trying.

Try eliminating the negative construction:

> We are still trying to equal the results of the old formula with our new one.

We are *not* advising you to lie or to stretch the truth—and we are certainly not advising you to use euphemisms and abstractions to cover up your project's shortcomings. We are, however, urging you to be as positive as you can be, *consistent with the truth*. (Of course, if the project is a hopeless disaster, your readers will need to know that.) The tone you want to strive for in your progress report is one of cautious optimism—*optimism*, because the main goal of the report is to reassure your readers; *cautious*, because, as we said before, you don't want to have your readers expecting more than you can deliver.

 ## Choosing a Format

Progress reports may be written as memos or letters (informal reports) or as formal reports with title pages, abstracts, tables of contents, appendixes, and the works. Generally, you will be told on the job which format to use. The format requirements given to you will be designed to fit your readers' needs. Thus, you may be asked only to fill out printed forms monthly or quarterly, or you may be asked to submit memoranda focusing on a few specific topics or tasks. A progress report on a research project funded by the government is shown in Figure 16.2. This format is typical of progress reports submitted to federal agencies.

If you are left to your own devices to choose a format, base your choice on your audience. If you are writing to a direct superior or to someone you know quite well, a memo or letter will probably be fine. If you are writing to the Board of Directors, on the other hand, a more formal presentation is in order.

SECTION IV

APPLICANT: REPEAT GRANT NUMBER SHOWN ON PAGE 1	GRANT NUMBER
SECTION IV—SUMMARY PROGRESS REPORT	AM 22032-03

PRINCIPAL INVESTIGATOR OR PROGRAM DIRECTOR (Last, First, Initial)	PERIOD COVERED BY THIS REPORT	
Adler, Robert A.	FROM	THROUGH
NAME OF ORGANIZATION	7/01/81	6/30/82
Dartmouth College		

TITLE (Repeat title shown in Item 1 on first page)
Prolactin As A Diabetogenic

1. List all publications, not previously reported, resulting from work supported by this grant (author(s), title, page numbers, year, journal or book). List manuscripts separately as submitted for publication or accepted for publication.
2. Provide two reprints of publications not previously submitted to the awarding unit.
3. Progress Report. (See instructions)

1. A. Adler, RA, Brown, SJ, and HW Sokol, Characteristics of prolactin secretion from anterior pituitary implants. In: Progress in reproductive biology: Advances in prolactin, vol. 6, pp. 24-30 (Karger, Basel 1980).

   Adler, RA, Effects of fasting and TRH in hyperprolactinemic and normal rats. Abstracts of the Endocrine Society Annual Meeting, June, 1981.

   Marceau, M. Adler, RA, and HW Sokol, Is prolactin involved in mammalian water balance? Abstracts of the Endocrine Society Annual Meeting, June, 1981.

   B. North, WG, Gellai, M, Sokol, HW and RA Adler, Demonstration that rat prolactin has no intrinsic antidiuretic activity in the rat, in press, Hormone Research.

   Sokol, HW, Marceau, M, and RA Adler, The effects of vasopressin and endogenous or exogenous prolactin on urinary flow and concentration in rats with hereditary hypothalamic diabetes insipidus (Brattleboro strain.) (submitted for publication)

   Adler, RA, and HW Sokol, The effects of elevated circulating prolactin levels on glucose tolerance in rats. (submitted for publication)

2. (Attached)

3. Progress Report -

   The major objective of this work is to determine if the anterior pituitary hormone, prolactin (PRL), has anti-insulin properties. We have studied glucose tolerance in rats with chronic endogenous hyperprolactinemia produced by implantation of extra anterior pituitary glands (AP) under the kidney capsule. We have submitted our surprising findings for publication: AP-implanted rats have no decrease in glucose tolerance. On the contrary, after dextrose injection the hyperprolactinemic rats had lower serum glucose levels than control rats. Serum insulin concentrations were similar in the two groups, suggesting that PRL may potentiate insulin's effect and/or have an insulin-like effect of its own. After fasting, AP-implanted rats had lower serum glucose levels than sham-operated control rats, again with similar insulin concentrations. For the next budget period we will continue these whole rat studies by observing the effects of insulin-induced hypoglycemia on PRL secretion in the rats, the effects of fasting on PRL secretion, the effects of hypophysectomy on glucose tolerance, and the effects of streptozotocin in AP-implanted and control rats.

PHS-2590-1
(Rev. 9/80)
PAGE 5   (Use Continuation Page as necessary)

FIGURE 16.2   Federal Research Progress Report.
(Reprinted with permission of the author.)

Adler, Robert A.

Grant number   AM 22032-03

Nonetheless, to determine the mechanisms behind our observations, new methods have had to be developed. Specifically we have worked with two methods of purification of rat PRL so that in vitro tests can be performed. Rat PRL (like ovine PRL) is contaminated with other substances including vasopressin (AVP) and catecholamines. In our laboratory polyacrylamide gel electrophoresis and isoelectric focusing have been used to remove the vasopressin--as measured by radioimmunoassay--and, theoretically, the other contaminants. At present we are refining our purification techniques and submitting the preparations to analysis by radioimmunoassay, radioreceptor assay, and bioassay (incorporation of ^{14}C-acetate into fatty acids in mouse mammary gland culture). Our studies on osmoregulatory effects of PRL have helped us assess these preparations because the contamination by AVP can be measured easily by radioimmunoassay and by biological testing in rats with hypothalamic diabetes insipidus. We have now shown (see publications above) that removing the AVP contamination of rat PRL preparations obliterates all the antidiuretic effects of the crude PRL extract.

Production of adequate amounts of purified PRL will enable us to expand our work on isolated adipocytes. Fat cells removed from hyperprolactinemic rats seem to oxidize more ^{14}C-glucose to $^{14}CO_2$, but this effect has not been as reproducible as first observed. Certainly we have not found decreased insulin-induced glucose uptake in these rat cells. Nonetheless it will be necessary to add purified PRL to adipocytes in vitro to be sure of these insulinlike effects.

We hypothesize that PRL will turn out to be similar to what Lewis has described for human growth hormone: a complex of peptides. We will need to look carefully at both the physical and biological properties of the PRL we purify in our attempt to identify the effects of this important hormone. Our work has implications for the basic question of what prolactin really does in mammals. Is this hormone truly "versatilin," a substance with many diverse actions? Are many of the reported actions of PRL attributable to the contaminants of the PRL preparations? Are there different forms of mammalian PRL with different biologic activities? We feel that steps are being made to answer these basic questions.

Review Questions

1. Who reads progress reports? What are they looking for as they read?
2. What are the goals of a progress report?
3. What information belongs in the introduction of a progress report?
4. Under what circumstances is chronological order most appropriate for organizing the present work section of a progress report? Under what circumstances might topical order be better? Combined order?
5. How and where do you indicate problems you might be having with your project?
6. What sort of detail do you include in a progress report?
7. Characterize the language most appropriate for discussing the progress you have made.

Assignments

1. Write a 750-word report to your instructor, outlining the progress you are making toward completing the final project of this course.
2. Write a memo to your advisor explaining the progress you have made in your academic career. Include such topics as the number of credits you have already taken, the number of required courses you still need to take, any special courses you have taken or plan to take, and so on.
3. Write a progress report for the instructor in one of your lab courses, explaining to him or her the status of an experiment you are conducting.

17

Analytical, Evaluation, and Recommendation Reports

Technical reports elude classification according to length, format, or style and tone: Each must be handcrafted to its particular audience and particular goal. Nevertheless, reports *can* be broadly classified according to their purpose as analytical reports, evaluation reports, and recommendation reports. All the known types of final reports for which "official" formats exist fall into one of those three categories. In this chapter, we will discuss the differences in organizational thrust among analytical reports, evaluation reports, and recommendation reports, and we will describe some specific types of final reports that you may be asked to produce on the job—inspection reports, laboratory reports, feasibility studies, and environmental impact statements.

 ## Analyzing, Evaluating, and Recommending

Analytical reports, evaluation reports, and recommendation reports have different goals. Analytical reports emphasize findings; evaluation reports emphasize conclusions; recommendation reports emphasize recommendations. When you are assigned the task of preparing a final report, you should find out first of all what the goal of the report is supposed to be. Who will be reading it and why? Are you being asked to explain a complex topic to uninformed readers by breaking it down into accessible subtopics, providing only information without judgments? If so, your report is analytical. Are you being asked to assess the advantages and disadvantages of a proposed action, to determine the feasibility of a project, to judge the positive and negative aspects of a plan, including your conclusions but not what you would do about them? If so, your report is evaluative. Are you being asked to provide specific recommendations for action on a topic and to bolster those recommendations with carefully selected supporting evidence? If so, yours is a recommendation report.

 ## Analytical Reports

The most straightforward of all final reports are analytical reports, which describe a complex topic by breaking it down into subtopics and addressing

371

those subtopics one at a time. What differentiates analytical reports from evaluation reports and recommendation reports is that analytical reports do not venture very far into the realm of opinion. They are confined to facts and inferences that may safely be drawn from those facts. Their goal is to convey those facts to readers for *their* use: Analytical reports are informative rather than persuasive and as strictly objective as any technical document can be made to be.

Routine analytical reports are part of the daily output of many technical professionals—engineers, chemists, physicians. In many organizations—in a police department, for instance—those routine reports are written on pre-printed forms. Short analytical reports intended for circulation within an organization are often written in memo form. Analytical reports intended for customers are frequently written as letters.

Organizing Analytical Reports

Because the purpose of an analytical report is to present readers with findings for *their* evaluation, analytical reports are organized so as to present those findings candidly and directly.

The Introduction. The introduction to an analytical report is purely informational. It does not need to attract readers' attention, ease them into the topic of the report, or entice them to keep reading. People do not read analytical reports at all, as a rule, unless they are interested in the analysis they contain. Therefore, all analytical reports, whether formal or informal, begin with a statement of purpose that announces the topic and situates the report within the topic:

> In this report, the causes of unwanted feedback and oscillation in public address systems and musical instrument amplifiers will be specified.

To put your statement of purpose in context, you will probably want to provide a brief overview of the specific technical problem being addressed. And if the statement of purpose isn't explicit about exactly what the analysis covers and what it excludes, you will want to include a statement of scope as well.

The introduction to an analytical report usually contains, in addition to the statement of purpose, background discussion, and statement of scope, a brief description of the methods used to conduct the analysis. The introduction should give the source or sources of the information used in the report, but it may or may not need a rationale; some analyses require justification and some do not. In general, the more routine the analysis, the less you need to explain why it was performed. Depending on its length and complexity, an analytical report may have in its introduction a description of the intended audience for the report, background on the need for the analysis, definitions of key terms, and a list of findings. The introduction to an analytical report may also include a description of the credentials of the people who performed the analysis, acknowledgments to people who assisted in the analysis, and a summary of the steps in the analysis.

The introduction to an analytical report customarily ends with a preview of the organization of the report as a whole. A fairly complex analytical report will have a detailed table of contents, of course, but some readers won't bother to study it. A paragraph or two announcing the major subsections can work wonders in helping readers understand the structure of the report. Those who intend to read the entire report will appreciate the road map; those who don't will use this section as a checklist for what to read and what to skip.

You might find it hard to weave all of these elements into a smooth-flowing series of paragraphs. If you do, consider dividing the introduction into subsections with headings and subheadings.

The Body of the Report. Virtually all analytical reports have the same type of introduction, but the organization of the body of an analytical report varies with its subject matter. Reports of the findings of a detailed physical inspection are quite often organized spatially: Moving the description from left to right, from bottom to top, or from inside to outside helps readers of the report visualize what is being inspected. Investigation reports, such as those written by police officers and insurance investigators, are most often organized chronologically so that they duplicate the sequence of the investigation itself. Laboratory reports also follow a chronological sequence. Readers of laboratory reports will presumably be most interested in the results and discussion sections, but tradition demands that those sections come last, after an introduction and a description of the procedure used in the experiment.

Because chronological organization places equal emphasis on everything, many analytical reports use topical organization instead to highlight some points and soft-pedal others. A chemical analysis of blood is organized by constituent elements; a causal analysis of equipment failure is organized by contributing factors. If the topic of an analytical report is not naturally divisible into subtopics, subtopics can be created for the report. The analytical report in Figure 17.1, reproduced in its entirety, is a good example of topical organization. Its purpose is to make clear to readers the design and projected performance of a solar energy system for an oil refinery. It accomplishes that purpose in five steps: discussing the baseline solar plant from which the system was devised, describing the results of trade-off studies on the baseline plant, describing the full system, describing the subsystems, and predicting the performance of the system when operational. These five subtopics represent five different ways of looking at the system, arranged in order of decreasing importance.

Because analytical reports are usually organized either spatially, chronologically, or topically in order of importance, most do not build inexorably to a conclusion. Often analytical reports simply stop rather than end. When they do "end," it is often with a summary, a discussion of especially significant

Central Receiver Solar Energy System for an Oil Refinery

Robert E. Sommerlad   Robert A. Pichnarcik
Ramjee Raghavan   Provident Energy Company, Inc.
Foster Wheeler Development   Phoenix, Arizona
    Corporation
Livingston, New Jersey   Russell T. Neher
                             McDonnell Douglas Astronautics Company
                             Huntington Beach, California

Abstract

The paper discusses the conceptual design of a central receiver solar energy
system that will provide practical and effective use of solar energy for an
oil refinery currently being designed by Foster Wheeler Energy Corporation
for the Provident Energy Company. The refinery will be built 25 miles
southwest of Phoenix, Arizona, and is scheduled to be in operation by 1983.
The system will be designed to make maximum use of existing solar thermal
technology consistent with existing refinery design and operating techniques,
will provide for the best possible economics for the application, and will
offer the best combination of solar and fossil-fuel energy.

The design is for a baseline system consisting of a tower-mounted,
natural-circulation water/steam receiver with an 18.4 kg/s (146,000 lb/h)
capacity. A flat-panel absorber generates saturated steam that is
superheated to the desired temperature in a separate, oil-fired superheater
prior to admission to the main refinery superheated steam headers. Solar
energy is concentrated on the receiver by a field of heliostats north of the
tower.

1. Introduction

Foster Wheeler Development Corporation (FWDC), McDonnell Douglas Astronautics
Company (MDAC), Foster Wheeler Energy Corporation (FWEC), and Provident
Energy Company, Inc. (PEC), are engaged in a conceptual design of a central
receiver solar energy system for the PEC refinery. The program objectives
are:

·   Provide 20 percent or more of the steam requirements of the refinery

·   Attain the best possible economics for the solar energy application,
    considering both capital and operating costs

·   Utilize existing solar thermal technology

·   Typify a solar energy retrofit for the refinery industry.

The 92.0 dm^3/s (50,000 bbl/day) refinery, being designed by FWEC for the
PEC, is to be built in Mobile, Arizona, which is 40 km (25 miles) southwest
of Phoenix, Arizona. The refinery is scheduled for operation by 1983.

The refinery peak steam requirement is 25.2 kg/s (200,000 lb/h) and the
average is 21.4 kg/s (170,000 lb/h) at 4.2 MPa gage (600 $lb/in^2 g$), $370^\circ C$

FIGURE 17.1   An Analytical Report.
(This paper was published by Foster Wheeler Development Corporation as an interim account
of work performed under contract to the Department of Energy. It is included only as an
example and not for its technical or engineering content. The authors are R. E. Sommerlad and
R. Raghaven of Foster Wheeler Development Corporation, R. A. Pichnarcik of Provident Energy
Company, Inc. and R. T. Neher of McDonnell Douglas Astronautics Company.)

(700°F). Steam is also utilized for process heat at two lower pressure
levels.

The solar facility will be located adjacent to the northeastern edge of the
refinery on a site that is ideal because the land is relatively flat and
insolation is very high. The annual average daily value of direct normal
insolation is 7.6 kWh/m^2 (24000 Btu/h·ft^2). Figure 1, the refinery plot
plan, shows the heliostat layout and the location of the utilities, where the
steam generators are located.

2. Baseline System

The baseline solar plant used for preliminary analysis and trade-off studies
consists of a tower-mounted, natural-circulation water/steam receiver with a
capacity of 18.4 kg/s (146,000 lb/h). The flat-panel receiver generates
saturated steam, superheated to the desired temperature in a separate,
fossil-fired superheater before admission to the main refinery steam
headers. Solar energy is concentrated on the receiver by a field of
heliostats north of the tower. There is no provision in the baseline system
for thermal storage.

3. Trade-off Studies

Trade-off studies and configuration optimization studies were conducted to
determine the most practical and cost-effective solar plant design. The
results of the major studies are summarized below.

·   Buffer storage is required to protect the refinery from cloud-induced
     solar transients.

·   No economic benefit accrues from thermal storage of 1-hour maximum steam
     flow rate or less.

·   The flat-panel receiver was selected over a cavity-type receiver, based
     largely on its simplicity.

·   A fossil-fired superheater was selected over a solar superheater.

·   A door over the receiver face to preserve heat during periods of nonsolar
     operation is not economically advantageous.

·   A free-standing steel tower is preferred over either guyed steel or
     concrete.

·   Natural circulation is preferred over forced circulation because of its
     simplicity.

4. System Description

A preferred system design (Figure 2) was developed using the results of the
trade-off, optimization, and refinery compatibility studies. This system is
comprised of five subsystems: heliostat, receiver, buffer storage,
fossil-fired superheater, and controls.

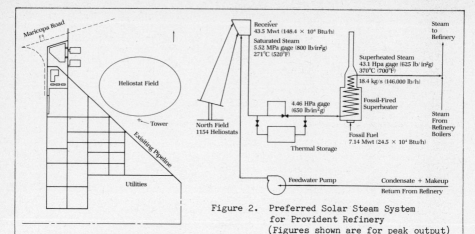

Figure 1. Refinery Plot Plan

Figure 2. Preferred Solar Steam System
for Provident Refinery
(Figures shown are for peak output)

A north field of heliostats directs reflected energy to the flat-panel
receiver. This natural-circulation boiler generates 5.51 MPa gage (800
lb/in²) saturated steam, which is conducted down the tower and fed into
the fossil-fired superheater. Superheated steam at 4.31 MPa gage (625
lb/in²) and 370°C (700°F) is combined with steam from the refinery's
own boilers to provide thermal energy for the refinery.

A pressurized-water buffer storage system is utilized to isolate the
refinery process from short-term solar transients. The addition of the
buffer storage necessitated increasing the receiver pressure to 5.52 MPa
gage (800 lb/in²). Steam delivery rates and refinery conditions are also
shown in Figure 2.

5. Subsystem Conceptual Design

5.1 Heliostat

The heliostat subsystem consists of a field of 1154 heliostats, related
controls, and necessary power for drive purposes. The preferred layout is a
north field elliptical arrangement. The total mirror area will be 6.5 hm²
(700,217 ft²) and the field will occupy 268 km² (67 acres) of land.

5.2 Receiver

The natural-circulation-type receiver shown in Figure 3 absorbs heat in an
exposed north-facing flat panel of tangential tubes. The exposed panel,
consisting of 210 0.55 dm (2 in.) O.D. tubes, is tilted 20 deg from the
vertical to face the heliostat field at an optimal angle. Water and steam
leaving the tubes of the panel are collected in the upper header, from which
they are carried by 20 riser tubes to the steam drum. In the drum the steam

is separated from the water and discharged as dry saturated steam. The
water recirculates through the 4 receiver downcomers, from which it is
distributed to the lower panel header by 10 feeder tubes. Feedwater enters
the steam drum below the water level.

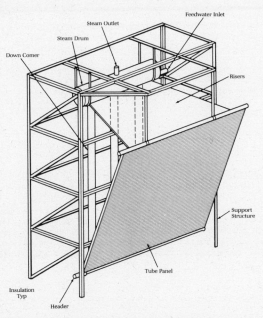

Figure 3. Natural-Circulation Receiver

The front face of the panel is painted with a high-absorptance paint, and
all pressure-part surfaces that do not absorb heat are insulated to minimize
heat loss. The receiver pressure parts are top-supported from a structural
steel framework and are free to expand downward. The receiver tubes are
welded to each other along their length with 6.4 mm (0.25 in.) fins to form
a MONO-WALL™.

5.3 Fossil-Fired Superheater

The superheater, a vertical, upfired steeple design with a cylindrical
furnace 3.6-m (11-ft 10-in.) O.D. and 7.92 m (26 ft) high, contains six
passes and is comprised of convection and radiant sections. It has an
absorption rate of 7.14 MWt (24.4 x 10^6 Btu/h) and can superheat 18.4 kg/s
(146,000 lb/h), 4.48 MPa gage (650 lb/in^2), 258°C (496°F) saturated
steam to 370°C (700°F), accompanied by a 103 kPa (15 lb/in^2) pressure
drop. The unit is designed for a 173 kPa (25 lb/in^2) pressure drop. The
superheater has a calculated efficiency of 84.5 percent.

5.4 Buffer Storage

A Buffer Storage Subsystem is provided to store a portion of the steam output from the receiver and subsequently provide steam to the refinery in times of cloud cover. The Buffer Storage Subsystem is a variable-pressure steam accumulator, designed to hold pressurized water, at 5.4 MPa gage (780 lb/in^{2}g), which will be flashed to 4.9 MPa gage (650 lb/in^{2}g) to provide 2.05 Mg (8500 lb) of steam. Since the solar plant's ability to generate will decline rapidly in times of cloud cover, the Buffer Storage Subsystem is provided with pressure controls and valves. These devices will enable the solar plant to maintain operating pressure and temperature at the refinery steam supply header in times of cloud cover, when the fossil boiler is ramping to make up for the decreasing steam output of the solar receiver.

5.5 Controls

The Control Subsystem is designed to provide stable performance in all operating modes including start-up, shutdown, full solar, intermittent cloudy, and nonsolar operation and during transition between modes.

As currently planned, the refinery's fossil boilers will be operating during solar operation--but at a very low output. The control system's function is to modulate the fossil boiler output in response to the solar steam production so that refinery header pressure is maintained at the required level. In times of cloud transient, the control system will modulate the steam output of the buffer storage system so as to maintain the refinery header pressure within the operating levels. The firing rate of the superheater will be varied in response to superheater outlet steam temperature to maintain the required temperature at the refinery steam header.

6.   Solar Performance

Approximately 105 GWht (359 x 10^9 Btu) of energy annually is generated as saturated steam with the conceptual design. Allowing for the fossil-fired superheater efficiency of 84 percent and the fossil boiler efficiency of 81 percent, the proposed solar plant displaces about 130,000 MWht (445 x 10^9 Btu) or about 11,350 m^3/yr (71,400 bbl/yr) of the salable residual fuel oil produced by the refinery. The annual refinery steam demand (expressed in energy units) is 506 GWht (1.73 x 10^{12} Btu) based on an average steam load of 21.4 kg/s (170,000 lb/h) throughout the year. The solar plant furnishes 20.8 percent of this demand on an annual basis.

details from the data presented in the body, or a list of conclusions to be drawn from the data. Many analytical reports do conclude with some weighing of the evidence, but such discussion should be strictly limited because the primary purpose of an analytical report is to present the evidence to readers so that *they* can make the necessary evaluation.

Types of Analytical Reports

Three types of analytical reports that technical writers are frequently asked to produce are inspection reports, investigation reports, and laboratory reports (sometimes called physical research reports). In this section, we will discuss these three types of reports briefly and show you examples taken from both government and business-world documents.

Inspection Reports. Figure 17.2 shows a full-length inspection report. In this example, the report is organized topically; its subtopics—the areas of investigation at the site—were mandated by law. Its author prepared it, like many other short, routine reports, as a memorandum for internal distribution within the state's Department of Environmental Protection.

Investigation Reports. Investigation reports differ from inspection reports in that they present data, but not necessarily physical data obtained by observation. The data may be quantitative or factual; they may have been collected by an individual or by an investigative team; the results of the investigation may be predictable or may come as a complete surprise to the investigator. Occasionally, investigation reports are organized chronologically, following the sequence of the investigation, but more often they are organized topically. The following excerpt from NOAA Technical Memorandum ERL MESA-6, "Contaminant Inputs to the New York Bight," exemplifies topical organization in an investigation report.

Here is the report's statement of purpose.	This report presents an estimate of the location and magnitude of contaminant inputs into the New York Bight, indicates their relative importance, and identifies data gaps.
The parameters used in the analysis are presented first. Each of the broad parameters is then analyzed into its constituents; much of the data in the report itself will be presented in terms of these constituents.	Four sources of contaminant inputs were evaluated in the study: barge dumps and atmospheric fallout as direct bight inputs, and wastewater and runoff as coastal inputs to water ultimately draining into the bight. The sources were further subdivided into their various constituents: dredge spoils, sewage sludge, acid wastes, chemical wastes, and rubble from the barge dumps; gauged stream flow; urban runoff and groundwater outflow for the runoff; and municipal and industrial wastewater inputs. The wastewater inputs were evaluated only downstream of the gauged stream stations because

Memo
DEPARTMENT OF ENVIRONMENTAL PROTECTION

To _____ Jeffrey S. Dent _____

From _____ Mary Johnson _____ Date _____ 29 March 1981 ____

Subject ___ Site Inspection of Edgewater Landfill _____

BACKGROUND

The Edgewater landfill is located on Cove Road in Ocean Township.
The facility has been closed since early 1979. The property is
owned by the Borough of Edgewater and is leased to a commercial
concern that operated the landfill.

DESCRIPTION OF MATERIAL DISPOSED OF AT THE SITE

During the period it was in operation, Edgewater landfill accepted
solid municipal waste from the borough, construction waste, junked
autos, tires, leaves, tree trunks, and agriculture and food
processing waste. Allegedly, no chemical wastes, either liquid or
solid, were accepted at the landfill.

DESCRIPTION OF THE SITE

Edgewater landfill is approximately 62 acres in size and 20 feet
in height. It lies in the floodplain of the Woodbury Creek, which
runs into the Patuxent River. The attached map shows the location
of the site in relation to roads, receiving waters and the
surrounding residential area.

The landfill consists of two mounds which together cover an area
of approximately ten acres. Both mounds have a clay cover and
patchy scrub grass growth. There is no evidence of drums or any
other waste present. On the day of the inspection there was no
leachate observed around the base of the fill area.

DESCRIPTION OF SURROUNDING AREAS

The residences closest to the landfill are in a development on the
northern edge. The only landfill-related problem experienced by
the residents of the development is some flooding and deposition
of silt in the back yards of some of the houses nearest to the
landfill. This occurs after a heavy rain and is due to the fact
that the landfill is higher than the developed area. I consulted
local engineers about this problem and it was suggested that the
area between the landfill and the development be filled in and a
drainage pipe run out to Cove Road. There is no indication that
the runoff poses a problem for any reason other than flooding.
Another problem was a citizen complaint of a "chemical odor" from
the landfill. A state inspector investigated and found that the
odor was attributed to freshly dug clay which had been brought in

FIGURE 17.2   An Inspection Report.

from construction on Ray Street in Cove City. The odor dissipated
once the clay had been spread and dried. There have been no
further complaints.

GEOLOGY AND GROUNDWATER

Clay soil underlies the landfill which is situated on the
Woodbury-Magothy aquifer. As a result, any leachate which could
be generated at this facility would most likely run into Woodbury
Creek rather than into the groundwater. However, as was
previously pointed out, there was no leachate observed on the day
of the inspection.

SAMPLING

No samples were taken at this site.

STATE, LOCAL INVOLVEMENT

The facility is inspected periodically by the state and a
registration permit is on file with the state.

DISCUSSION OF IMMINENT HAZARD ASPECTS OF SITE

Edgewater Landfill appeared to be well managed, and no hazards to
vegetation, receiving waters, or nearby residents were evident on
the day of the inspection. Overall, at this time this site can be
classified as a low priority.

Here is the description of the methodology used in the analysis.

The source of the data is given next.

The description of the methodology continues with a discussion of the ways in which the data were analyzed.

As in many technical reports, raw data have been relegated to the appendix.

The report begins with a discussion of the location and magnitude of contaminant inputs, the first topic given in the statement of purpose.

This is the first of four subsections discussing, in turn, the input of heavy metals, suspended solids and organic matter, nutrients, and microbes. Their relative importance is discussed in each subsection.

all inputs above these points are reflected in the gauged runoff values. In addition to flow or volume for each source, twenty-three separate contaminants were investigated to estimate the inputs of solids, organic matter, nutrients, heavy metals, and microbes.

A large quantity of recent (1970–1974) data was obtained from numerous agencies, including federal, state, municipal, and academic. These data were rigorously analyzed on a flow-weighted basis to obtain the average mass loads into the New York Bight for each contaminant. A summary of the contribution by source and location is presented as well as an estimate of the annual or seasonal variability for selected sources. To evaluate the significance of the mass loads generated in the study, they were compared to an estimate of the background mass loads entering the bight as a result of a net current across the ocean boundaries.

Because of their usefulness in water quality modeling, the raw dredge spoil, wastewater discharge, and gauged runoff data for each source are included as appendixes to this report.

1. The major contaminant inputs to the New York Bight originate from the Transect Zone, which includes the New York metropolitan area and the northern New Jersey and Hudson River drainage basins. Wastewater, runoff, and barge discharges constitute the major waste sources from this zone. The inputs from both the Long Island and the New Jersey coastal zones are small, generally contributing less than 6 percent of total bight inputs. Groundwater inputs are insignificant for all of the waste parameters considered in the study.

2. Heavy metals, especially lead and chromium, comprise the most significant manmade inputs into the New York Bight when compared to background loads from the flow entering the bight across its ocean boundaries. Dredge spoils contribute the major portion (24 to 80 percent) of the heavy metal input, with the exception of mercury for which wastewater contributes 70 percent. Municipal wastewater and runoff each contribute from 5 to 22 percent of the heavy metal loads, whereas sewage sludge from barge dumps accounts for less than 6 percent. Atmospheric inputs constitute 9 and 18 percent of the lead and zinc, respectively, while separate industrial wastewater inputs are significant only for copper. . . .

This subsection fulfills the third objective given in the statement of purpose.

The data are now evaluated, and conclusions are drawn.

This closing analysis of the value of the data on which the report is based helps to put its conclusions into perspective and suggests what the report does and does not contribute to the practical problem of learning the exact extent and sources of contaminant inputs to the New York Bight.

7. Significant data gaps exist with respect to atmospheric and urban runoff inputs.

8. The various sources of mass loads into the New York Bight are highly interrelated. The poor quality of the dredge spoils is caused by settling of contaminants from poorly treated wastewater and urban runoff. Increasing degrees of wastewater treatment produce a greater mass of municipal and industrial sludge for ultimate disposal. For conservative substances such as heavy metals, various control measures may cause a redistribution of the load among the sources but no net decrease in the total load.

9. The actual contaminant load reaching the New York Bight overlying waters and sediments is not known. The load that ultimately reaches the New York Bight from the coastal zones may be significantly altered because of chemical reactions, bacterial decay, sedimentation, adsorption and leaching of sediments, and the growth cycles of biological organisms. The distribution of the dredge spoil and sewage sludge mass loads between sediment and overlying bight waters is also unknown.

Laboratory Reports. Reports of laboratory research follow a more-or-less standard organizational plan. They have seven sections.

Introduction. The introduction to a laboratory report begins with the purpose of the experiment. There are several ways of approaching the statement of purpose in a laboratory report. Often it is phrased as an answer to one of the following questions: Why was this experiment performed? What hypothesis was being tested in this experiment? What was this experiment designed to determine?

Literature Review. The literature review section describes work done by other scientists that is relevant to the work being described in the report. Normally, this means that similar or antecedent experiments must be cited and their results summarized. If any procedures in the experiment have been adapted from published literature, this is the place to say so. If the literature review for your experiment is very short, you can incorporate it into the introduction.

Materials and Methods. In the materials and methods section of a laboratory report, you give a detailed account of exactly how the experiment was carried out and under what conditions. Central to this section is, of course, the experimental procedure itself, but it should also include the types of glassware used, the names and amounts of various chemicals used, and such variables as time and temperature. If you used experimental animals in your procedure, you should include their scientific (Latin) names.

Results. The results section that appears in all laboratory reports gives the data garnered through the experimental procedure. Often these data are summarized in charts, tables, or graphs. The results section should also include statistical information about the data. It should not, however, contain any interpretation or other discussion of the results. This section ends, customarily, with a summary of the most significant results.

Discussion. In the discussion section, you present interpretations of your results and state the conclusions that you reached on the basis of those results. You may wish to compare those conclusions with the hypotheses you advanced in the introduction or to compare your results with the results of others who have performed similar experiments. (If you do mention other people's work in this section, you should cite it just as you would in the literature review.)

Sources of Error. A section discussing sources of error is optional. (It may also be incorporated into the discussion section.) Its purpose is to give you an opportunity to redeem yourself intellectually and account to your superiors if your experiment went awry. If you obtained totally inexplicable results, you may wish to discuss where you think the experiment went wrong: poor technique, inaccurate measurements, stale chemicals, incorrect preparation of solutions, and so on. Even if your results are usable, you may use this section to discuss their limitations.

References. You should end a laboratory report with a list of standard bibliographic references for the literature you have cited in your report. We have described how to prepare such references, and how to cite them, in Chapter 13, "Technical Report Format."

▶ Evaluation Reports

Evaluation reports differ from analytical reports in that they present not only facts but also educated opinions, and in that they synthesize those facts and opinions into a coherent framework with a persuasive thrust. Analytical reports have a statement of purpose, but no real thesis statement. Evaluation reports *must* have a thesis: X caused Y; X requires Y; X is better than Y. Because of their persuasive thrust, evaluation reports are often organized to answer a specific question: What caused Y? What is necessary for X to function effectively in the future? Which is better—X or Y? The answer to the question is the report's principal conclusion—its thesis. An evaluation report is designed to apprise readers first of that principal conclusion, next of the secondary conclusions that support the principal conclusion, and finally of the data on which all of those conclusions are based.

Organizing Evaluation Reports

Because they are persuasive, evaluation reports must be organized in such a way that they induce readers to accept the conclusions being proffered by the author. Analytical reports may be fragmented, with no clear coherence among their subsections, because the purpose of an analysis is to break down a subject into its parts and then discuss the parts individually. In an evaluation report, all the findings must instead be arranged to support and drive home the report's thesis.

The Introduction. The introduction to an evaluation report generally contains not only a statement of purpose but also a statement of the problem and a discussion of the background of the problem. Usually, to provide a context for the statement of purpose, this background discussion comes first:

Background

The report begins with two nontechnical assumptions that even lay readers can easily understand.

The source of the problem is stated.

The statement of the problem immediately follows the delineation of its source.

The problem is elaborated for nontechnical readers: The economic theory underlying the evaluation is explained briefly, and operational definitions are provided.

The conventional approach to water pricing has been premised on two assumptions: (1) that water is endlessly abundant, and (2) that, because water is necessary for life, it should be provided almost regardless of cost.

Conventional water supply management has addressed itself only to filling maximum projected demand by studying the technical feasibility of increasing supplies.

The technical orientation of conventional water supply management has caused both the rising supply cost of water and its diminishing marginal value to be ignored. As more water is used by a single consumer, the marginal benefit derived from each additional unit will decrease while marginal supply costs increase. In order for optimal social welfare to occur, the marginal benefit and the marginal cost of a unit produced must be equal: This is the point where all resources will be used at their highest value. The economic principle of allocative efficiency requires that water services be used at their highest value so that no waste is induced, whereas the principle of technical efficiency requires adoption of the best available production techniques and maximum labor/management pro-

Here is the report's thesis statement.	ductivity. Economic theory can be used to adapt conventional water supply management in the direction of optimal social welfare.
	Purpose
The assumptions described in the first section provide a context for this statement of historical fact. Definitions of "efficiency" in the background section make this use of the term clearer. The excerpt ends with the report's statement of content.	As water rate levels have increased, they have become more controversial, and many new concepts of water management through pricing have been developed. These new pricing structures attempt to direct water to its most efficient use. This report will evaluate the following rate structures: flat charge, uniform rate, step rate, declining block rate, inverted block rate, true cost approach, demand rate, peak load pricing, and social pricing.

As you can see from this example, the introduction to an evaluation report customarily contains the same elements as the introduction to an analytical report, but it is likely to be longer and more elaborate. Background information is virtually indispensable, and a rationale for the evaluation may also be required.

In this particular example, the author has not yet specified which of these rate structures she prefers. She has left herself two choices as to where in the report she can commit herself on that issue—at the end of the introduction or in the conclusion. In most evaluation reports, the thesis statement comes either at the beginning or at the end of the introduction. Persuasion theory tells us that readers prefer to be told up front what they are being asked to endorse or approve, provided they expect you to have an opinion already. In those rare cases where you, as the report writer, know that your readers will disagree with your views and will not expect you to disagree with theirs, you can place your thesis statement at the end of your report and build up to it gradually, presenting all the relevant findings first. This writer has no reason to suspect that her readers will distrust her preference in water rate structures; therefore, she can express that preference at the end of her introduction, immediately before launching into its justification.

One element normally contained in the introduction to an evaluation report that is *not* usually in the introduction to an analytical report is a description of the evaluation criteria and the standards of judgment applied in the evaluation. The sorts of criteria commonly used in evaluation reports include cost, safety, durability, reliability, and efficiency. However, the criteria in any given report may be a great deal more specific: Metal alloys might be evaluated for elasticity and tensile strength; hybrid tomato strains, for yield and hardiness. In the preceding example, rate structures are to be evaluated for the extent to which they make marginal cost equal to marginal benefit, and thus lead to allocative efficiency and optimum social welfare.

The Body of the Report. The organization of evaluation reports varies with the number of things being evaluated. If your report addresses only one subject—the advantages and disadvantages of a single proposed departmental reorganization, the strengths and weaknesses of a particular new product design, the on-the-job performance of an individual colleague or subordinate—it may be organized in either of two ways: You can take up the criteria or aspects one at a time, usually in order of importance. Or you can group all the advantages into one section and all the disadvantages into another.

If your evaluation report must weigh two or more alternatives against one another instead of examining one alternative in depth, use comparative order (see Chapter 2). Again, you have two choices—one alternative at a time or one criterion at a time.

If you are considering numerous alternatives in your evaluation or applying a large number of criteria, it may be advisable for you to group and summarize either alternatives or criteria. Also, you may not wish to discuss all the alternatives or criteria in the same order each time. You may wish to put the best alternative first under each criterion, showcasing each alternative's strengths. Or, if you group by alternatives, you may want to subdivide by performance, listing first under each alternative the criterion on which it performed best, and so on down the line. Whatever organizational matrix you choose, however, you will use either the alternatives being evaluated or the criteria being applied for your major divisions and then use the other for your major subdivisions.

Evaluation reports that are essentially comparative are particularly well suited to tabular organization. As you know from Chapter 11, tables can show several alternatives and several criteria simultaneously, and they permit readers to see easily how the report writer arrived at his or her conclusions.

Types of Evaluation Reports

Two of the most common types of evaluation reports are feasibility studies and environmental impact statements. Feasibility studies are rapidly becoming one of the most common forms of technical writing, and environmental impact statements are multiplying so fast that the demand for them cannot be met exclusively by specialists in the genre.

Feasibility Studies. A feasibility study is a report that evaluates a proposed or contemplated action to see if it is (1) physically possible, (2) economically possible, or (3) desirable, or any combination of these. If your employer is considering starting up a van pool for employees, a feasibility study will be made before any decision is reached. If a city is planning to close a school because of declining enrollments, the school board will first commission a feasibility study. Feasibility studies—both those prepared in-house and those prepared by teams of management consultants brought in from outside—are almost always written by technical writers.

The feasibility study in Figure 17.3 shows how a table and an accompanying discussion can be used together to weigh the advantages and disadvantages of four alternative locations for a day-care center.

Feasibility studies may be either evaluation reports or recommendation reports, depending on their goal and audience. When a corporate or governmental organization hires a consulting firm to prepare a feasibility study, it usually asks for recommendations. When a superior assigns a subordinate to prepare such a study to aid the superior in making an important decision, he or she will probably *not* expect recommendations to be made. The decision-maker will want to see all the relevant evidence, not an edited version that makes a strong case for only one alternative. When you are asked to prepare a feasibility study, be sure to find out whether or not you are expected to offer recommendations.

Environmental Impact Statements. Another special type of evaluation report that you may need to be familiar with as a technical writer is the environmental impact statement. Environmental impact assessments are required by the National Environmental Policy Act for all federally funded or federally regulated projects "significantly affecting the quality of the human environment."[1] By law, these assessments must address the following topics:

1. The environmental impact of the proposed action.
2. Any adverse environmental effects that cannot be avoided should the proposal be implemented.
3. Alternatives to the proposed action.
4. The relationship between local short-term uses of the human environment and the maintenance and enhancement of long-term productivity.
5. Any irreversible and irretrievable commitments of resources that would be involved in the proposed action should it be implemented.

The Council on Environmental Quality guidelines for the preparation of environmental impact statements[2] do not specify a format for such reports. They range in length from a few pages to many volumes—depending on the project's controversiality as well as its complexity. An environmental impact statement is usually written after the decision has been made to go ahead with a project, not as part of the decision-making process. The key to writing a good environmental impact statement is to be as complete and as scrupulously honest as possible. Projects don't often get stopped for bad environmental impacts; as long as the environmental impact statement says that a bad effect may happen, the legal requirement is fulfilled, and the writer is protected. But projects can and do get stopped for bad environmental impact statements— statements that fail to weigh all the data and consider all the alternatives. The summary paragraph on page 392, from an environmental impact statement dealing with offshore oil and natural gas production in Texas, provides a balanced assessment of the risks to the environment posed by offshore drilling in the event of a hurricane.

DAYSCHOOL STEERING COMMITTEE REPORT

On March 1, 1981, Dayschool was notified by Arlington Borough that
federal funds had been allocated to improve the intersection adjacent to
Dayschool and that 60% of the land that now makes up Dayschool's play yard
would be appropriated by the borough for the improvement. Since state
standards require that child care centers provide at least 100 square feet of
outdoor play space per child, Dayschool will be unable to conform to state
standards if the school remains at its present site. Dayschool must move to a
new location by October 1, 1981 if it is to remain in operation. The steering
committee suggests that the Dayschool Board of Directors consider the following
alternative locations:

1. 166 Hermitage Avenue - The property located at 166 Hermitage Avenue
is a two-story house with large rooms, ample yard space, a sheltered porch that
would be suitable as a play area on rainy days, kitchen facilities, and laundry
facilities. The property could be rented for $325 a month, but state
regulations forbid the use of the second floor for child care. The second
floor could, however, be used for offices and storage space.

2. 1126 Renfrew Street - The property located at 1126 Renfrew Street is
a single-story house on a two-acre lot with kitchen and laundry facilities and
a full basement. The house has only 1200 square feet of floor space, including
basement; while Dayschool could operate on the site on a temporary basis, it
would be necessary to build an addition to the property if Dayschool were to

FIGURE 17.3 A Feasibility Study.

Page 2

remain there permanently. The property is for sale for $35,000; the addition, which could be designed specifically as a child care center, would cost approximately $15,000.

3. Greenville Shopping Center - Space could be rented for Dayschool at Greenville Shopping Center for $550 per month; the owner of the property is willing to let Dayschool make whatever structural modifications are necessary in order for the property to conform to state standards for child care centers. A contractor would need to be hired to remove two interior walls and put in two fire doors, but the Dayschool staff could install interior shelving and cabinets and fence in a portion of the land behind the shopping center for a play yard. There are no kitchen or laundry facilities.

4. RFD #4, Box 457 - The property located on County Road 518 is a single-story house on a little less than ten acres of land. Kitchen and laundry facilities are adequate and there are 1800 feet of floor space; no additional floor space would be required in the foreseeable future. No modifications to the property would be necessary beyond the installation of cabinets and shelving. The property is for sale at $45,000; it is, however, eight miles from Dayschool's present location. If the parents of children currently enrolled in Dayschool are unwilling to drive the additional distance to keep their children in Dayschool, it might be necessary for the school to furnish transportation from the center of Arlington. Transportation costs, including the cost of additional insurance, would probably have to be included in tuition fees.

Page 3

All the sites described above conform to state fire and building codes and possess sufficient floor space per child if Dayschool's enrollment remains at 35 children. It will be necessary for Dayschool to obtain a zoning variance before opening at any of the locations, but no objection from residents is anticipated.

The table below summarizes the pros and cons of each site.

Site	Cost	Floor Space	Facilities	Modifications Required	Distance from Arlington
166 Hermitage Avenue	$325/mo.	2200 sq. ft.	Kitchen/laundry	None	1 mile
1126 Renfrew Street	$40,000	1200 sq. ft.	Kitchen/laundry	Major	2 miles
Greenville Shopping Center	$550/mo.	2000 sq. ft.	None	Major	1/2 mile
RFD #4, Box 457	$45,000	2800 sq. ft.	Kitchen/laundry	None	8 miles

Of the available sites, the house on County Road 518 offers the fewest drawbacks. The price is not too high, and extensive remodeling would not be required. The added cost of providing transportation from the center of Arlington would be offset by the savings in capital expenditures, since no additional floor space would be required for Dayschool in the foreseeable future. Dayschool would be purchasing the house, so there would be no danger that the school would be asked to relocate, and the ten-acre lot provides ample space for any future expansion.

Most possible adverse effects on the environment that could not be avoided seem to be related to some major natural disaster. The most common natural disaster along the Texas coast is the hurricane. Up to this time, however, hurricanes have not caused any major pollution incidents resulting in serious damage to water-based oil and gas facilities and pipelines. If state and federal regulations are followed and routine maintenance is performed on the wells and platforms, the chances of significant pollution incidents resulting from hurricanes are slight. The wells are equipped with safety devices that should substantially reduce the chance of major oil spillage in the event of severe damage to the well structure. It is recognized that, if some unusual damage did occur, a major pollution problem could result that could create a hazard to public health, recreational areas, and the Gulf and estuarine biota. The greatest potential threat would be to the shallow, semienclosed coastal bays.

Style and Tone in Evaluation Reports

Because evaluations are judgmental, they require an exceptionally dispassionate tone if they are to avoid sounding biased. When you prepare an evaluation report, you must take special care not to use loaded language— words with emotional associations or strongly negative or positive connotations. You can express your preferences in an evaluation by pointing to favorable facts such as a low price, high-quality construction, or convenience, but you aren't supposed to communicate your preferences subliminally through your choice of words. Another way in which writers of evaluation reports evade charges of linguistic bias is by making their judgments relative rather than absolute. They don't say, "Site A is perfect"; instead, they say, "Site A is far superior to other available sites" and then provide the evidence that proves the assertion. Negative evaluations are often expressed in positive terms—not "the Renfrew Street house has insufficient floor space" but rather "the Renfrew Street house would require additional floor space." The fact in itself is damaging enough; there is no need to present it in a disparaging tone.

▶ Recommendation Reports

The difference between evaluation reports and recommendation reports is essentially one of degree. In a recommendation report, you do not simply imply your recommendations; you state them outright. Recommendation reports are not only informative but also persuasive, often strongly persuasive. Sometimes their persuasive function even overrides their informative function: Readers of recommendation reports have been known to read only the recommendations and skim the rest of the report, especially if they are sympathetic to the recommendations being set forth.

The goal of a recommendation report is to urge that some specific action or actions be taken. That is not the goal of evaluation reports: Their goal is to draw conclusions from a set of facts—pro and con—that are pertinent to a debatable issue. Like evaluation reports, recommendation reports draw conclusions, but they go one step further; they do not leave it to readers to infer what action should be taken. The following paragraph, if it were added to the Dayschool Steering Committee Report, would turn it into a recommendation report:

> We recommend that the Dayschool Board of Directors extend an offer of $45,000 to the owners of the property at RFD #4, Box 457. If the property is not available, we recommend that the board make arrangements to rent the property at 166 Hermitage Avenue until such time as a similar property becomes available for purchase.

Not only evaluation reports but analytical reports as well can turn into recommendation reports if a list of recommendations is added to them. The following recommendations closed the technical memorandum "Contaminant Inputs to the New York Bight," which we discussed on pages 379–383.

> 1. Obtain data on dredge spoil and sewage sludge leaching rates to determine if a significant portion of the contaminants from these sources reach overlying waters. Continually analyze dredge spoils and sewage sludge samples to determine the quality of changes occurring with time. Obtain the interrelation between the discharges to harbor waters and the quality of the dredge spoils.
> 2. Define more accurately the net contaminant load to the New York Bight by continuing the measurement of the net flux of contaminants across the Sandy Hook-Rockaway Point transect and modeling of the Transect Zone inland waters. Additional data are also required on atmospheric and urban runoff inputs, the least reliable estimates of this study. The effects of the contaminant inputs on water quality and on biological life should be continually evaluated.
> 3. Construct a mathematical model of the New York Bight to evaluate the significance and interrelations of the various contaminant inputs and to serve as a guide for bight sampling and research requirements. Initially, a relatively rough model should be constructed and modified with time as additional data become available. After adequate verification, the model will provide a sound technical base for evaluating alternative management decisions with respect to use of the New York Bight as a water resource.

Organizing Recommendation Reports

Recommendation reports are organized to highlight the recommendations they contain. Most readers of recommendation reports are sympathetic and perfectly willing to accept at face value the evaluation on which the recommendations are based even if they are not willing to go along with every single one of the recommendations. But they may be too busy to study all the evidence. Recommendation reports are organized to guide readers to the essential parts of the report and help them skim or skip through the rest.

The introduction to a recommendation report normally contains a statement of the problem (and background, if necessary), followed immediately by a list of findings, conclusions, and recommendations. Usually, this introduction can be blunt and to the point; readers do not need to be eased into the problem because they already know about it and they are expecting the report writer to recommend a solution.

The body of the report contains the evidence, usually arranged either by alternative—with the recommended alternative coming first—or by criterion—with the recommended alternative set apart from the also-rans:

   I. Introduction
  II. Criterion A
     A. Recommended alternative
     B. All others summarized
 III. Criterion B
     A. Recommended alternative
     B. All others summarized

Sometimes, recommendation reports do not even have a body. The following organizational pattern is becoming more and more common as the amount of information that technical professionals are expected to take in continues to increase exponentially:

   I. Introduction
     A. Statement of the problem
     B. Statement of purpose
     C. Statement of scope
     D. Evaluation criteria
     E. List of alternatives evaluated
  II. Conclusions and Recommendations
     A. List of recommendations
     B. Rationale for recommendations
       1. Summary of conclusions
       2. Description of selected findings
 III. Description of How Recommended Alternative Satisfies
     Evaluation Criteria
 Appendixes: Complete Findings

Notice that, in this organizational pattern, the "body" of the report is in the appendixes.

Offering Unsolicited Recommendations

Suppose *you* want to make recommendations in a report, even though you may not have been asked for them. Reports whose recommendations are unsolicited are organized in virtually the same way as evaluation reports because they must persuade readers that the recommendations they contain are worthy of serious consideration. You will have to write such reports when you think something should be done—or *not* done—in order to solve a problem, but you suspect that your readers either won't agree with you or won't think it is your job to suggest what to do. Because unsolicited recommendations come as a surprise, your tone should be subdued, and you should present your evidence in the body rather than in an attachment at the end.

The more unwelcome the recommendations, the stronger the necessity for *inductive* organization—easing the readers into the problem, weighing all evidence carefully and evenly, discussing advantages and disadvantages frankly and fully, eliminating undesirable alternatives with regret, and building to a strong conclusion in which the best possible case is made for the alternative you recommend. The case for the recommended alternative should be the last thing your audience reads; even if you decide to introduce your recommendations briefly at the beginning of the report, you should give them their most complete and most formal elaboration at the end.

 ## Writing Analytical, Evaluation, and Recommendation Reports

We have spent this entire chapter building a case that there are clear organizational differences among reports designed to analyze, with an emphasis on findings; reports designed to evaluate, with an emphasis on conclusions; and reports designed to recommend. In reality, of course, distinctions in form among these three types of final reports are, at best, blurry; their style and organizational strategies overlap and, in fact, they represent a progression from disinterested neutrality to partisan advocacy.

No organizational formula can substitute for a thorough audience profile and analysis of goals. All technical report writers must find ways to organize their reports so that the most important information is in sections that the audience is certain to read. The organization of technical reports is—above all—practical. If none of the patterns we have suggested to you here or in earlier chapters works for your particular report, create your own structure.

Review Questions

1. Where is the primary emphasis in an analytical report? In an evaluation report? In a recommendation report?
2. What is the difference between an inspection report and an investigation report?
3. What are the major parts of a laboratory report?
4. What is a feasibility study? An environmental impact statement?
5. What is the major difference between analytical reports and evaluation reports?

6. Where should you place the recommendations in a recommendation report?

7. How objective should you be when you prepare an evaluation report?

Assignments

1. Reread assignment 2 in "Graphics," Chapter 11, in which you were asked to gather information on who takes technical writing and to generate graphics based on that information. Now write the report. Include a description of how you obtained your data as well as the data themselves.

2. You are a building inspector at your school. Examine either a dorm room or a common room, looking for damaged furniture, broken and cracked windows, chipped or peeling walls and ceilings, faulty electrical switches, etc. Present your analysis in the form of a memo to your supervisor.

3. Write a short report in which you analyze a campus problem and recommend a solution to the appropriate authority. Write a follow-up report in which you evaluate several other possible solutions to the same problem.

4. The company you work for is considering the purchase of two new staff cars, and your supervisor has asked you to conduct the preliminary investigation. Because the cars will be used to carry equipment as well as people, the company is especially interested in station wagons. (Some of the equipment they will be required to carry measures $4' \times 4' \times 2'$.) The cars will be used primarily in town, but may occasionally be needed for long hauls and field work. They will not be used for off-road driving, however. Although the company would like to buy American cars, foreign cars are not to be ruled out. You are to evaluate station wagons on the basis of price, gas mileage, storage capacity, seating capacity, and ease of parking. Do not include recommendations. (You can gather information for this assignment from popular magazines—*Motor Trend, Consumer Reports*, etc.—or from car dealers in your area.)

5. The situation is similar to that described in the last assignment, except that this time you have been asked to recommend a specific make and model and, of course, to justify your recommendation.

6. Your school is considering making computer literacy a prerequisite for graduation. The administration has asked the Student Government Association, of which you are a member, to prepare a report focusing on the effects that this requirement would have on students. Your report should evaluate the pros and cons, and conclude either in favor of or against the proposal. If you decide in favor, feel free to offer recommendations on how a computer literacy course might be implemented.

Notes

1. *National Environmental Policy Act*, Public Law 91-190, 91st Congress S. 1075, January 1, 1970.

2. The original Council on Environmental Quality guidelines may be found in the *Federal Register*, vol. 38, no. 147, pp. 20550-20562, August 1, 1973.

18

Oral Reports and Group Meetings

You can expect to spend upwards of half your time on the job engaged in some form of technical communication—writing proposals and reports, composing letters, dashing off memos, and so on. Not all that time, however, will be spent writing. Technical people do a good deal of talking as well. They talk to each other while they work, of course, and they discuss work over coffee in the cafeteria, but they also have occasion to do more formal talking: They present oral reports to their supervisors or to the company's board of directors, make presentations to potential customers, address groups of their colleagues at conferences, speak to the public and to local officials, lead group meetings with their staff, attend group meetings with other people from the firm, and so on. Recent studies suggest that technical people, especially researchers, devote nearly a quarter of their time to preparing and delivering oral communications.

In this chapter we discuss the two basic types of oral communication that will be required of you: the oral report and the group discussion. The distinction between them is that in an oral report the speaker is the sole focus of attention, delivering information to an audience; in group discussions, on the other hand, everyone present is expected to participate.

 Oral Reports

The basic process of preparing an oral report is like the process of preparing a written one: You must establish your goals, analyze the audience, gather information, and organize the material so that it has a clear and effective beginning, middle, and end. The big difference is that oral reports are presented to live audiences, so you can adjust your presentation as you are making it. If listeners look confused or bored, or start whispering to each other instead of listening to you, you can make changes right on the spot. A written report, of course, admits no such flexibility. As a speaker you can even stop your presentation and solicit responses from your audience if you think you aren't getting across, and you can end with a question-and-answer period to cover any loose ends. In other words, if you are attentive to the feedback you receive from your audience, an oral report allows for maximum clarity of the material presented.

The limitations of an oral report are also audience-related. The biggest one is that listening is a much harder activity than reading. A reader can take occasional breaks from a written report—get up and stretch, fix a cup of coffee, walk around the desk—and then come back to the report refreshed. He or she can reread complex passages, skim over sections that don't seem relevant, and study especially relevant ones in great detail. Listeners, on the other hand, cannot set their own pace: They hear an oral report only once and run the risk of missing something important if they let their concentration lapse even for a minute. This need to concentrate takes its toll. As any experienced speaker (or listener) knows, an audience's attention span is fairly short, especially when the subject matter is abstract or difficult—as it usually is in technical presentations. Even if your listeners pay close attention to every word you speak, they can easily get lost in a complex argument that would give them no trouble at all if they were reading it and had section headings, paragraph indentations, section summaries, or other visual cues to keep them oriented and to provide graphic transitions.

What all these limitations suggest in practical terms is that an oral presentation must be simpler than a written one: Its scope must be narrower; its goals, more modest. It must be put together so that listeners can follow the organization easily. And it must take special care to keep listeners oriented—usually through liberal use of repetition and transition.

Goal and Audience

Analyzing your audience usually takes precedence in oral reports even over defining your goals. With a little effort you can usually find out exactly who is going to be filing into the conference room or auditorium to hear you. You may have invited them yourself; more likely, they invited you. If you are making an in-house presentation or one to a client organization that you have worked with in the past, you will probably know most of the people attending by name. If you are addressing colleagues at a conference, you will probably know some of them, if only by reputation. Even if you don't know the names and faces of the audience you will be addressing, a few phone calls or letters of inquiry will give you all you need to know to size them up as a group.

Assessing Listeners' Interests and Knowledge. As you analyze your audience, two essential points must be borne in mind: Your oral report must speak to your audience's interests, and it must speak at your audience's level. Ask yourself first off *why* these people are coming to hear you speak. What do they want to know? What do they need to know? Often in oral presentations the answers to these two questions are quite different, and your report will have to devote some time to convincing your listeners that they *should* want to know what you are telling them. In other words, you will have to tie your report securely to their interests.

As best you can, moreover, you should determine the attitude toward your subject that the audience will bring to your presentation. Will they have strong feelings or opinions about your subject, about you or the company you work for, about technology or science in general? You may need to use the opening moments of your speech to gain their confidence or to defuse their hostility.

Analyzing their level of understanding is just as important. You need to determine how much your listeners already know about the subject you will be reporting on. Because listening is such a difficult—and, unfortunately, underdeveloped—skill for most people, even experts in your field may have trouble following an oral report filled with technical abstractions, quickly spoken equations, and the like. People who aren't experts will, of course, find it hopeless. They will let the point slip away and perhaps miss something crucial in your presentation. To be on the safe side, try to keep your language a little simpler than you would if you were writing to the same audience.

Limiting Your Scope and Goals. Like written reports, oral reports try to inform, to convince, or to motivate their audience. Their goals, however, are not identical with those of written reports. Evidence overwhelmingly shows that oral reports are much better than written ones at motivation and emotional persuasion, much worse at information and rational persuasion. As conveyers of pure information, they surpass written reports only when an actual demonstration can be part of the presentation. In other words, oral presentations are best suited for showing, not telling; for inspiring, not instructing; for explaining why, not how.

Listener attention span being what it is, you will have to accept the fact that your report can cover only a few major points—its scope must be modest. As a result, your goals, too, will have to be modest; one presentation is not going to change long-standing policies or values. As you formulate your goals, make certain they are suitably narrow:

Long-range goal:	to change company policy on research
Immediate goal:	to convince audience to appoint a committee to reconsider policy

General goal:	to inform audience about my research
Narrow goal:	to summarize results of study series 3

In sum, you should determine what you can reasonably expect your listeners to come away with, given what they came in with, and make that your goal.

By limiting yourself to a few major talking points, you allow the time necessary to ensure that what you do cover gets through to your listeners.

You will use this time to reinforce your points. One of the cardinal rules of oral reporting is to support every point you make with concrete examples. The more abstract the point, the more examples you should use.

Clusters of examples all leading to the same point may strike you as inefficient and repetitive. But they are necessary. Another cardinal rule of oral reporting—perhaps THE cardinal rule—is to *repeat yourself*. Don't just state a conclusion once in your presentation; mention it again and again, whenever you talk about the evidence that supports it or the recommendations that follow from it. Don't just state the purpose of your research once in the introduction to your presentation; keep referring to it when you discuss the methods you used to achieve it or the findings that illuminate it. Pepper your talk with summaries and recapitulations, with transitions and restatements—all to orient your listeners, to clarify meaning, and to increase the chances that your listeners will hear and retain what you say.

Types of Oral Reports

Oral reports on technical topics are usually read from manuscript or presented extemporaneously from notes. Memorized or impromptu presentations are seldom wise in technical fields; without a manuscript or notes you are too vulnerable to factual error.

The Read-from-Manuscript Report. The great disadvantage of the read-from-manuscript report is that unskilled speech readers may drone through a whole speech without ever lifting their eyes off the page to acknowledge the audience's presence. A more skilled speech reader pauses to check audience response and to answer questions and then easily picks up where he or she left off.

Although a read-aloud speech isn't as spontaneous-sounding as one in which the speaker frames the sentences on the spot, it can be effective so long as you remember as you write it that it is to be *spoken*—it should include lots of examples and repetition, and its sentences and vocabulary should be simple. Because all of the details are written out, this is usually a highly accurate and credible type of oral report, especially popular among academics delivering papers at conventions and scholarly meetings.

The Extemporaneous Report. From the point of view of the person delivering the report, the read-from-manuscript presentation has one overwhelming advantage: The speaker need not worry about fumbling for the right word. Nonetheless, the extemporaneous report is preferable because of its flexibility. Because you make up the actual phrases and sentences as you deliver the report, the presentation usually sounds less wooden and more spontaneous. Moreover, the extemporaneous report allows maximum feedback from the audience. With no finished text to glue your eyes to, you can look out at your audience and watch for signs of attention or inattention, agreement or disagreement, comprehension or befuddlement. You can pause, paraphrase what you just said or expand on it, or even entertain questions from the house—and then easily pick up the thread of your speech from the outline. If the feedback suggests that you are spending too much time explaining the obvious, you can pick up the pace; if you are running long, you can skip some minor points or an example or two. It is much harder to do this kind of on-the-spot editing with a written-out text because all the points are nestled into complete sentences and paragraphs and thus are hard to pick out at a glance.

You should keep in mind that the extemporaneous report is *not* impromptu—it should be carefully rehearsed. Once you have established your outline of talking points, you should practice delivering your report in front of friends or in front of a mirror, trying different ways of saying the same thing, sharpening your delivery until you feel confident that you can get through all your points adequately in the allotted time. The parts that make you wish you had a manuscript—key statements, controversial ideas, particularly effective transitions, particularly clever turns of phrase—you can write out word for word.

Organizing an Oral Report

Developing an outline for an oral report is much the same as developing one for any other type of report. In fact, because oral reports tend to be limited in scope, they are generally fairly easy to organize. Start by organizing the body of the report.

The Body. Once you have limited yourself to four or five speaking points, your only problem will be to settle on the most suitable order of presenting them. All the options we discussed in Chapter 2, "Gathering and Organizing Information," are available—chronological, spatial, order of importance, and so on—depending, of course, on the nature of your material.

For material that lends itself to order of importance—and that, in fact, is most material—we suggest a slight variation that specifically takes into account the listening behavior of a live audience. Experienced speakers have long noticed that audiences listen most efficiently at the beginning of a presentation; that as the presentation proceeds, their ability to concentrate and to absorb information diminishes; but that it increases somewhat when they know the end of the presentation is approaching. An oral report arranged in strict order of decreasing importance, then, would reinforce an audience's tendency to pay less and less attention to the middle sections; moreover, it would present listeners with only the least important points at the end of the report and thus not make good use of their renewed attention. To counteract this problem, many seasoned speakers prefer to begin with a strong point—perhaps the most difficult one, if not the most important one—then "downshift" to a minor point and build up from there, ending with a very strong one. This arrangement is designed to hold the audience's attention and also to put the points you most want heard in positions where they are most likely to be heard.

The Introduction. The introduction of your report deserves special care. You want your audience to know from the outset that you are knowledgeable, comfortable with yourself and your material, and as friendly and approachable as the occasion permits. A number of attention-getting devices are worth considering. You might, for instance, begin your presentation with an unusual or surprising fact or statistic, or with a brief, relevant quotation. Or you might say something about the occasion that has brought you and the audience together—something to help establish a mutual interest or concern. Or you might want to start with a telling anecdote that ties directly into your subject. Sometimes the statement of purpose alone will be sufficient to grab your listeners' attention: "I want to talk to you today about the future of your jobs. To be frank, they are in jeopardy."

In addition to securing attention, the introduction should provide the purpose of your talk, the background and context of your talk, and a brief summary of the points you will cover. The statement of purpose should tell your listeners clearly and precisely why you are addressing them: "I want to share with you a technique that can cut production time by as much as 25 percent." To establish a context for your report, you might want to give your listeners a brief history of the subject under discussion or of your involvement with it. Or you might want to explain the importance to your listeners of the subject or to refresh their memories about the kind of work done by people in your field—to provide, in other words, the background that will enable your audience to see where your report fits in the general scheme of things.

Most introductions end with a brief summary of the major points you intend to cover in the report. Such a summary isn't usually necessary in the introduction of a written report because the abstract and table of contents already provide it. But because an oral report has neither abstract nor table of contents, you should take extra care to orient your listeners in the introduction. This summary provides a simple and effective transition into your first point.

The Conclusion. If a good introduction ends by telling listeners what you are going to tell them, a good conclusion begins by telling them what you have just told them. It again summarizes the major points covered in depth in your presentation. Though this may seem repetitious, it is useful and effective—it is the equivalent of *rereading*, which, of course, is something that listeners cannot do. An oral report whose goal is to inform may end with just the summary and a brief "thank you and good day." If the goal is to motivate or persuade, the summary should lead into a call for action, a direct appeal, a challenge, or even a rhetorical question ("Having examined these problems in detail, shouldn't we now do such and such?").

Preparing for an Oral Report

Assuming that you are going to make an extemporaneous presentation, your next step after writing the outline is to prepare a performance copy—

the written document that you will refer to as you speak. Some speakers prefer to use a formal outline with Roman numerals, letters, and the rest. Others prefer a series of note cards; and still others, a series of notes on loose-leaf paper held in a binder.

The real problem isn't choosing between note cards and loose-leaf paper. It is deciding on the kind and amount of information to put on the cards or paper. If your notes are too extensive—with every supporting idea, illustration, definition, and so on written out—the tendency will be to read them rather than to talk, and you will end up with a disguised read-from-manuscript presentation. In any event, it is easier to pick out your main points if they are surrounded by white space rather than by words. Your notes, therefore, should be as brief as possible. They should remind you of the points you need to cover and give you only enough information to support those points. Obviously, the more you have practiced, the more familiar you will be with your material, and the less you will have to write down. Often a few key words will be all you need for a whole section of your talk. If, for instance, you are talking to a general audience about an experimental geothermal power plant, your notes for the section on environmental effects might look like this:

Environmental effects minimal to nonexistent:

seismic   hydrologic   noise   emissions

*****Use California example*****

Of course, if you have quotations or statistics to present, you should write them out and not trust your memory.

The important thing to remember about your notes is that they are for your eyes alone. Thus, they don't have to be uniformly brief or extensive throughout. If you know some parts of your speech less well than others, you can make the notes to those parts more complete. Feel free, moreover, to include any sort of "stage direction" that you think might help you. You might, for instance, want to remind yourself to turn on the overhead projector or to change the display board, or you might want to bracket points or examples that you can cut if you start running long. You can put all your main points in all capitals or red ink or whatever so that they can easily be distinguished from supporting ones. You can mark places where you think you might like to pause and field questions from the house. Some speakers even put approximate speaking times for each section in the margin and keep a watch handy to control their pacing. Anything that will help smooth out your talk is fine.

Presenting an Oral Report

Three components make up an oral presentation: your language (the words themselves), your delivery (voice, body language, and so on), and, usually, the visual aids that augment your talk.

Language. Almost everything we said in Chapter 12 on "Technical Report Style" applies to speaking as well as to writing. Clarity and precision are probably even more important in oral reports, because listeners cannot spend time poring over your words trying to figure out what you must have meant. Though you should for the most part stick to standard usage and accepted grammar, the fact that you are speaking and can therefore modulate the pitch and volume of your voice for emphasis means that you can sometimes use sentence fragments and other substandard grammatical devices effectively— in small doses. Oral reports are generally a good deal less formal than written ones, so standard contractions and common idioms are perfectly acceptable (in a speech we would have said "OK"). And feel free to use "I"—after all, because you are physically present, it is only natural to acknowledge that you exist. Here are three additional suggestions.

Use Simple, Concrete Language. Stay away from the kind of inflated, "grand" language that often passes for oratorical style: "It is indeed an honor, and it is certainly a distinct pleasure, to stand before you today, on this most auspicious of occasions. . . ." Watch out, too, for abstract language; if you need abstractions, be sure to back them up with concrete explanations. And avoid unnecessary jargon. If you absolutely need a technical term, be sure to define it—and the second time you use it, define it again, just in case.

Avoid Complex Sentence Structure. Keep your sentences brief and simple, the way you naturally do when you are speaking to someone. Sentences filled with dependent clauses are going to lose listeners at every turn. People can *read* complex sentence structures without much difficulty, but the ear is much less efficient than the eye; if a long dependent clause separates subject and verb, the subject is soon forgotten—and, of course, the ear can't scan back for it. Run-on sentences, on the other hand, work fine for the ear. Listeners have no trouble following a string of short sentences that begin with conjunctions like "and," "but," "yet," or "so."

Provide Transitions. As we mentioned at the beginning of the chapter, guiding your listeners from idea to idea is especially difficult because they have none of the cues that readers do—no headings, no paragraph indentations, no white spaces. Thus you have to give them verbal equivalents to signal that you have come to the end of one point and are beginning another, that you are continuing in the same vein, or whatever. Even within a sentence or section of your report, you will probably need more transitional words and phrases than you would if the report were being written.

Between major points or between sections of your report, you may need to devote a sentence or even a paragraph to making the transition.

Having discussed the problem in detail, let us now consider our options for dealing with it.

To summarize briefly, the proposed project makes a good deal of sense from a purely scientific viewpoint: It will demonstrate *X*, *Y*, and *Z*. The next question we must consider, then, is whether the project is economically feasible. There are two points to keep in mind here. First, . . .

Delivery. You learn effective gesturing, voice projection, articulation, and pitch by making speeches and practicing. You already know that you should speak clearly and not mumble; that you should stand comfortably and relaxed, not slouch over the podium and shift nervously from leg to leg; that you should use hand gestures to emphasize your points; that you want to sound expressive and not speak in a monotone. All these points are obvious, and, obviously, they are easier to recommend than to do.

Preparation is crucial. Try to find out the size of the room in which you are going to speak, and, if possible, try to get there before the audience so that you can practice projecting your voice. Keep in mind, too, that the larger the room, the more slowly you will have to speak to be audible in the back rows; between the ambient noise and the echo of your own voice, your words can sound blurry as they travel. Speaking too fast is a problem even in a small room; nervous speakers tend to talk faster than people can listen. A speech timed at home at exactly forty-five minutes can come in at around thirty in front of an audience. (This is why putting times in the margin of your notes is not a bad idea. If you see yourself rushing through your material, force yourself to talk at what you think is a very slow pace—it will probably be about right.)

Before you get in front of your audience, try tape-recording your talk, and note places where you can add emphasis by modulating your voice. Raising or lowering your voice, changing pitch, changing speed of delivery, even creating momentary silences—all these make a talk sound more alive.

Finally, it is important to make eye contact with people in the audience. For one thing, it will emphasize the fact that you are speaking to them and that they matter. But more important, it will allow you to see how well you are getting across.

Visual Aids. Maps, charts, tables, graphs, photos, drawings, and models are all useful visual aids. Even a list of key words or phrases from your talk, boldly displayed, can help to reinforce your points.

Advantages of Visual Aids. Visuals can augment your description of an experimental design or procedure. For example, listeners who would have a hard time following all the intricate steps of an elaborate procedure will benefit enormously from a flowchart. They will be able to see where each step you take up fits into the overall design. Similarly, a good photograph or drawing of a piece of equipment will enhance your verbal description; an actual working model will, of course, enhance it even more.

Visuals are also very useful for presenting numerical data. If you try to read numbers, percentages, equations, and the like to your listeners, you are bound to swamp them. Instead, put your data into a table or graph that they can study as you talk about it. And keep such visuals clean, clear, and simple—

to be useful, they have to be easy to read. Bar charts, pie charts, and line graphs (pp. 250–254) are especially good because they require little explanation.

Tables (pp. 231–240) can be helpful, too, but you have to take special care to explain them and keep them simple. To be useful for an oral presentation, your tables should have no more than a few columns or rows of numbers, the numbers themselves should be large enough so that people in the back can make them out, and the tabulated data should be arranged to demonstrate a discernible comparison, pattern, or trend—don't just present raw data. A good rule to remember in preparing tables for oral reports is this: Never put any piece of data into a table unless you plan to discuss it in your presentation.

Types of Visual Aids. Visuals can be presented in various ways. If you can get into the room before the presentation, you can put your data on the blackboard, and then cover the blackboard until that part of the speech. (It is usually a bad idea to draw your graphs and tables during the presentation; you don't want to interrupt yourself or turn your back on the audience.) A second way to present visuals is with display boards, which you prepare before the presentation. You can set these up in order on an easel and remove them one at a time after you have discussed them. Best for presenting information to a small group of people, display boards are much less useful if you are addressing a large audience—they are usually too small to be visible in the back rows.

Thus, for large audiences, most speakers prefer to project their graphics on a screen, by means of either an overhead projector or a slide projector. Slides give the sharpest, most accurate image, and they are especially good for showing photographs. But they have three disadvantages: They are expensive, they must be prepared weeks in advance (so that you have time to get them processed), and they require that you darken the room during your presentation. (You are bound to lose some control over the audience when you become a disembodied voice.) Overhead projector transparencies, on the other hand, are cheaper, easier to prepare, and much more controllable. You can draw your graphics on regular paper and then transfer them onto transparencies on almost any office duplicator. And you don't need to darken the room to show them. Moreover, you can set the projector right next to you so that you never have to relinquish your position as center of attention—you can point to something on the transparency without turning your back to the audience.

Don't use handouts to augment your talk. Once they have been passed out, they are out of your control, and there is no way to prevent your audience from rustling through them instead of listening to you. (You might, however, consider distributing handouts *after* you have finished your presentation.)

Whether you use display boards, slides, or transparencies, be sure that they are in the right order before you begin your talk. And remember that they can't serve as aids unless you use them—put cues in your notes to remind you to "Go to Table 5."

▶ Group Meetings

The opportunity to make an oral presentation does not arise every day. Especially early in your career, when you are more likely to be part of a team than its director or supervisor, you may have occasion to deliver an oral report only rarely. What you will have occasion to do—and often—is participate in group meetings.

Although the variations on the activities of a group meeting are limitless, most meetings are called to solve problems. Because group meetings draw on the resources of a number of interested and involved people, they can be an effective means of solving problems that would overwhelm one person. But good meetings are unfortunately rare. All too often, meetings are inefficient, ineffectual, and emotionally draining: Everyone talks at once, trying to shout down opposing viewpoints, or no one talks at all; one loudmouth acts as though he or she were the only one whose opinions mattered; the discussion goes off on a tangent and never returns; people stick to their guns no matter what, going over the same ground again and again until they have exhausted, if not convinced, the other members. In such meetings, problems usually end up getting tabled or left for the group leader to resolve, and everyone leaves with the feeling that time has been frittered away.

Culpability for this kind of horror rests to a large extent with the leader, whose job it should be to keep the discussion on target and under control. But individual participants also determine the effectiveness and efficiency of a group meeting. In this section we discuss things that you can do both as a participant and as a leader to make the meetings you attend worthwhile and productive. Because early in your career you will be attending more meetings than you will be leading, we will start there.

Participating in a Meeting

To make the meetings you attend effective and efficient, you must recognize that showing up is not enough: You should take an active role in the process.

Do Your Homework. Well in advance of the meeting, you should find out what problem or problems are going to be discussed so that you can prepare. If the group leader has drawn up and distributed an agenda, read it over carefully, and then make notes to yourself of things you want to say at the meeting. Naturally, you won't have an equal amount to say about every item on the agenda, but you should at least think about each one. Jot down points you think need to be discussed or clarified, questions that need to be raised. For items that are especially important to you or that are in your specific area of expertise, spend some time gathering relevant information so that you can cite exact facts and figures at the meeting. Having hard data to contribute will probably make any position you take stronger; it will certainly improve the group's ability to reach a decision.

Even if no formal agenda has been circulated, you can still prepare for the meeting. Talk to the leader, and see what she or he has planned. If you have a particular item that you want discussed at the meeting, you can lobby for it with the leader and other members so that they will come prepared to consider it.

Clarify Goals; Define Problems. An able and efficient group leader will usually open a meeting with an overview of the agenda, whether or not it has been previously circulated, making certain that everyone in the group knows what is to be discussed and what the group's responsibilities are. (Some leaders may invite participants to propose other agenda items as well.) Without direction-setting of this sort, discussions tend to wander off the mark. If the leader doesn't perform this important function, therefore, somebody else should—you, for instance. Once it is clear that things are wandering, you might suggest that the group pause to enumerate the problems facing it, the scope and dimensions of each, and the relationships among them, so that a logical assault can be devised. Furthermore, you might ask the group to delineate its goals: Are you meeting to suggest a solution to a problem or to make a firm decision? Will the group implement the decision or will it leave that to someone else? Is the group's decision binding, or is it subject to revision by a higher authority?

Don't Be Afraid to Talk. It is easy for a member, especially a shy one or a newcomer, to defer to those members who talk voluminously—easier than fighting for the floor, much easier than contradicting a facile talker. What can you do to overcome the reluctance to participate? First, having your notes in front of you should help you get over the fear that you won't be able to express yourself adequately. Second, you can *learn* how to get the floor. Sensitive leaders will notice signals that indicate that you want to speak—a knowing glance, a hand gesture, a sudden body movement—and they will clear a path for you. With less sensitive leaders, you will have to run your own interference. A good way to begin is to make a transition from the last speaker's remarks to your own:

❙ "John, you just mentioned such and such, and I'd like to amplify. . . ."

❙ "Susan's observation about such and such raises a related problem. . . ."

Or even:

❙ "Mary. Mary!! Let me interrupt you for a second to expand on the excellent point you just made."

Or to the leader:

I "Tom, may I interrupt Mary for a moment?"

Keep in mind, too, that not everything you say has to be a startling fact or a staunchly defended opinion. You should feel free to ask questions about other members' ideas and how they arrived at them, to ask for clarifications of points you don't understand or for definitions of terms you are unsure of, and so on. And to ensure that both you and the other group members are hearing the same things, you might want to try summarizing or paraphrasing other people's points, and then asking the group if you have gotten it right.

Don't Be Afraid to Listen. Meetings are really about conflict and conflict resolution. If all the members of the group agreed on every point, if there were no problems to thrash out, there would be no need to talk or to listen to each other; all decisions would be affirmed by acclamation, and the meeting would become a kind of quaint ceremony. Although this happens occasionally, much more often meetings uncover genuine divisions of opinion that require lengthy debate to resolve. Thus, how well you listen to opposing viewpoints is just as important to the success of the meeting as how well you put forth your own. For one thing, you have to listen to make sure that the other members understand what you have said so that you can clarify your position if they have gotten it wrong. And you have to listen to what they say about your position so that you can defend it rationally. Notice that *clarifying* and *defending* are different activities. Other members may understand exactly what you mean and still disagree with you. To defend your position entails showing how and why you arrived at it; merely to restate your position once it is clear to everyone will not resolve the debate. Conversely, defending it *before* it is clear to everyone is pointless; clarification must come first.

Listening is not merely a strategy for countering opposition. You should listen to others' opinions because they are likely to have some merit. This seems an obvious point, but many people act as though their ideas were unassailable; they will assert and reassert their position and never allow themselves to hear the opposition's case. Indeed, what they are really defending is not their position, but their *ego*—and ego is the greatest subverter of meetings. If you ever find yourself going over the same ground again and again, try to face the possibility that your own desire to be right may be impeding the progress of the meeting.

You should listen to everything that is said at a meeting even if your own ideas and opinions aren't on the line. Maintaining a constant, high level of attention is difficult, especially when group discussion moves into an area in which you are only peripherally interested. Yet your participation in such a discussion can be critical: Precisely *because* you are not emotionally involved with the issue being discussed, you might be able to provide the kind of candid, unbiased insight that leads to a happy resolution. At the very least, you will be able to provide an outsider's viewpoint—and that will give the major combatants some new information to assimilate.

Watch for Hidden Agendas. The desire to have others think we are smart and capable is just one item on a secret list that most of us keep—our hidden agenda. To demonstrate our power, to intimidate a rival, to impress a superior—these are likely to be other items on the same list. The trouble with hidden agendas is that they can interfere with the decision-making process of a meeting. Here is a common example of a hidden agenda in operation: One member of the group might agree to support a resolution sponsored by another member in exchange for support on a separate and unrelated issue. Meetings in which the participants give priority to their hidden agendas can result in distrust, hostility, and bad decisions. When a meeting goes awry, asking yourself whose hidden agendas are at work can help you diagnose the problem and decide how to deal with it.

Be Attentive to the Feelings of the Other Members. In the heat of an intense debate, it is easy to forget that the people you are arguing with are probably as emotionally committed to their side as you are to yours. And although we are all supposed to know that an attack on one of our ideas is not necessarily intended personally, the fact remains that egos bruise, and that bruised egos make bad meetings. Therefore, you should do your part to maintain good feelings among the group members. After observing the group's dynamics for a while, you will know when it is time to reduce tension, to encourage one member to discontinue his or her attack on another member's argument, to rush to someone's defense, to relent a bit in pressing your own argument, and so on. It may take no more than a word of encouragement to a beleaguered member to clear the air, or it may take something more—some joking around, some lighthearted irrelevance, maybe even a total change of subject or a brief recess. But do not cave in on a point you feel strongly about merely to avoid hurting someone's feelings; on the contrary, if you can maintain a feeling of good fellowship among the members of the group, you will actually be increasing the likelihood of resolving disputes to everyone's satisfaction.

Running a Meeting

As you can see, many of our suggestions about participating in a meeting concern how to act in the absence of effective leadership—when the leader fails to prepare adequately for the meeting, fails to give every member an opportunity to be heard, fails to keep the discussion calm and relevant, and so on. In a well-run meeting, group members can usually concentrate on matters of substance and leave these other concerns mostly to the leader. The leader's primary responsibility is not substantive, but administrative: to open and close the meeting, to decide when to move on to a new topic, to control the flow of the discussion. Here are some practical suggestions for running a meeting.

Set the Agenda. Establishing the list of items to be discussed at the meeting should be your first concern, well before the day of the meeting itself. If the problem to be discussed is complex, you need more than just a list. Consider the following problems:

> What should we do about reorganizing the department?
> How can we increase production efficiency?
> What is our position on the director's proposal?
> How can we modify our design to suit our customers' most recent demands?

To avoid too freewheeling and inefficient a discussion, you need to plan an orderly attack on such problems. Break the problem down into discussable parts. Define the scope of the problem. (How much of an "increase" are you aiming for? What parts of the design do the customers want modified? How extensive can the modifications be?) Determine how the problem affects the members of the group, noting particularly those members with special expertise in dealing with this particular type of problem. Decide whether you want them to brainstorm freely, to list pros and cons of the problem so that you can make a final decision, to come to a decision themselves, or to legitimize a decision you have already reached. Establish possible criteria for evaluating solutions proposed by the group. Your main objective in this analysis is to provide a framework for discussion; it is not to solve each problem before the group meets.

If you have more than a few items on your agenda, give some thought to the order in which you want them discussed. A common mistake in agenda-setting is to begin with the least important items, saving the most complex ones for last. What invariably happens is that the group ends up wrangling over the easy problems, exhausting itself and leaving too little time to do justice to the big problems. Better to launch into the most problematical item first when everyone is still fresh and eager. If the group has only a few minutes left at the end of the meeting for the trivia, participants are unlikely to inflate them out of proportion. It is wise to allot a certain amount of time to each item, to guarantee that all the items have a chance to be covered.

Once you have set the agenda, you should distribute it—or at least a simplified version of it—to the members of the group. If possible, give it to them in advance of the meeting, so that they have time to study and prepare; otherwise, give it to them at the start of the meeting. Also consider alerting group members with special interest or expertise in particular agenda items that you will be relying on their participation. If knowledgeable members come prepared to share their knowledge, you will have an easier time getting the ball rolling. Finally, bear in mind that you can solicit additions to your agenda from participants, either in advance or at the meeting.

Open the Meeting. At the beginning of the meeting you should probably run through the agenda, briefly summarizing your analysis of each item. In addition, you should make sure that everyone understands the function of the group: to solve a problem, to implement an action—whatever.

The transition from this overview of the agenda to a discussion of the first item is often awkward, with each member reluctant to be the first to volunteer. If you have lined up some experts, now you can call on them to break the ice. If you haven't, resist the temptation to play the role yourself; it sets a bad precedent when the leader asks a question and then answers it, a precedent that reinforces group members' silence. Instead, coax someone likely to have good things to say to speak, or suggest that everyone make a short analysis of the problem one at a time. Once people begin to talk, they usually loosen up, and the discussion proceeds less awkwardly.

Be aware, however, that members have a tendency to leap to solutions before the problem itself has been thoroughly examined—and once the group has begun to debate solutions, it is difficult to go back and pick up any overlooked pieces of background information. Thus, if early in the discussion participants begin to offer solutions, cut them off as quickly and as graciously as you can.

Control the Discussion. A major part of the leader's job is to control discussion—to keep it moving and to keep it on target. Keeping it moving involves, among other things, encouraging timid people to talk and imposing silence tactfully on the garrulous.

Controlling the discussion also involves leading the participants back to the topic after they have gotten embroiled in a complex and interesting digression. And when discussion trails off—either because the group has reached a consensus or because it has simply run out of things to say—it will be your job to lead the way to the next item. It is often useful to make a summary transition between segments: to summarize the discussion of the problem before moving on to the discussion of the solution; then to summarize both before moving on to the next item on the agenda. The summary rounds off the discussion nicely. It says, in effect, "Here is what we have accomplished so far."

Mediate Conflicts. Dealing with conflicts is the group leader's most important function. When a serious conflict brings the group's progress to a halt, you must intercede. Conflict is inevitable; the hostility and emotional fireworks that often accompany it are probably not. Unfortunately, many leaders seem to lump them together and try to avoid hostility by avoiding conflict. The phrase "agree to disagree" exemplifies this sort of conflict-avoidance. If two people agree to disagree on whether they should paint their house green or blue, they may be able to avoid conflict, but they won't be able to paint their house.

As the leader of a meeting, you have two basic options when a conflict arises: You can *dictate* a resolution, or you can *facilitate* one. Dictating is efficient and tidy; with a single sentence you can erase a conflict. ("Having heard both sides, I must agree with Helen.") But it works only under certain conditions: First, the group must trust and respect you enough to defer to your opinion without resentment; second, you must firmly believe in your decision and be willing to accept sole responsibility for the consequences. If you try to resolve a conflict *merely* to resolve it, without meeting these two conditions, you will come off as arbitrary, and you will probably antagonize everyone in the group, including the person you sided with. Worse yet, you may be wrong, or you may miss a better solution that is waiting to be found.

If you can't really dictate a resolution, you should instead facilitate it, acting as a referee among members whose positions are in conflict. As facilitator, you want to keep the argument focused and clean. Make them defend—not merely restate—their positions; probe until you find out what they would need to alter their positions. In other words, help the opposing sides search for common ground. And encourage the other participants to join in the search; more often than not, the source of the final solution is not one of the original combatants.

At the same time, work to diminish hostility. Be on the lookout for signs of trouble: curt or sarcastic responses to honest questions, insistent repetition of a single point, aggressive actions like slamming a pen onto the table or collapsing heavily into the chair with a dismissive facial or hand gesture, and so on. If you are paying attention, you should be able to jump in before tempers flare. Let both sides know that you are genuinely concerned, and encourage them to work it out. If tension continues to build, you might try to vent it by changing the format of the meeting: Have each of the arguers summarize and justify his or her position uninterrupted, or have each member of the group comment on how he or she sees the conflict, or call a five-minute recess to allow nerves to settle.

Obviously, the best way to resolve a conflict is to keep plugging away until a true consensus is reached. A decision reached by consensus will have something in it to please every member and nothing in it that deeply offends anyone. And because decisions reached by consensus call for rigorous discussion and argument, they are usually well thought-out and effective. For the very same reasons, however, they take a long time to reach. And the plain fact is that you may not always have enough time to achieve a unanimous decision, especially if you are pushing up against a hard deadline. Thus, once you have allowed as much time as you possibly can, you may have to settle for majority rule rather than consensus. Because voting produces winners and losers, it is likely to leave some members unsatisfied. And because it is much more subject to hidden agendas than consensus-building, the decisions it produces are often less good than consensus decisions. It is, nevertheless, the next best way to resolve conflicts—and it is certainly more efficient. If the vote is heavily one-sided, you might be able to appease the losers a bit by noting their opposition in the draft of the resolution. And if the vote is close, you should probably allow the losers to draft a separate "minority report."

Close the Meeting. When all the items on the agenda have been dealt with— or when time runs out—review the group's progress by recapping the decisions you have reached together. And be sure to specify who is to do what about each item after the meeting: "John will check with marketing on this one and work up a draft policy by next Thursday." In addition, outline the work that still needs to be done, especially the problems that will be taken up at the next meeting—and announce when the next meeting will be.

Follow Through. Many a motion that has been rigorously debated, revised, brought to a vote, and passed has then quietly slipped from view, never to be heard from again. Part of the group leader's job is to follow through on the decisions reached by the group. You should keep track of the decisions and make sure both that the members of the group carry out the things they have agreed to and that the people to whom you have submitted your decisions are implementing them. To this end, you should appoint someone to keep and distribute the minutes of the meeting so that all concerned parties have a written record of what you have done and what they should do. It will probably be your responsibility, moreover, to write up the reports of your group's decisions and to write the letters and memos necessary to make sure that those decisions were not reached in vain.

Review Questions

1. Why must the scope of an oral report be narrower than that of a written report on the same topic?
2. Define the role of repetition in an oral report.
3. What are the advantages and disadvantages of the read-from-manuscript report? Of the extemporaneous report?
4. What is audience feedback?
5. What is the disadvantage of organizing the body of your report in strict order of decreasing importance? How can you overcome this disadvantage?
6. What should you include in the introduction to your report?
7. Characterize the notes you would prepare for an extemporaneous report.
8. Describe the language most appropriate to an oral report.
9. What are the characteristics of an effective visual aid?
10. What is the primary function of most group meetings?
11. How can a group member prepare for a meeting?
12. What is the difference between clarifying a position and defending it? What is the difference between clarifying a position and restating it?

13. What is an agenda? What is a hidden agenda?
14. Characterize the major functions of a group leader.
15. What options does a leader have when a conflict arises?
16. What is the difference between consensus and majority rule? What can a leader do to help the group reach consensus?

Assignments

1. Attend one of the guest lectures presented at your school. During the presentation, make notes about the organization and presentation of the material: Does the speaker limit discussion to four or five main points? Does the introduction set the context, state the purpose, and preview the talk? Does the speaker use examples and repetition effectively? Is the vocabulary understandable? Are visuals used wisely? Write a two-page memo to your technical writing instructor, evaluating this presentation.

2. Turn one of the written reports you have done for this class into a five- to ten-minute oral report, and present it. Then write a brief analysis of the changes you made when you switched from written to oral reporting.

3. Present a five-minute report to your technical writing class on a topic in your field. Topics might include a laboratory or field experiment that you have done, a natural phenomenon of interest to you (for example, the northern lights), a technical procedure or mechanism, and so on.

4. Suppose you were going to present the subject matter in the preceding assignment to students in your major. Develop a list of talking points, then write a brief analysis of the changes you would make to accommodate this change in audience.

5. Attend a committee meeting of a campus organization that you do *not* belong to. Observe the performance of the leader and the interactions among the members, especially their methods of dealing with conflict and making decisions. Write a memo to your instructor, analyzing this meeting.

6. Break into groups of about eight people, and appoint a group leader. Discuss this question: How much, if at all, should class participation count in determining a technical writing student's final grade? The leader should be responsible for ensuring that all the implications of the question are examined before a decision is reached and that everyone in the group participates in the decision. Submit a report to your instructor, explaining your decision—if you don't reach a consensus, write minority reports as well. Submit a separate evaluation of the group discussion process you went through.

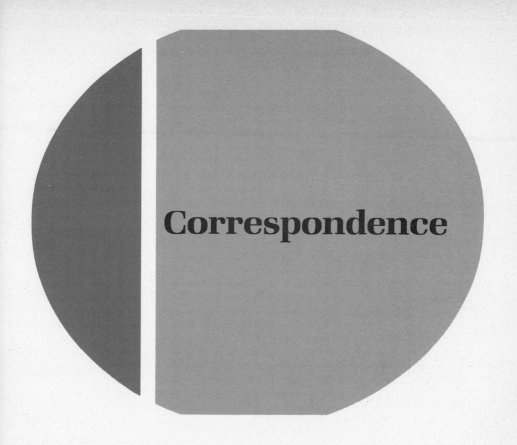

Correspondence

PART 4

19 Correspondence

Of all the forms of technical writing, correspondence is probably the easiest to master. Letters and memos are short and highly structured; they have a single overt goal and a limited readership.

Correspondence is also enormously important to master. A continual exchange of information—much of it in the form of official correspondence—keeps technology and business humming. Convention requires a memorandum or letter in many situations—to extend an invitation, to register a complaint, to transmit a report, to respond to a letter or memo you have just received, to request information. Moreover, effective writers use correspondence as a simple and powerful tool to build new contacts and reinforce old ones: to confirm an appointment or suggestion made over the phone, to summarize a conversation so as to avoid any misunderstanding, to compliment the author of an especially interesting report or article. Each good letter you write is a strand in the web of your professional contacts. That web is one of your most important assets to your employer.

We begin this chapter with a summary of basic memo and letter formats. Then we outline strategies for organizing any type of correspondence. We end with a detailed discussion of one variety of correspondence sure to be of immediate interest to you—the job application letter and résumé.

 ## Memo and Letter Formats

Corresponding with colleagues, subordinates, superiors, and strangers is part of the daily life of every technical professional. Whether you correspond by letter or by memo depends on your intended audience:

If you are communicating with an individual outside your organization or work group, write a letter.

If you are communicating with many people outside your organization or work group, write a letter and mass-produce it so that each recipient gets an original.

If you are communicating with many people within your organization or work group, write a memo.

If you are communicating with an individual within your organization or group and do not regard the communication as privileged, write a memo.

If you are communicating with an individual within your organization or group but want the communication to be a private one, write a letter.

Memo Formats

The format of a memorandum is always the same and always different— that is, every memo has a heading that says who sent it, who is to receive it, when it is being sent, and what it is about. Those four obligatory elements may be arranged in any number of ways to create a heading, and you will learn when you start a job what particular arrangement your employer insists on. In some organizations, memos are typed on preprinted forms. In others, they are typed on letterhead stationery; in still others, they are typed on plain white paper. The name of the addressee and the date usually appear first; the name of the writer and the subject follow. The elements may be lined up along the left margin or balanced—with some on the left and some on the right. Here are three examples:

TO: Corporate Sales Staff
FROM: Robert E. Blandings,
            Director, Promotions Department
SUBJECT: Arrival of 1985 Sales Manual
DATE: June 24, 1984

To: Patricia Smithers   Date: 18 March 1984
From: Justin Washington
Subject: Proposed change in Unit 3/A-23 testing procedure

INTERDEPARTMENTAL MEMORANDUM   May 9, 1984
          TO: Harold Worthy
     FROM: Margaret Szernic
       CC: John Aaron
             Susan Kravitz
             Christine O'Leary
       RE: QUARTERLY PROGRESS REPORT

You must decide whether or not to include titles in the "To" and "From" sections; it is wise to do this if the occasion is very formal or if the memo is being sent to numerous readers who don't know each other—or you. Moreover, if you have multiple addressees, you must decide whether to list them in alphabetical order or in order of decreasing status or seniority. Your organization may have a policy on this, so check. You must also decide who should be an actual addressee of the memo and who should be only a "CC" recipient. ("CC" or "cc" stands for "carbon copy" and is still used despite the fact that most copies are no longer made with carbon paper.) Copy recipients are being sent the memo for their information only, so include people among your addressees only if the memo is asking them to *do* something; "CC" everyone else.

Letter Formats

As you might expect, letter formats are more detailed and more personalized than memo formats. A memo has just a heading and a body, but a letter has no fewer than six obligatory format elements, as well as five optional ones. These are the obligatory elements:

1. Return address
2. Dateline
3. Internal address
4. Salutation
5. Complimentary close
6. Signature

The optional elements are as follows:

1. Attention line
2. Subject line
3. Stenographic reference
4. Carbon copy notation
5. Enclosure notation

Unlike memo formats, letter formats do not vary from organization to organization. The format conventions displayed in Figures 19.1 and 19.2 are correct for any kind of technical correspondence. We have annotated these letters to call your attention to their use of these conventions. Some formatting conventions, however, need further explanation.

Internal Address. Also called the inside address, the internal address should be the complete address of the person to whom you are writing, including his or her name and title and the zip code. Don't use abbreviations, with this exception: Reproduce the company name exactly as it appears on the company letterhead, abbreviations and all.

Salutation. If you are writing to someone you know fairly well, use his or her first name: "Dear Eric." Otherwise use a title and the person's surname. (Ex-

If you are not using letterhead stationery, write your full address without abbreviations; use two or more lines with no punctuation between the lines.

The salutation line ends with a colon.

The subject line is useful if you are writing "blind" to an organization and have no idea who is in charge of your type of letter.

The format known as "blocked style" lines up the return address, dateline, complimentary close, and signature on the right side of the page.

If you are sending a copy of your letter to anyone other than the person to whom it is addressed, a carbon copy notation lets the addressee know. Include titles unless you are sure the addressee knows who they are.

```
                                        104 First Avenue
                                        Rideout, Oklahoma 76239
                                        February 17, 1985


         Algorithm, Inc.
         2200 Industrial Boulevard
         University Park, Michigan 48112

         Gentlemen and Ladies:

         Subject:  Word Processing Equipment

              I am writing on behalf of the Committee on
         Technology at Harrover Corporation.  Our company is in
         the process of installing microcomputers for word
         processing and routine office management, and this
         committee has been asked to select a microcomputer to
         be placed at all work stations.

              We are much impressed with your TMI-442, but your
         Scripto software does not have underlining or
         mathematical symbols--both essential for the technical
         reports and specifications we prepare.  We are leaning
         toward Micromatic's Wordmaster.  Is it possible to run
         that software on your machine?

              Please let us know by mid-March if you can supply
         a hardware/software package that meets our needs.  We
         must make our decision by April 1.

              Thank you for your help.

                                        Sincerely,

                                        William G. Hardcastle

                                        William G. Hardcastle

         cc:  R. Jennings Hopgood, Chair, Committee on Technology
              J. W. Harrover, President, Harrover Corporation
```

FIGURE 19.1   A Letter in Blocked Style.

cept for the abbreviations "Mr.," "Ms.," "Mrs.," and "Dr.," titles should be given in full: "Professor Winitski," "President McNamara," and so on.) Choose the title your correspondent prefers if you know what it is; otherwise, Mr. and Ms. are safest. If you don't know the sex of the person to whom you are writing, don't assume it is a man. "Dear Sir or Madam" is customary when writing to an anonymous individual; "Gentlemen and Ladies" or "Ladies and Gentlemen," when writing to a department or organization. Both sound a little stilted, but no consensus substitute has been found yet. The salutation ends with a colon.

The date should never be abbreviated in a business letter.

If you use an attention line, put the same notation in the lower left-hand corner of your envelope.

The format known as "full blocked style" lines up every line in the letter on the left.

Always type your full name and title. If you know your correspondent well, sign your first name only.

This letter contains only one enclosure. If you have several enclosures, give the number in the enclosure line.

ALGORITHM, INC.
2200 Industrial Boulevard
University Park, Michigan 48112

February 27, 1985

Committee on Technology
The Harrover Corporation
104 First Avenue
Rideout, Oklahoma 76239

Attention: William G. Hardcastle

Dear Committee Members:

We are delighted to learn that you are considering the TMI-442 for use by your office staff. It can easily be adapted to accept Wordmaster if you purchase our Offix Softcard, which is available for a 20% discount when you purchase the complete TMI-442 system.

The Offix softcard gives you great flexibility in programming. It enables you to use not only Wordmaster but all other programs based on the CP/M operating system, as well as those that use TDOS, the system with which the TMI-442 is equipped.

I am enclosing a brochure that describes Scripto II, our updated word processing system. This advanced software includes many features not incorporated into Scripto, including both underlining and mathematical symbols, and is fully competitive with Wordmaster.

Please call collect if you have more questions about the TMI-442. I would be happy to meet with you to discuss how Algorithm's products can meet your needs.

Cordially,

Glenn Cereghino

Glenn Cereghino
Member, Technical Staff

GC/nr
Encl.

FIGURE 19.2   A Letter in Full Blocked Style.

Attention Line. If you think you know who should receive your letter but you are not sure, you can use a general internal address and salutation, with an attention line between them. You also use an attention line when you want to send your letter to a "contact person," whose job it is to channel letters to people who can best respond to them. If you use an attention line, put the same line in the lower left-hand corner of your envelope.

Subject Line. A subject line is useful when you are writing "blind" to an organization; if you don't know the name or even the department of the person who should read your letter, supplying a subject line will simplify the task of getting your letter to the right person. The subject line usually goes between the salutation and the body:

> Dear Sir or Madam:
>
> Subject: Seminar in marine acoustics
>
> Super Acoustics, Inc., is pleased to announce . . .

Carbon Copy Notation. The abbreviation for the carbon copy notation is not standardized: "C.C.," "CC.," "CC:," "c.c.," "cc.," and "cc:" are all common. If copies are going to several people, list their names alphabetically or in order of rank, directly beneath each other, right after the stenographic reference, if any. Include titles unless you are sure the addressee knows who they are. It is acceptable to send someone a copy of your letter without telling the addressee; this is called a "blind copy," and obviously the letter includes no reference to its existence.

Other Conventions. Two approaches are available for arranging the elements on the page: blocked style and full blocked style. Blocked style lines up the return address, dateline, complimentary close, and signature on the right side of the page. It may or may not indent the first line of each paragraph in the body. Full blocked style lines up everything on the left and never indents. Unless the letter is being written on letterhead stationery, full blocked style often looks unbalanced; however, on letterhead, it usually looks fine, and secretaries prefer it because it requires less tabbing.

Whichever style you choose, use letterhead stationery if possible; otherwise, use white, heavyweight 8-1/2-by-11-inch typing paper—no onion skin, no strange colors or sizes or shapes. Avoid corrasable bond, too; it smears, and it is rarely used by anyone but students. If you make minor typing errors, white them out, and then correct them on the typewriter. If you make major errors—or many minor ones—retype the letter: A letter should look "letter-perfect."

If your letter runs more than one page, each additional page should carry a heading with the name of the addressee, the date, and the page number:

-2-

Joseph Torborg   7 January 1985

or

Joseph Torborg
7 January 1985
Page 2

Organizing Memos and Letters

Despite their differences in formatting, memos and letters are organized identically. Both record momentary and immediate responses to a particular situation. Both are brief, concise, narrow in scope, concrete, and highly specific. A memo may consist of only a sentence or two, or it may run to several pages, depending on how complex its subject is. A letter is usually at least a paragraph long, because it must greet its audience cordially and establish goodwill as well as convey its message. In both memos and letters, several short paragraphs are preferable to one long one, and each of the paragraphs should contain only one idea—or perhaps two at the most if they're closely related and easily grasped. No one likes to read single-spaced print, so if you make your paragraphs too long, people will tend to read the first two or three sentences and the last sentence and skim the rest. If you want to avoid that, use lots of paragraphs and keep them brief.

Depending on your topic and purpose, the paragraphs in your letter or memo may build on one another in some way, or they may each relate independently to the main point. Three techniques will help you make these relationships clear to your readers. First, organize each paragraph like a mini-letter, with its main point in the first sentence and the supporting points immediately following. Second, if your paragraphs build on each other, arrange them in logical order, usually in order of decreasing importance if they justify a main point you have already discussed and in order of increasing importance if they build to a main point you haven't mentioned yet. If all the paragraphs relate independently to the main point of the memo or letter, number them to make their lack of logical sequence explicit. Third, use lots of transitions to indicate how each paragraph relates to the preceding paragraph ("however," "therefore," "first," "finally," and so on).

For relatively routine correspondence, the following structure is standard:

1. State your main point, usually your goal in writing the letter. If appropriate, devote a few sentences to establishing a common interest with the reader; then link that to your main point.
2. Present your first supporting point, and tie it to your main point.
3. Present your second supporting point, and tie it to both your main point and your first supporting point.
4. Present additional supporting points as necessary, and tie each one to your main point and the previous supporting point.
5. Restate your main point, and say what response you want from the reader.
6. Close with a brief expression of goodwill.

This basic structure works just fine for both letters and memos if they bring good news or pleasing information. If they bring unwelcome information, then you will need to use a variant of this structure that is more persuasive:

1. Establish a common ground with your reader. State a fact that will please your reader ("We have been buying Tinkos from your company for nearly twenty years") or say something on which you and your reader can agree.

2. Present background information that will lead you up to your main point, stating the main point only indirectly. Present reasons that will prepare your reader for the unpleasant information to come. ("SuperTread Tires are not guaranteed against damage resulting from abuse.") Help your reader understand why you must be the bearer of bad news.
3. State your main point as kindly as you can, and tie it to the supporting points you have already made.
4. Present the rest of your supporting points, tying each to the main point.
5. Restate your main point (perhaps in a subordinate clause), and say what response you want from the reader.
6. Close with a brief expression of goodwill.

Every memo or letter you write as a technical professional will use a version of one of these two basic structures.

 # Job Application Letters and Résumés

One of the most important pieces of technical writing you are likely to undertake in the near future is the letter of application for a job. Like most correspondence, the job application letter has the advantage of a clear-cut goal and a one-person audience. This greatly simplifies the task of designing the letter (though, of course, it does not make that task any less important). The résumé that accompanies many job applications presents tougher design problems: You want it to fit the format, yet stand out in a crowd; you want it to work for the widest range of potential employers you can manage; and, above all, you want it to showcase your employment assets.

Goal and Audience

When you write a job application letter and résumé, your goal is to start the process that will end with a job offer. And who is your audience? Ask yourself these questions:

• What sort of person will be reading this letter?
• What does that person want from me?
• What do I want that person to know?
• What do I want that person to do?

Once you have answered these questions as truthfully and as completely as possible, you can devise a strategy for creating a letter that accomplishes your goal.

What sort of person will be reading this letter? Someone who is busy, who doesn't know you, and who would rather be doing something else.

What does that person want from me? Clear evidence that you may be right for the job—or, failing that, clear evidence that you are wrong for it and don't need to be interviewed. "Right for the job," by the way, means different things to different employers. They all want the appropriate technical qualifications, but some are looking for innovators with new ideas while others are looking for diligent toilers who won't rock the corporate boat. Find out which by reading company brochures, talking to employees and former employees, consulting other professionals in the field, and browsing through trade journals for news of the company. This research is as important as memorizing the ad or job description in helping you decide how to sound "right for the job."

What do I want that person to know? That you are right for the job, of course. Most important are your qualifications, personal and professional. But you have to assume many applicants will be qualified. In the face of stiff competition, you will need something extra—an experience, credential, or idea that makes you more qualified than most for this particular job. Knowing a bit about the company's current activities will help here, and a good job letter should show that you know.

What do I want that person to do? Hire you on the spot—but that is hardly realistic. An invitation to interview is a more feasible first step.

With all this in mind, look at the two letters that follow. (We have omitted the addresses and such.) It is easy to see that the second letter is better, but *why* is it better? Contrast the two word for word, sentence for sentence, and paragraph for paragraph.

A. Dear Ms. Smith:

During the spring semester this year I took an advanced seminar on industrial and commercial applications of solar technology for use in heating without conventional energy sources, or with reduced use of conventional energy sources. The course, which was very interesting to me, was taught by Professor R. H. Kidder of the Industrial Engineering Department of New Mexico Tech, where I was then a junior.

Now that I am a senior and nearing graduation, I am interested in considering the possibility of future employment in solar technology. Enclosed please find for your information a typed copy of my résumé. I would be interested in working with the Solar Systems Corp. and would appreciate receiving application materials at your earliest convenience. I've been really enthusiastic about working in the Southwest ever since my folks sent me to summer camp in New Mexico when I was a kid, and I think I'm the man for the job opening you have available.

I read about the opening in your classified ad in last week's Sunday *New York Times*, which said you were looking for someone who was familiar with industrial and commercial solar technology and could serve in the function of a public and technical information specialist, which I assume means writing brochures on your products as well as instruction manuals and the like. I am not really trained as an information specialist, and I guess I might not be completely qualified for the opening, but as I hope you can tell, I am really interested. I'm a hard worker, and I'm told I'm a pretty good writer when I try hard. That's what my boss told me when I worked as a junior technical editor last summer at school. Anyhow I thought the opening was worth a letter.

The undersigned awaits a reply.

Sincerely yours,

George T. Fenwick

B. Dear Ms. Smith:

I would like to apply for the position as public and technical information specialist that you advertised in the May 12 *New York Times*. The job at Solar Systems Corporation sounds like an excellent opportunity for me to combine my interests in solar energy and technical writing.

Solar Systems Corporation's demonstration project testing the cost-effectiveness of industrial solar panels is especially interesting to me. Although several companies have accepted federal grants to test solar equipment, you are one of the few companies that have taken the initiative to finance such a project themselves. I have sent a separate letter to your public information department requesting a copy of the final report on the project when it is completed. This sounds like the type of project that would be exciting to explain to potential customers and the concerned public. Although I have been aware of the project for some time, I have not seen any news coverage on it. Are you thinking of inviting reporters to inspect the plant and discuss the project with company engineers?

Your work in industrial solar heating was one of several projects discussed in an advanced seminar at New Mexico Tech on the industrial and commercial applications of solar technology, one of three courses I took there on energy technology. As the enclosed résumé indicates, I worked last summer as a junior technical editor and took a course in technical writing this year. I will graduate from New Mexico Tech's Industrial Engineering Department in June and will be ready for employment July 1.

I would be glad to interview with you any time between now and June 9. Tuesdays and Thursdays are the best days for me to come to your office. Please write if you would like to set up an appointment. I look forward to hearing from you.

Sincerely,

Frank T. Fenwick

Frank's letter answers our list of questions more effectively than George's does. By discussing a specific project and suggesting a way to publicize it, Frank shows that he knows and cares about the company's interests. George, on the other hand, seems to know little and care less about Solar Systems

Corporation; he could have sent the same letter to any other company with a similar opening. Frank's goal is clear from the first sentence. George never quite states his goal, though by the third paragraph you can figure out which job he wants. Frank ends by telling Ms. Smith exactly what he wants her to do—write to set up an interview. George doesn't say; perhaps he expects to be hired sight unseen.

Being Persuasive

The main point of thinking through your goal and audience is to devise a persuasion strategy that links the two. The essence of being persuasive is to convince the reader that doing what you want is in his or her best interests.

If persuasive communication has a "first principle," this is it: Communication changes people by reinforcing a value or need they already possess that predisposes them in the direction of the desired change. In practice, this means establishing a common bond with your reader (a shared interest), tying your goal to that shared interest, then suggesting an action (preferably something easy) that will forward both your reader's interests and your goal. Instead of pointing to differences and disputing them, persuasion theory tells us to point to similarities and mobilize them. Look again at Frank's second paragraph, the one about the solar demonstration project. It grabs Ms. Smith's attention because it deals with a topic *she* cares about. It shows her that Frank knows something about the company and shares its concerns, and it demonstrates that he can help advance its interests—*if* he gets the job. The paragraph neatly ties one of the reader's interests (the demonstration project) to Frank's goal (proving he is the person for the job).

Research in persuasion theory also shows us that readers are more influenced by an appeal from a high-credibility source than by one from a low-credibility source. Neither Frank nor George has any credibility with Ms. Smith to begin with: They are strangers, students applying for a job. It is up to them, in their letters, to build their credibility to the point where Ms. Smith will think, "This candidate certainly knows what he's talking about." George misses his opportunity. He as much as says, "I'm interested, but not especially qualified." Frank, on the other hand, builds credibility by stressing his particular interest and experience in solar energy and technical writing, and then follows up with his "official" credentials (degree, university, and so on).

Notice Frank's last paragraph too. He doesn't leave it to Ms. Smith to decide how to respond to his plea, nor does he ask for a huge commitment (a job offer) right off the bat. Instead he suggests a specific and relatively easy response—write to set up an interview.

Here, then, are four key characteristics of a persuasive job application letter:

1. State your goal clearly and early. Include the title of the position you are applying for and where you heard about it.
2. Learn something about the organization to which you are applying. Ad-

dress its concerns in your letter and show how they relate to your goal and background.

3. Build your credibility by pointing out highlights from your résumé that relate directly to the particular job you are applying for.

4. Tell the reader what response you want (a call, a letter), and make it easy to comply (certainly by supplying a telephone number and perhaps by suggesting times for an appointment).

Being Personal

When you apply for your first professional job, you will understandably want to stress the credentials you have spent the past several years accumulating. However, your reader may not be quite as interested in those credentials *per se* as you are. In your eagerness to prove how well qualified you are, don't forget that most other serious candidates for the job probably have approximately the same credentials. Your reader will be perusing dozens, maybe hundreds, of job application letters and résumés, and they are all going to sound pretty much the same. You want yours to sound special because you want yours to be chosen.

To sound special, your job application letter must complement, not simply reiterate, your résumé. (We will tell you what you should put in your résumé a little later in the chapter.) Don't just list your strongest credential; say what makes that particular credential especially relevant to this job. Draw connections among different facets of your background that are in different parts of your résumé. If you want to work for a particular firm because it is in Boston and you have always wanted to live there—by all means, say so. George Fenwick has the right idea, though the reason he gives is trivial. Any detail that gives your reader a personal glimpse into what you are like will help give your letter a touch of individuality and will encourage the impression that you might be an interesting person to interview.

Another way to interject the personal touch into a job application letter is by using the word "I." But use it *sparingly*, or you will sound as if you are self-interested and see everything only in terms of how it relates to you personally. Instead, try to use the word "you"—to readers, that is the most welcome word in the English language. Rather than talking about how working for Solar Systems Corporation would benefit *you*, tell how having you on the payroll would benefit the *company*. Notice how many times Frank Fenwick uses the word "you" in his letter, especially in the second paragraph. Notice, too, how he addresses Ms. Smith directly at the end of that second paragraph. He sounds like a real person writing a letter to another real person. Ms. Smith is bound to find him friendly, warm, and courteous. Friendliness and amiability are desirable attributes in any employee, but they are particularly important in a public information specialist.

Achieving the Correct Tone

As part of your attempt to be personal, you should use approximately the same tone in your job application letter that you would use in person or on the telephone. Achieving a natural, unaffected tone in a job application letter is much harder than you think. Overcome by lack of confidence, most first-timers write job application letters that sound anything but natural. In fact, five specific tone traps are so common that they deserve special mention. The first ensnares business letter writers everywhere; the other four are especially likely to catch newcomers.

The Stuffy Tone. Concerned that speaking in their own voice might make them sound clumsy and amateurish, many inexperienced letter writers reason that if they string together stock phrases and formulaic expressions, they will not say something inane. Unfortunately, the opposite is true. Those stock phrases and formulaic expressions are now well over a hundred years old, and to a modern reader they sound stilted and meaningless. A stuffy letter always impresses its writer more than its reader; such a letter says very little, does not speak directly to readers, and carries none of the writer's personality. In a job application letter, whose goal is to make its author sound competent and interesting, the stuffy tone is not only inappropriate but counterproductive as well.

The following phrases—and others like them—lend a certain quaintness to nineteenth-century business correspondence, but in a contemporary business letter they merely sound stuffy.

Allow me to
Enclosed please find
Herewith/hereby/hereto
I remain
Kindly (for "please")
Permit me to say
Rest assured that
Thanking you in advance
The undersigned

The Chatty Tone. In an attempt to sound relaxed, many letter writers go too far and use a tone that is more appropriate for a personal letter than a business letter. "I did my first two years of college at Colorado State. Attending school away from home really developed my interest in psychology. I rubbed elbows with people with the most unbelievable life-styles, values, and attitudes." Slang and colloquial expressions, contractions like "I'd" and "you'll," and abbreviations such as "ad" for "advertisement" have no place in formal business correspondence between strangers, not only because they are presumptuous and lack dignity but also because they are frequently unclear. The chatty tone is all-too-often vague, cliché-ridden, and very long-winded.

The Mixed Tone. Perhaps the strangest tone, and one of the most prevalent, is a curdled mixture of stuffy and chatty. Whatever your tone, at least it should be consistent. "I'd really like this job because I've admired your company for I don't know how long" just doesn't belong in the same letter with "The undersigned awaits your reply."

The Timid Tone. Unsure of themselves and uncertain about their qualifications, some people write letters in a tone that almost begs for rejection: "I almost received my degree in electrical engineering, but my grades were poor last semester because of family problems. I need a job badly, so I thought I would apply for the job as electronics engineer that you advertised in last Sunday's *New York Times*." Every newcomer to the job market feels scared and inadequate. Don't worry about it, and don't let it creep into your business letters. Instead of apologizing for your qualifications, present them in as positive a light as honesty and honor permit: "I have been studying electrical engineering for the past three years." Readers are more likely to respond favorably to a letter if it is direct and confident.

The Aggressive Tone. When we say "direct and confident," we do not mean "pushy and obnoxious." You may be "just the right person for the job" and "fit the qualifications to a T," but that is for the personnel director, not you, to decide. Be assertive—but assert facts, not self-evaluations. And try not to sound high-handed or to press too hard. There is all the difference in the world between "I will call your office next week to arrange an interview" and "May I call your office next week to discuss the possibility of an interview?" Often the difference between "direct and confident" and "pushy and obnoxious" is in the judicious use of the word "you." Don't close your letter with "If you wish to arrange an interview, I can be reached at (609) 555–3456 after 5 p.m." Instead, say: "I would be happy to come to Chicago for an interview at any time; you may reach me at (609) 555–3456 any day after 5 p.m."

Let's look again at the Fenwick letters. What tone does George use? We would call it mixed, and it seems quite likely to leave Ms. Smith with the impression that George is confused and unreliable. George also alternates between timidity and overconfidence. "I think I'm the man for the job," he says, then adds in the next paragraph: "I guess I might not be completely qualified." Frank, on the other hand, keeps his letter safely in the middle range instead of oscillating among the various extremes. "I would be glad to interview," he writes, instead of "I'd sure like an interview" or "permit me to suggest an interview" or "I'm certain you will want to interview me" or "I know you're probably too busy to interview me."

Organizing a Job Application Letter

Whatever their tone, job application letters—like all business letters—should be tightly structured. A rambling letter not only risks that the reader may miss your main point; it also communicates that you are either too confused or too uninterested to do your own organizing and are thus forcing the reader to do it for you. By making its main point impossible to miss, a well-organized letter tells the reader both that you know what you're doing and that you respect the value of his or her time.

The main point of a job application letter is usually your reason for writing it. Revealing it in the first sentence helps your reader know what to think about as he or she reads on. "Please consider me for the position of . . ." is thus an excellent way to begin a job application letter. Compare the Fenwick letters again. Frank states his purpose in the first sentence whereas George waits until the middle of the second paragraph—and then he doesn't come right out and say that he wants to apply for the job. Nor does George's introductory paragraph serve the persuasive purpose of building a bond with the reader. Ms. Smith probably spent the time it took her to read it wondering why a total stranger was writing to her about his course in solar technology.

The body of your letter backs up its main point by offering evidence that you are the best candidate for the job. Arrange your central paragraph (or paragraphs) so that the ideas build on one another. Above all, make sure that each supporting point in your letter relates clearly to your main point. Any time you suspect that your reader may be losing track of the relationship, state it explicitly: "This experience should prove useful on your hardened silo project."

Now back to the Fenwick letters. Look at the chaotic organization of George's third paragraph, for example. What he stresses in its important first sentence is that he read about the opening in a classified ad in the *Times*— hardly a major point in his effort to sell his services to Solar Systems. In a string of clauses he then repeats and interprets the job description, information Ms. Smith does not need to hear again. George admits in the next sentence that he may not be qualified, then spends three sentences explaining that he is nonetheless interested and a hard worker. In the final sentence he shrugs, having talked himself out of a job.

The body of Frank's letter is much more tightly organized. He begins by stating that he wants to apply for a particular job, stressing how it suits his interests and experience (main point). The second paragraph demonstrates his knowledge of the company and shows how he would handle a specific situation (supporting point). The third gives his qualifications (supporting point), and the fourth restates his interest and suggests a specific action. Frank's transitions between paragraphs are subtle, but they work. The last sentence of the first paragraph focuses on his twin interests in solar energy and technical writing. The second paragraph shows how the two would work together in the demonstration project. The project then becomes a lead-in to the third paragraph on his qualifications. Strictly speaking, Frank's second paragraph— the one about the demonstration project—doesn't stem directly from his main point. But even though the paragraph digresses a bit from Frank's qualifications for the job, it is an essential part of his persuasion strategy. Ms. Smith will easily understand why it is there.

The end of a letter should restate its main point, in slightly different words and with perhaps a different emphasis. A good way to end a job application letter is to restate indirectly that you want the job by giving a reason why you want it. People tend to skim letters, paying most attention to the first and last paragraphs, so you want to make them count. (For similar reasons, the first and last sentences of each paragraph are the most important; never hide a vital fact or idea in the middle of a paragraph.) Your restatement of your main point should be tied to the specific action you want from the reader:

This means that you should end a job application letter by requesting an interview, as Frank does. You should then close the letter with an expression of goodwill—"I look forward to hearing from you," "I appreciate your interest in my qualifications," "I hope to see you on the 21st." The complimentary close above your signature says good-bye.

Organizing a Résumé

A résumé, which you attach to every letter of application for a job, is a short summary of facts that an employer might wish to know about you. Because its goal is to encourage an employer to hire you, we recommend that a résumé be strategically organized so that it presents your qualifications in order of decreasing importance *to the employer*. There is no standard form for a good résumé; in fact, the more standardized you make yours look, the more it will look like everyone else's. Although résumés should never include any falsehoods, they are not totally objective either. Their purpose is to present your credentials as persuasively as possible—stressing your strong points, playing down the minor details, and ignoring catastrophes.

If, for example, your last job had a prestigious title, you might want to put your employment history first in the résumé and type the job titles in full caps. If, on the other hand, you have had little relevant work experience, you will probably want to describe your education first and, when you get to your jobs, stress the responsibilities instead of the titles. If your best experience has come from extracurricular activities and volunteer jobs, you might want to abandon the conventional "Employment" category altogether. Under a broader heading like "Experience" you can organize your qualifications according to skill areas—"Laboratory Work," "Technical Writing," "Management and Administration"—listing under each skill the relevant jobs, activities, and even courses. Ask yourself: "If I were hiring, what would I be looking for in a candidate?" Then showcase those skills in your résumé.

The typical résumé has four almost mandatory categories and a large number of optional ones. We'll consider the key categories first, then describe the major options.

Personal Data. The personal data section is almost always the first section of the résumé, usually without a heading. It includes, of course, your full name, home address, and home telephone number (including the area code). If you have a job already, include your business telephone number too—but only if you will be available to take calls and your boss won't mind. Your social security number is optional but seldom useful. Some résumé writers list their sex, date of birth, place of birth, height, weight, health, marital status, and the like. This used to be customary, but some of the information is useless, and some of it (such as sex, age, and marital status) can become a basis for discrimination if it is included. Of course, you should not include your race or religion.

Education. A professional with years of experience usually condenses the education section into a list of degrees, universities, and dates, working backwards from the last school attended. But for a student seeking a first real job, education is an important category. Among the information you might want to include, consider your major, other concentrations of course work, special projects or internships, and your grade point average. Also include special courses, workshops, and the like that are relevant to your job skills even if they didn't lead to any sort of degree. Don't go back any further than high school.

Employment. Employment is usually the most important category (which is why you may want to try an untraditional format if you haven't got much to say here). List each job you have held, starting with the most recent—you should leave out *very* minor or ancient jobs unless they are relevant. For each job you list, name your employer, the month and year you began and left, your job title, and your responsibilities. Don't include the complete address of your employer (the city is enough), the name of your supervisor, or your salary. The main stress belongs on the responsibilities; be as specific as you can, and emphasize the ones that are most relevant to the jobs you are applying for now. If you were promoted on the job, say so. It is a good idea to include recent summer jobs even if they are not especially relevant; employers want to know you didn't loaf for three months. Technically, employment must be paid, but if you change the title to "Work Experience" you can incorporate your most important volunteer work as well.

References. Your references are the names, titles, business addresses, and possibly telephone numbers of at least three people who can testify to your ability, industry, and charm. Choose your references carefully. You want the highest-status people you can get, but an eminent professional you met once briefly isn't as valuable a reference as an unknown who really knows and likes your work. Ex-bosses and major professors are the usual references for students and recent graduates. Get their permission before you include their names. Many experts recommend putting a phrase like "References available upon request" on your résumé and waiting till you're asked for them; others think it's better to include the references. Either way, this section comes at the end of the résumé.

Job Objective. The job objective category, which is optional, does focus attention on what you want and what you offer, but it may eliminate you from consideration for a job you would settle for. If you decide to include it, hold it to a sentence or two near the top of the résumé. Make it specific enough to be useful but not so specific that it rules you out from too many jobs: "I am looking for employment as a chemical engineer with a large company, preferably in the oil industry."

Extracurricular Activities. If your extracurricular activities at college have taught you important job skills and you can't find anywhere else to include them, add this optional category. Don't forget that leadership is a relevant skill for every job; if you were vice-president of the senior class, work it in elsewhere or put it here, but do include it somewhere.

Professional Skills. Ultimately what most interests a prospective employer is your relevant skills. If a few additional abilities don't show up anywhere else on your résumé, such as a foreign language you handle fluently, use this optional category to list them. If *many* of your most relevant skills are hidden by the traditional format, think about reducing the "Education" and "Employment" sections to quick lists, organizing the details in an extended skills section.

Organizations. Most successful professionals belong to a number of professional organizations. If you want to look like a go-getter, join the same groups yourself and list them under the heading "Organizations" or "Memberships." You can include service and civic groups as well, or devote a separate category if you like to these nonprofessional organizations.

Honors, Awards, and Publications. If you have honors and awards to boast about, by all means include them—the name of the award, the organization that presented it, the year, and a brief description if necessary. Include academic scholarships and election to honorary societies in this category. The same goes for publications, including reports, trade journal articles, and pamphlets. Give the complete citation for each publication. If you've got more than one (including material about to be published), make "Publications" a separate category.

Other. Anything else that will help you land the job is worth including on your résumé, in a special category if you need one. Two common items that we would not include, however, are the month you are available for work and the salary you expect. We think the former dates your résumé and belongs in the cover letter instead. The latter may price you out of some jobs and cost you money in others; deal with it at your interview.

Writing Effective Résumés

Working within these categories, you still have quite a bit of flexibility in writing your résumé. You can select among optional categories or make up new ones; you can decide which activities to stress and which to downplay; you can arrange everything on the page as you see fit. You do all this, of course, in terms of your goal and audience, but whatever decisions you make, keep in mind four concepts: clarity, conciseness, liveliness, and perfection.

Clarity. No supervisor or personnel manager with fifty résumés to read is going to stop to puzzle out your credentials. So whatever categories you use—especially if you use untraditional ones—label them clearly and arrange them logically and consistently. Use indentation, underlining, capitalization, and white space to help guide the reader to the most important facts. Consider having the résumé professionally printed; if you are applying for a job that requires artistic skill, letter it. Use good taste, however; you want the résumé to look elegant, not garish.

Conciseness. The best résumés are short—one page if you can manage it, two at the most. That puts a heavy premium on conciseness; there is *no* room in a résumé for flab or padding. There may be no room for complete sentences either; it is quite acceptable to use fragments or to eliminate pronouns and articles as long as your meaning is clear. "Managed part-time staff of three in study of lakeshore septic tank seepage" is a good résumé "sentence."

Liveliness. All the factual material in a résumé inevitably gives it a stagnant, passive flavor. Yet selling yourself means sounding dynamic and forceful. Concrete verbs are your only hope. You didn't "participate in an analysis," you analyzed. You didn't "work on evaluating," you evaluated. Other useful verbs include administered, created, designed, developed, directed, edited, expanded, improved, invented, maintained, managed, operated, organized, oversaw, planned, prepared, presented, produced, reorganized, sold, supervised, trained, and wrote.

Perfection. If you think a *letter* has to be letter-perfect, consider the résumé. In a tight job market, employers are looking for reasons to eliminate you, and a single typo can do the trick. Never send a résumé with a misspelling, a dangling modifier, or a penciled correction. Edit and proofread your résumé a dozen times; then get your friends to edit and proofread it; then (if possible) have it professionally typed and reproduced. And if you find an error, throw it out and start again.

Finally, unless you are striving for an offbeat reputation, stick to 8-1/2-by-11-inch, high-quality white paper. Strange sizes, shapes, and colors stand out, but ultimately they do more harm than good.

Many people make up only one résumé for all prospective employers, but if you are one of the growing number of people with expertise in more than one field, you may decide to write several résumés, each stressing a different strength. Our old friend George Fenwick, for example, might want one résumé stressing his engineering skills and another stressing his technical writing skills. Each time he applied for a job he would pick the appropriate résumé.

In Figures 19.3 and 19.4, we have written George's two résumés for him. Both use acceptable formats, though they are quite different, and we hope you will see from them how very differently the same information can be arranged. Use either as a starting point for designing your own résumé, and you will be well on your way to a good job.

George T. Fenwick

Personal Data
    Home address: 14907 Shady Lane, Santa Barbara, CA 94207
    Home telephone: (408) 272-4958
    Business telephone: (408) 931-7319
    Born January 4, 1962; Portland, Oregon

Education
    B.S., New Mexico Institute of Technology, 1983
        Graduated magna cum laude in Industrial Engineering, with special
        concentration (6 courses) in design and application of solar
        technology.
    San Diego Institute, summer seminar on industrial heating, 1981
    Jefferson High School, Santa Barbara, CA 1979

Employment
    6/83-date   Information Assistant, California Solar Information Center,
                Santa Barbara, CA. Summarize literature and interpret
                technical data for Center publication. Brief staff writers on
                solar engineering developments.

    9/82-5/83   Research Assistant (to Prof. R. H. Kidder), New Mexico
                Institute of Technology. Analyzed energy efficiency of solar,
                gas, and oil installations in local industry.

    6/82-8/82   Assistant Technical Editor, Publications Department, New
                Mexico Institute of Technology. Edited manuals and brochures
                on new electrial engineering lab, energy efficiency research,
                microprocessing, and other topics.

    6/81-8/81   Range Guide, Sunshine Dude Ranch, Gillette, Wyoming. Planned
                and conducted 10-day hunting trips.

    7/80-8/80   Counselor, Camp Wyandot, Livingston, CA. Ran "science shack"
    7/79-8/79   and assisted on waterfront.

Publications
    "Comparative Costs and Benefits of Four Heating Systems for Small
    Businesses," with R. H. Kidder and Jamie Hall, Extension Bulletin #T-48,
    New Mexico Institute of Technology, in press.

References
    Mr. Ralph Whitty, Director, California Solar Information Center, 370 Price
        Avenue, Santa Barbara, CA 94202

    Dr. R. H. Kidder, Department of Industrial Engineering, New Mexico
        Institute of Technology, Alarum, NM 73907

    Ms. Kathy Wilder, Extension Publications Editor, Cooperative Extension
        Service, New Mexico Institute of Technology, Alarum, NM 73907

FIGURE 19.3   **One Résumé Format.**

Preparing a Follow-up Letter

If your résumé and job application letter are well crafted, they will probably accomplish their immediate goal—to get you an interview. Even in a

George T. Fenwick
14907 Shady Lane, Santa Barbara, CA 94207
(408) 272-4958

JOB OBJECTIVE: A technical writing position using my knowledge of industrial
engineering, and, if possible, my special training and
experience in solar technology.

PROFESSIONAL SKILLS:

Technical Communication. Wrote and edited manuals and brochures on a variety
of technical topics. Prepared technical reports on a funded research
project and wrote newspaper articles on the results.

Engineering Research. Studied industrial energy efficiency on a project with
Prof. R. H. Kidder. Interpreted research findings for popular writers.

Management Leadership. Chaired Student Engineering Society for two years.
Conducted 10-day hunting trips for tourists as sole guide.

EDUCATION:

B.S., New Mexico Institute of Technology, Industrial Engineering, 1983
(magna cum laude; concentration in solar technology)
San Diego Institute, summer seminar in industrial heating, 1981
Jefferson High School, Santa Barbara, CA, 1979

EMPLOYMENT:

Current   Information Assistant, California Solar Information Center,
                    Santa Barbara, CA. Summarize solar developments in
                    nontechnical language, resulting in more than 40 brochures,
                    pamphlets, news releases, and displays so far.

1982-1983   Research Assistant to Prof. R. H. Kidder. Surveyed local
                    industry on energy efficiency. Drafted progress reports,
                    final report, extension bulletin, and newspaper summary.

Summer 1982   Assistant Technical Editor, Publications Department, New
                    Mexico Institute of Technology. Edited user and repair
                    manuals, wrote publicity brochures, revised faculty research
                    reports, supervised layout and production.

Summer 1981   Range Guide, Sunshine Dude Ranch, Gillette, Wyoming.

Summer 1980   Counselor, Camp Wyandot, Livingston, CA. Ran "science shack."
Summer 1979   Revised camp publicity and drafted form letters to parents.

PUBLICATIONS:

"Comparative Costs and Benefits of Four Heating Systems for Small Businesses,"
with R. H. Kidder and Jamie Hall, Extension Bulletin #T-48, New Mexico
Institute of Technology, in press.

"Industry Looks at Solar Energy," Vigilante (student newspaper), June 3, 1982.

Other reports, pamphlets, news releases, etc. available on request.

REFERENCES: Available on request.

FIGURE 19.4   An Alternative Résumé Format.

buyer's market, a clear, concise, lively, and perfect résumé is rare enough that
employers usually want to meet its author personally. Though it would be
outside the scope of this book for us to discuss how you should conduct
yourself at the interview (the people in your college placement office can give

you some good suggestions), we would like to discuss one more written aspect of the job application process—sending a thank-you letter after a job interview. Not only is it courteous to write such a letter, but it is also a sound piece of persuasion strategy because it keeps your name in the mind of everyone who interviewed you. The organizational pattern that we gave you for the job application letter works just fine for this letter, too. The following letter is organized according to that outline, and it achieves the correct tone of restrained enthusiasm.

Dear Mr. Burbot:

Thank you for taking the time yesterday to interview me and introduce me to the other Marine Water Center personnel. I know you have a busy schedule, and I appreciated the interview, the introductions, and, above all, the personal tour you gave me of the hatcheries.

Throughout the day, I was very impressed by the camaraderie I saw among your staff. Having worked for two summers in a large, impersonal lab, I was delighted by the professional yet friendly atmosphere. This was especially apparent in late afternoon, when the staff joined our discussion on aquaculture.

After our discussion, I recalled a reference that may interest you. "The Effects of Motorboats on Offshore Clam Hatcheries" (*Journal of Shellfish Management*, vol. 17, no. 2) points out some of the hazards involved in the type of hatchery we talked about. You may find this article useful in your feasibility study.

Thanks again for your time and interest. If I can supply you with any further background materials or references, please call.

Sincerely,

Morris K. Tuttle

Review Questions

1. What are the primary differences between a memo and a letter?
2. For what audiences is a memo appropriate? For what audiences is a letter appropriate?
3. How should you organize a letter that brings neutral or pleasant news? How is the organization different if the letter brings unwelcome news?
4. What is the immediate goal of a job application letter and résumé?
5. How do you make a job letter persuasive?
6. What tone is appropriate in a job letter?
7. How is a job letter organized?
8. What are the major parts of a résumé?
9. How should the résumé be organized?
10. What is the value of a follow-up letter after a job interview?

Assignments

1. You work in the research department of Alpha Enterprises. Albert Sampson, vice-president, Export Division, has asked you to send him information on American production of lemons and limes since 1970. The most recent information you can find is that presented in the table on page 258. Write your response in the form of a memo.

2. Assume that Mr. Sampson writes back to you that your information is too old to be useful to him, that he wants current information, and that he wants it within a week. You must write him that you do not have the current figures, but that you are sending to the Department of Agriculture to get them. (Your response, in other words, is one that contains unwelcome news.)

3. Send a letter of inquiry to the Department of Agriculture requesting the information. (Use a subject line or an attention line if you can't find the specific person to whom the letter should be addressed.)

4. Select a job advertisement from a publication in your field, and write a letter applying for the job.

5. Draft your résumé.

Index